Distribution System Modeling and Analysis with MATLAB® and WindMil®

This Fifth Edition includes new sections on electric vehicle loads and the impact they have on voltage drop and transformers in distribution systems. A new and improved tape-shield cable model has been developed to produce more accurate impedance modeling of underground cables. In addition, the book uses state-of-the-art software, including the power distribution simulation software Milsoft WindMil® and the programming language Mathworks MATLAB®. MATLAB scripts have been developed for all examples in the text, in addition to new MATLAB-based problems at the end of the chapters.

This book illustrates methods that ensure the most accurate results in computational modeling for electric power distribution systems. It clearly explains the principles and mathematics behind system models and discusses the smart grid concept and its special benefits. Including numerous models of components and several practical examples, the chapters demonstrate how engineers can apply and customize computer programs to help them plan and operate systems.

The book also covers approximation methods to help users interpret computer program results and includes references and assignments that help users apply MATLAB and WindMil programs to put their new learning into practice.

Distribution System Modeling and Analysis with MATLAB® and WindMil®

Edited by

William H. Kersting
Robert J. Kerestes

CRC Press
Taylor & Francis Group
Boca Raton London New York

CRC Press is an imprint of the
Taylor & Francis Group, an **informa** business

MATLAB® is a trademark of The MathWorks, Inc. and is used with permission. The MathWorks does not warrant the accuracy of the text or exercises in this book. This book's use or discussion of MATLAB® software or related products does not constitute endorsement or sponsorship by The MathWorks of a particular pedagogical approach or particular use of the MATLAB® software.

Fifth edition published 2023
by CRC Press
6000 Broken Sound Parkway NW, Suite 300, Boca Raton, FL 33487-2742

and by CRC Press
4 Park Square, Milton Park, Abingdon, Oxon, OX14 4RN

CRC Press is an imprint of Taylor & Francis Group, LLC

© 2023 William H. Kersting, Robert J. Kerestes

First edition published by CRC Press 2001
Fourth edition published by Routledge 2017

Reasonable efforts have been made to publish reliable data and information, but the author and publisher cannot assume responsibility for the validity of all materials or the consequences of their use. The authors and publishers have attempted to trace the copyright holders of all material reproduced in this publication and apologize to copyright holders if permission to publish in this form has not been obtained. If any copyright material has not been acknowledged please write and let us know so we may rectify in any future reprint.

Except as permitted under U.S. Copyright Law, no part of this book may be reprinted, reproduced, transmitted, or utilized in any form by any electronic, mechanical, or other means, now known or hereafter invented, including photocopying, microfilming, and recording, or in any information storage or retrieval system, without written permission from the publishers.

For permission to photocopy or use material electronically from this work, access www.copyright.com or contact the Copyright Clearance Center, Inc. (CCC), 222 Rosewood Drive, Danvers, MA 01923, 978-750-8400. For works that are not available on CCC please contact mpkbookspermissions@tandf.co.uk

Trademark notice: Product or corporate names may be trademarks or registered trademarks and are used only for identification and explanation without intent to infringe.

ISBN: 978-1-032-19836-1 (hbk)
ISBN: 978-1-032-19838-5 (pbk)
ISBN: 978-1-003-26109-4 (ebk)

DOI: 10.1201/9781003261094

Typeset in Times
by SPi Technologies India Pvt Ltd (Straive)

Contents

Preface xi
Acknowledgments xvii
Authors xix

Chapter 1   Introduction to Distribution Systems 1

           1.1   The Distribution System 2
           1.2   Distribution Substations 2
           1.3   Radial Feeders 4
           1.4   Distribution Feeder Map 6
           1.5   Distribution Feeder Electrical Characteristics 6
           1.6   Summary 8
           References 8

Chapter 2   The Nature of Loads 9

           2.1   Definitions 9
           2.2   Individual Customer Load 10
                  2.2.1   Demand 11
                  2.2.2   Maximum Demand 11
                  2.2.3   Average Demand 11
                  2.2.4   Load Factor 12
           2.3   Distribution Transformer Loading 12
                  2.3.1   Diversified Demand 13
                  2.3.2   Maximum Diversified Demand 14
                  2.3.3   Load Duration Curve 14
                  2.3.4   Maximum Non-coincident Demand 14
                  2.3.5   Diversity Factor 15
                  2.3.6   Demand Factor 16
                  2.3.7   Utilization Factor 17
                  2.3.8   Load Diversity 17
           2.4   Feeder Load 17
                  2.4.1   Load Allocation 18
                         2.4.1.1   Application of Diversity Factors 18
                         2.4.1.2   Load Survey 18
                         2.4.1.3   Transformer Load Management 22
                         2.4.1.4   Metered Feeder Maximum Demand 22
                         2.4.1.5   What Method to Use? 24
                  2.4.2   Voltage Drop Calculations Using Allocated Loads 24
                         2.4.2.1   Application of Diversity Factors 24
                         2.4.2.2   Load Allocation Based Upon
                                     Transformer Ratings 24

v

	2.5	Individual Customer Loads with Electric Vehicles.................29
	2.6	Summary..31
	Problems...32
	References..36

Chapter 3 Balanced System Method of Analysis 37

    3.1 Line Impedance 37
    3.2 Voltage Drop...38
    3.3 The LIT...42
        3.3.1 Linear Network...42
        3.3.2 Non-linear Network..46
    3.4 Summary...52
    Problems..52
    References...53

Chapter 4 Series Impedance of Overhead and Underground Lines 55

    4.1 Series Impedance of Overhead Lines 55
        4.1.1 Transposed and Balanced Three-Phase Lines 56
        4.1.2 Untransposed Distribution Lines...............................56
        4.1.3 Carson's Equations..59
        4.1.4 Modified Carson's Equations 61
        4.1.5 Primitive Impedance Matrix for Overhead Lines........63
        4.1.6. Phase Impedance of Matrix for Overhead Lines........64
        4.1.7 Parallel Overhead Distribution Lines 71
    4.2 Series Impedance of Underground Lines 73
        4.2.1 Concentric Neutral Cable..74
        4.2.2 Tape Shielded Cables 79
        4.2.3 Parallel Underground Distribution Lines 83
    4.3 Summary...87
    Problems..87
    WindMil Assignment..92
    References...92

Chapter 5 Shunt Admittance of Overhead and Underground Lines 93

    5.1 General Voltage Drop Equation..94
    5.2 Overhead Lines...94
        5.2.1 The Shunt Admittance of Overhead Parallel Lines 99
    5.3 Concentric Neutral Cable Underground Lines 102
    5.4 Tape Shielded Cable Underground Lines 106
    5.5 The Shunt Admittance of Parallel Underground Lines..........107
    5.6 Summary...109
    Problems...109
    WindMil Assignment...110
    References..110

Chapter 6 Distribution System Line Models 111

6.1 Exact Line Segment Model 111
6.2 The Modified Line Model 120
    6.2.1 The Three-Wire Line 121
    6.2.2 The Computation of Neutral and Ground Currents.. 122
6.3 Source Impedances 126
6.4 The LIT... 128
6.5 The General Matrices for Parallel Lines............................. 131
    6.5.1 Physically Parallel Lines 135
    6.5.2 Electrically Parallel Lines 140
6.6 Summary... 145
Problems... 146
WindMil Assignment.. 150
References 150

Chapter 7 Voltage Regulation 151

7.1 Standard Voltage Ratings 151
7.2 Two-Winding Transformer Theory.................................... 153
7.3 Two-Winding Autotransformer.. 158
    7.3.1 Autotransformer Ratings 162
    7.3.2 Per-Unit Impedance.. 165
7.4 Step-Voltage Regulators 169
    7.4.1 Single-Phase, Step-Voltage Regulators 171
        7.4.1.1 Type A Step-Voltage Regulator 171
        7.4.1.2 Type B Step-Voltage Regulator 172
        7.4.1.3 Generalized Constants 175
        7.4.1.4 The Line-Drop Compensator.................... 175
    7.4.2 Three-Phase, Step-Voltage Regulators 182
        7.4.2.1 Wye-Connected Regulators 182
        7.4.2.2 Closed Delta-Connected Regulators......... 194
        7.4.2.3 Open Delta-Connected Regulators 197
7.5 Summary... 210
Problems... 211
WindMil Assignment.. 216
References 216

Chapter 8 Three-Phase Transformer Models.................................... 217

8.1 Introduction 217
8.2 Generalized Matrices... 218
8.3 The Delta-Grounded Wye Step-Down Connection 219
    8.3.1 Voltages 219
    8.3.2 Currents 224
8.4 The Delta-Grounded Wye Step-Up Connection 234

	8.5	The Ungrounded Wye-Delta Step-Down Connection	236
	8.6	The Ungrounded Wye-Delta Step-Up Connection	247
	8.7	The Grounded Wye-Delta Step-Down Connection	248
	8.8	Open Wye – Open Delta	255
	8.9	The Grounded Wye – Grounded Wye Connection	261
	8.10	The Delta – Delta Connection	264
	8.11	Open Delta – Open Delta	273
	8.12	Thevenin Equivalent Circuit	278
	8.13	Summary	281
	Problems		282
	WindMil Assignment		286

Chapter 9 Load Models 287

- 9.1 Wye-Connected Loads 287
  - 9.1.1 Constant Real and Reactive Power Loads 288
  - 9.1.2 Constant Impedance Loads 288
  - 9.1.3 Constant Current Loads 289
  - 9.1.4 Combination Loads 289
- 9.2 Delta-Connected Loads 293
  - 9.2.1 Constant Real and Reactive Power Loads 294
  - 9.2.2 Constant Impedance Loads 294
  - 9.2.3 Constant Current Loads 295
  - 9.2.4 Combination Loads 295
  - 9.2.5 Line Currents Serving a Delta-Connected Load ...295
- 9.3 Two-Phase and Single-Phase Loads 295
- 9.4 Shunt Capacitors 295
  - 9.4.1 Wye-Connected Capacitor Bank 296
  - 9.4.2 Delta-Connected Capacitor Bank 296
- 9.5 Three-Phase Induction Machine 297
  - 9.5.1 Induction Machine Model 298
  - 9.5.2 Symmetrical Component Analysis of a Motor 301
  - 9.5.3 Phase Analysis of an Induction Motor 306
  - 9.5.4 Voltage and Current Unbalance 314
  - 9.5.5 Motor Starting Current 314
  - 9.5.6 The Equivalent T Circuit 314
  - 9.5.7 Computation of Slip 321
  - 9.5.8 Induction Generator 325
  - 9.5.9 Induction Machine Thevenin Equivalent Circuit ...325
  - 9.5.10 The Ungrounded Wye – Delta Transformer Bank with an Induction Motor 329
- 9.6 Electric Vehicle (EV) Chargers 335
- 9.7 Summary 340
- Problems 340
- References 344

Contents ix

Chapter 10 Distribution Feeder Analysis...345
    10.1 Power-Flow Analysis..345
        10.1.1 General Feeder 345
        10.1.2 Uniformly Distributed Loads 346
        10.1.3 Series Feeder 349
        10.1.4 The Unbalanced Three-Phase Distribution Feeder....350
                10.1.4.1 Shunt Components......................................351
        10.1.5 Applying the Iterative Technique 351
        10.1.6 Let's Put It All Together 352
        10.1.7 Load Allocation 360
        10.1.8 Loop Flow 362
                10.1.8.1 Single-Phase Feeder....................................362
                10.1.8.2 IEEE 13 Bus Test Feeder 365
                10.1.8.3 Summary of Loop Flow 374
        10.1.9 Summary of Power-Flow Studies..............................375
    10.2 Short-Circuit Studies 375
        10.2.1 General Short-Circuit Theory....................................376
        10.2.2 Specific Short Circuits...379
        10.2.3 Back-Feed, Ground-Fault Currents 384
                10.2.3.1 One Downstream Transformer Bank 384
                10.2.3.2 Complete Three-Phase Circuit Analysis...387
                10.2.3.3 Back-Feed Currents Summary...................392
    10.3 Summary..397
    Problems...398
    WindMil Assignment...404
    References..407

Chapter 11 Center-Tapped Transformers and Secondaries..................................409
    11.1 Center-Tapped, Single-Phase Transformer Model 410
        11.1.1 Matrix Equations 413
        11.1.2 Center-Tapped Transformer Serving Loads
                through a Triplex Secondary 419
    11.2 Ungrounded Wye – Delta Transformer Bank with Center-
          Tapped Transformer..425
        11.2.1 Basic Transformer Equations 425
        11.2.2 Forward Sweep..427
        11.2.3 Backward Sweep 433
        11.2.4 Summary 434
    11.3 Open Wye – Open Delta Transformer Connections 441
        11.3.1 The Leading Open Wye – Open Delta Connection..442
        11.3.2 The Lagging Open Wye – Open Delta Connection..442
        11.3.3 Forward Sweep..443
        11.3.4 Backward Sweep 447
    11.4 Four-Wire Secondary..451

11.5 Putting it All Together 454
       11.5.1 Ungrounded Wye – Delta Connection 454
       11.5.2 Open Wye – Delta Connections 459
       11.5.3 Comparisons of Voltage and Current Unbalances 460
11.6 Summary 464
Problems 465
WindMil Assignment 467
References 467

Appendix A Conductor Data 469

Appendix B Concentric Neutral 15 kV Cable 473

Index 475

Preface

We are now in the beginning stages of the smart grid, and distribution systems are getting smarter as automation becomes more prevalent. This is possible due to smart meters, otherwise known as advanced metering infrastructure. The US Energy Information Administration reported that 102.9 million smart meters have been deployed by 2020 and that 88% of these smart meters are for residential customers in distribution circuits [1]. At the very start, we want to emphasize that this text is intended to only develop and demonstrate the computer models of all the physical components of a distribution system. As the text develops the component models it will become clear that this thing we called "load" is the weak link in the overall analysis of a distribution system. Smart meters give access to much more data than was traditionally captured; however, they still have not been utilized to their full capacity. At the present time, the only true information available for every customer remains energy in kilowatt-hours consumed during a specified period. This topic is addressed in Chapter 2. The problem with load is that it is constantly changing. Computer programs can and have been developed that will very accurately model the components, but without real load data, the results of the studies are only as good as the load data used. As the smart grid is developed, more accurate load data will become available, which will provide for a much more accurate analysis of the operating conditions of the distribution system. What needs to be emphasized is the smart grid must have computer programs that will very accurately model all the physical components of the system. The purpose of this text is to develop very accurate models of the physical components of a distribution system.

In the model developments, it is very important to accurately model the unbalanced nature of the components. Programs used in the modeling of a transmission system assume that the system is a balanced three-phase system. This makes it possible to model only one phase. That is not the case in the modeling of a distribution system. The unbalanced nature of the distribution system must be modeled. This requirement is made possible by modeling all three phases of every component of the distribution system.

The distribution system computer program for power-flow studies can be run to simulate present loading conditions and for long-range planning of new facilities. For example, the tools provide an opportunity for the distribution engineer to optimize capacitor placement to minimize power losses. Different switching scenarios for normal and emergency conditions can be simulated. Short-circuit studies provide the necessary data for the development of a reliable coordinated protection plan for fuses, recloser, and relay/circuit breakers.

So, what is the problem? Garbage in, garbage out is the answer. Armed with a commercially available computer program it is possible for the user to prepare incorrect data and as a result, the program outputs are not correct. Without an understanding of the models and a general "feel" for the operating characteristics of a distribution system, serious design errors and operational procedures may result. The user must fully understand the models and analysis techniques of the program. Without this knowledge the garbage in, garbage out problem becomes very real.

The purpose of this text is to present the reader a general overall feeling for the operating characteristics of a distribution system and the modeling of each component. Before using the computer program, it is extremely important for the student/engineer to have a "feel" for what the answers should be. Engineers once used a slide rule for engineering calculations prior to the advent of hand calculators. The beauty in using a slide rule was you were forced to know what the "ballpark" answer should be. We have lost that ability thanks to hand calculators and computers but understanding the ballpark answer is still a necessity.

It has been very interesting to receive many questions and comments about previous editions of the text from undergraduate and graduate students in addition to practicing engineers from around the world. That gets back to the need for the "feel" of the correct answer. New students need to study the early chapters of the book to develop this "feel". Practicing engineers will already have the "feel" and perhaps will not need the early chapters (1, 2, and 3). In developing the fifth edition of the book, we have retained most of the contents of Chapters 1, 5, 7, 8, and 11. We have added the concept of electric vehicles as loads and their impact on the grid. This is visited in Chapter 2 at a high level and in a more detailed analysis-based treatment again in Chapter 9. We modified Chapters 3, 6, and 10 to give the book a quicker and more direct treatment of the concept of iterative power flow. Lastly, in Chapter 4, we developed a new model for tape-shielded underground cables, which we believe to be more accurate than the previous model.

This textbook assumes that the reader has a basic understanding of transformers, electric machines, and transmission lines. At many universities, all these topics are crammed into a one-semester course. For that reason, a quick review of the needed theory is presented as needed.

There are many example problems throughout the text. These examples are intended to not only demonstrate the application of the models but also to teach a "feel" for what the answers should be. The example problems should be studied very carefully since they demonstrate the application of the theory just presented. Each chapter has a series of homework problems that will assist the student in applying the models and developing a better understanding of the operating characteristics of the component being modeled. Most of the example and homework problems are very number intensive. In previous versions, all the example problems had used a software package called "Mathcad" [2]. Since distribution engineers will soon need the skills in both computer programming and data science in addition to traditional electrical engineering, we have elected to convert all example problems to MATLAB [2]. All example code is provided for instructors, students, and practicing engineers. We have decided to keep the Mathcad iterative routines in the textbook, as Mathcad provides a fantastic visual flow of how a program works.

As more and more components are developed, and the feeder becomes more complicated, it becomes necessary to use a sophisticated distribution analysis program. Milsoft Utility Solutions has made a student version of "WindMil" [4] available along with a user's manual. The user's manual includes instructions and illustrations on how to get started using the program. Starting in Chapter 4, there is a WindMil assignment at the end of the homework problems. A very simple system utilizing all the major components of the system will evolve as each chapter assignment is completed. In Chapter 10, the data for a small system is given that will allow the student/engineer to match operating criteria. The student version of WindMil and the user's manual can be downloaded from the Milsoft Utility Solutions website homepage. The address is:

Milsoft Utility Solutions
P.O. Box 7526
Abilene, TX 79608
Email: support@milsoft.com
Home page: www.milsoft.com

Unfortunately, there is a tendency on the part of the student/engineer to believe the results of a computer program. While computer programs are wonderful tools, it is still the responsibility of the users to study the results and confirm whether the results make sense. That is a major concern and one that is addressed throughout the text.

Chapter 1 presents a quick overview of the major components of a distribution system. This is the only section in the text that will present the components inside a substation, along with two connection schemes. The importance of the distribution feeder map and the data required is presented.

Chapter 2 addresses the important question, What is the "load" on the system? This chapter defines the common terms associated with the load. In the past, there was limited knowledge of the load and many assumptions had to be made. With the coming of the smart grid, there will be ample real-time data to assist in defining the load for a given study. Even with better load data, there will still be a concern on whether the computer results make sense. Electric vehicles are described as loads in Chapter 2 as well.

Chapter 3 introduces voltage drop and introduces the iterative power-flow program. This is done by introducing the "ladder" (forward/backward sweep) iterative method used by many commercial programs. This chapter covers the simplified approach to distribution systems analysis, assuming that distribution systems are balanced like that of their transmission level counterpart.

The major requirement of a distribution system is to supply safe reliable energy to every customer at a voltage within the American National Standards Institute (ANSI) standard, which is addressed in Chapters 4 and 5. The major goal of planning is to simulate the distribution system under different conditions now and into the future and assure that all customer voltages are within the acceptable ANSI range. Since voltage drop is a major concern, it is extremely important that the impedances of the system components be as accurate as possible. In particular, the impedances of the overhead and underground distribution lines must be computed as accurately as possible. The importance of a detailed feeder map that includes the phase positions for both overhead and underground lines is emphasized.

Chapter 6 develops the models for overhead and underground lines using the impedances and admittance computed in earlier chapters. Included will be the "exact" model along with an approximate model. Introduced are the matrices required for the application of the ladder analysis method. Included in the chapter are methods of modeling parallel distribution lines.

Chapter 7 addresses the important concept of voltage regulation. How is it possible to maintain every customer's voltage within ANSI standards when the load is varying all the time? The step-voltage regulator is presented as one answer to the question. A model is developed for the application in the ladder technique.

Chapter 8 is one of the most important chapters in the text. Models for most three-phase (closed and open) transformer connections in use today are developed. Again, the models use matrices that are used in the ladder iterative technique. The importance of phasing once again is emphasized.

Chapter 9 develops the models for all types of loads on the system. A new term is introduced that helps define the types of static load models. The term is "ZIP". Most static models in a distribution system can be modeled as constant impedance (Z), constant current (I), constant complex power (P), or a combination of the three. These models are developed for wye and delta connections. A very important model developed is that of an induction machine. The induction motor is the workhorse of the power system and needs, once again, to be modeled as accurately as possible. Several new sections have been included in this chapter that develop models of the induction machine and associated transformer connection that are useful for power-flow and short-circuit studies. Induction generators are becoming a major source of distributed generation. Chapter 9 shows that an induction machine can be modeled either as a motor or a generator. Lastly, a thorough ZIP model is developed for Level 2 electric vehicle chargers. Electric vehicles are being deployed at a steady rate, and adding this energy requirement to the electrical sector from the previous fossil fuel transportation sector is sure to have an impact. Accurate electric vehicle modeling will be very important to future distribution engineers.

Chapter 10 puts everything in the text together for steady-state, power-flow, and short-circuit studies. The "ladder" iterative technique is introduced in Chapter 3 and then again in Chapter 6. This chapter goes into detail on the development of the ladder technique starting with the analysis of a linear ladder network that is introduced in most early circuit analysis courses. This moves onto the non-linear nature of the three-phase unbalanced distribution feeder. The ladder technique is used for power-flow studies. Introduced in this chapter is a method used for the analysis of short-circuit conditions on a feeder.

Chapter 11 introduces the center-tapped transformer that is used for providing the three-wire service to customers. Models for the various connections are introduced that are used in the ladder iterative technique and short-circuit analysis. The WindMil assignments at the end of Chapters 10 and 11 allow the student/engineer to build, study, and fix the operating characteristics of a small distribution feeder.

MATLAB® is a registered trademark of The Math Works, Inc. For product information, please contact:

The Math Works, Inc.
3 Apple Hill Drive
Natick, MA 01760-2098

Tel: 508-647-7000
Fax: 508-647-7001
E-mail: info@mathworks.com
Web: http://www.mathworks.com

REFERENCES

1. US EIA Frequently Asked Questions (FAQs): https://www.eia.gov/tools/faqs/faq.php?id=108&t=3
2. Mathcad: www.ptc.com
3. MATLAB: www.mathworks.com
4. WindMil: www.milsoft.com

Acknowledgments

Bill would like to thank the many students and engineers who have communicated with him via email their questions about some of the contents of the fourth edition. It has been a pleasure to work with these individuals in helping them to better understand some of the models and applications in the text. Since Bill is retired, it has been a real pleasure to have the opportunity to work with many graduate students working on their research involving distribution systems. Bill hopes that students and practicing engineers will continue to feel free to contact him at bjkersting@zianet.com.

As always, Bill would like to thank his wife, Joanne, who has been very supportive of him for over 50 years. She has been very patient with him as he worked on the fifth edition. Without a doubt, it has been a wonderful experience working with Bob on this fifth edition.

Bob would like to thank his better half Amanda, and their three children, Noah, Amelia, and Emmy for their incredible patience throughout the process of writing the fifth edition of this textbook. In addition, Bob would like to thank his home institution, the University of Pittsburgh, for its constant support in achieving his dreams and to thank the wonderful students at Pitt for keeping him motivated and inspired. Bob would like to also thank one of his mentors, Tom McDermott, for introducing him to power distribution engineering, which ended up being his life's passion. Finally, Bob would like to thank Bill Kersting for trusting him with the amazing work that he has dedicated his career to. Bob is very honored to be a part of this endeavor. Bob still actively works in teaching and research in power distribution engineering and can be contacted at rjk39@pitt.edu for questions or possible collaborations.

Special thanks to Wayne Carr the CEO of Milsoft Utility Solutions, Inc., for allowing us to make WindMil a major part of the third, fourth, and fifth editions. Thanks also to the many support engineers at Milsoft who have guided us in developing the special WindMil assignments.

Authors

William H. Kersting received his BSEE from New Mexico State University (NMSU), Las Cruces, New Mexico, and his MSEE from the Illinois Institute of Technology. Prior to attending graduate school and for a year after graduate school, he was employed by El Paso Electric Company as a distribution engineer. He joined the faculty at NMSU in 1962 and served as professor of electrical engineering and from 1968 the director of the Electric Utility Management Program until his retirement in 2002. He is currently a consultant for Milsoft Utility Solutions.

Professor Kersting is a life fellow of the Institute of Electrical and Electronics Engineers (IEEE). He received the NMSU Westhafer Award for Excellence in Teaching in 1977 and the Edison Electric Institutes' Power Engineering Education award in 1979. Professor Kersting has been an active member of the IEEE Power Engineering Education Committee and the Distribution Systems Analysis Subcommittee.

Robert J. Kerestes, PhD, is an assistant professor of electrical and computer engineering at the University of Pittsburgh's Swanson School of Engineering. Robert was born in the Mount Washington neighborhood of Pittsburgh, Pennsylvania. He got his BS (2010), his MS (2012), and his PhD (2014) from the University of Pittsburgh, all with a concentration in electric power systems. His areas of interest are in modeling and analysis of electric power distribution systems, smart grid technology, and the integration of distributed energy resources and electric vehicles. Robert has worked as a physical system simulation modeler for Emerson Process Management, working on electric power applications for Emerson's Ovation Embedded Simulator. Robert also served in the US Navy as an interior communications electrician from 1998 to 2002 on active duty and from 2002 to 2006 in the US Naval Reserves.

1 Introduction to Distribution Systems

The major components of an electric power system are shown in Figure 1.1.

Of these components, the distribution system has traditionally been characterized as the most unglamorous component. In the last half of the twentieth century, the design and operation of the generation and transmission components presented many challenges to practicing engineers and researchers. Power plants became larger and larger; transmission lines crisscrossed the land forming large, interconnected networks. The operation of the large, interconnected networks required the development of new analysis and operational techniques. Meanwhile, the distribution systems continued to deliver power to the end user's meter with little or no analysis. As a direct result, distribution systems were typically overdesigned.

Times have changed. It has become very important and necessary to operate a distribution system at its maximum capacity, distributed energy resources such as solar power and energy storage are becoming more common, and new loads such as electric vehicles continue to reshape distribution systems. Some of the questions that need to be answered are as follows:

1. What is the maximum capacity?
2. How do we determine this capacity?
3. What are the operating limits that must be satisfied?
4. What can be done to operate the distribution system within the operating limits?
5. What can be done to make the distribution system operate more efficiently?

All of these questions can be answered only if the distribution system can be modeled very accurately.

The purpose of this text is to develop accurate models for the major components of a distribution system. Once the models have been developed, analysis techniques for steady-state and short-circuit conditions will be developed.

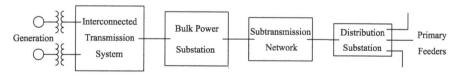

FIGURE 1.1 Major power system components.

1.1 THE DISTRIBUTION SYSTEM

The distribution system typically starts with the distribution substation that is fed by one or more subtransmission lines. In some cases, the distribution substation is fed directly from a high voltage transmission line, in which case, most likely, there is not a subtransmission system. This varies from company to company. Each distribution substation will serve one or more primary feeders. With the exception of dense metropolitan areas, the feeders are radial, which means that there is only one path for power to flow from the distribution substation to the user.

1.2 DISTRIBUTION SUBSTATIONS

A one-line diagram of a very simple distribution substation is shown in Figure 1.2.

Although Figure 1.2 displays the simplest of distribution substations, it does illustrate the major components that will be found in all substations.

1. High side and low side switching: In Figure 1.2, the high voltage switching is done with a simple switch. More extensive substations may use high voltage circuit breakers in a variety of high voltage bus designs. The low voltage switching in Figure 1.2 is accomplished with relay-controlled circuit breakers. In many cases, reclosers will be used in place of the relay/circuit breaker combination. Some substation designs will include a low voltage bus circuit breaker in addition to the circuit breakers for each feeder. As is the case with the high voltage bus, the low voltage bus can take on a variety of designs.
2. Voltage transformation: The primary function of a distribution substation is to reduce the voltage down to the distribution voltage level. In Figure 1.2, only one transformer is shown. Other substation designs will call for two or more three-phase transformers. The substation transformers can be

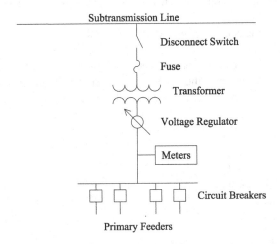

FIGURE 1.2 Simple distribution substation.

three-phase units or three single-phase units connected in a standard connection. There are many "standard" distribution voltage levels. Some of the common ones are 34.5 kV, 23.9 kV, 14.4 kV, 13.2 kV, 12.47 kV, and in older systems 4.16 kV.

3. Voltage regulation: As the load on the feeders varies, the voltage drop between the substation and the user will vary. To maintain the user's voltages within an acceptable range, the voltage at the substation needs to vary as the load varies. In Figure 1.2, the voltage is regulated by a "step-type" regulator that will vary the voltage plus or minus 10% on the low side bus. Sometimes this function is accomplished with a "load tap changing" (LTC) transformer. The LTC changes the taps on the low voltage windings of the transformer as the load varies. Many substation transformers will have "fixed taps" on the high voltage winding. These are used when the source voltage is always either above or below the nominal voltage. The fixed tap settings can vary the voltage plus or minus 5%. Many times, instead of a bus regulator, each feeder will have its own regulator. This can be in the form of a three-phase, gang-operated regulator or individual-phase regulators that operate independently.

4. Protection: The substation must be protected against the occurrence of short circuits. In the simple design of Figure 1.2, the only automatic protection against short circuits inside the substation is by way of the high side fuses on the transformer. As the substation designs become more complex, more extensive protective schemes will be employed to protect the transformer, the high and low voltage buses, and any other piece of equipment. Individual feeder circuit breakers or reclosers are used to provide interruption of short circuits that occur outside the substation. Protection has gotten more difficult due to distributed energy resources. Classic means of protection in many cases do not work for low fault currents supplied by these devices. One task for smart grid solutions is to employ alternative protection schemes [1].

5. Metering: Every substation has some form of metering. This may be as simple as an analog ammeter displaying the present value of substation current, as well as the minimum and maximum currents that have occurred over a specific period. Digital recording meters are becoming very common. These meters record the minimum, average, and maximum values of current, voltage, power, power factor, etc., over a specified time range. Typical time ranges are 15 minutes, 30 minutes, and one hour. The digital meters may monitor the output of each substation transformer and/or the output of each feeder.

A more comprehensive substation layout is shown in Figure 1.3.

The substation of Figure 1.3 has two LTC transformers, serves four distribution feeders, and is fed from two substransmission lines. Under normal conditions, the circuit breakers are in the following positions:

Circuit Breakers Closed: X, Y, 1, 3, 4, 6
Circuit Breakers Open: Z, 2, 5

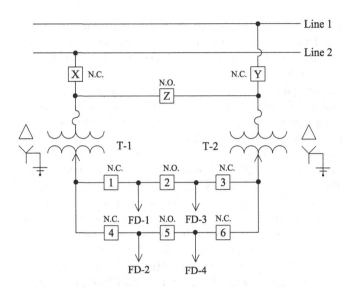

FIGURE 1.3 Two transformer substation with breaker-and-a-half scheme.

With the breakers in their normal positions, each transformer is served from a different subtransmission line and serves two feeders. Should one of the subtransmission lines go out of service, then breaker X or Y is opened and breaker Z is closed. Now both transformers are served from the same subtransmission line. The transformers are sized such that each transformer can supply all four feeders under an emergency operating condition. For example, if Transformer T-1 is out of service, then breakers X, 1, and 4 are opened and breakers 2 and 5 are closed. With that breaker arrangement, all four feeders are served by transformer T-2. The low voltage bus arrangement is referred to as a "breaker-and-a-half scheme" since three breakers are required to serve two feeders.

There are an unlimited number of substation configurations possible. It is up to the substation design engineer to create a design that provides the five basic functions and provides the most reliable service economically possible.

1.3 RADIAL FEEDERS

Radial distribution feeders are characterized by having only one path for power to flow from the source ("distribution substation") to each customer. A typical distribution system will consist of one or more distribution substations consisting of one or more "feeders". Components of the feeder may consist of the following:

1. Three-phase primary "main" feeder
2. Three-phase, two-phase ("V" phase) and single-phase laterals
3. Step-type voltage regulators
4. In-line transformers
5. Shunt capacitor banks
6. Distribution transformers

Introduction to Distribution Systems

7. Secondaries
8. Three-phase, two-phase, and single-phase loads

The loading of a distribution feeder is inherently unbalanced because of the large number of unequal single-phase loads that must be served. An additional unbalance is introduced by the non-equilateral conductor spacings of the three-phase overhead and underground line segments.

Because of the nature of the distribution system, conventional power-flow and short-circuit programs used for transmission system studies are not adequate. Such programs display poor convergence characteristics for radial systems. The programs also assume a perfectly balanced system so that a single-phase equivalent system is used.

If a distribution engineer is to be able to perform accurate power-flow and short-circuit studies, it is imperative that the distribution feeder be modeled as accurately as possible. This means that three-phase models of the major components must be utilized. Three-phase models for the major components will be developed in the following chapters. The models will be developed in the "phase frame" rather than applying the method of symmetrical components.

Figure 1.4 shows a simple "one-line" diagram of a three-phase feeder.

Figure 1.4 illustrates the major components of a distribution system. The connecting points of the components will be referred to as "nodes". Note in the figure that the phasing of the line segments is shown. This is important if the most accurate models are to be developed.

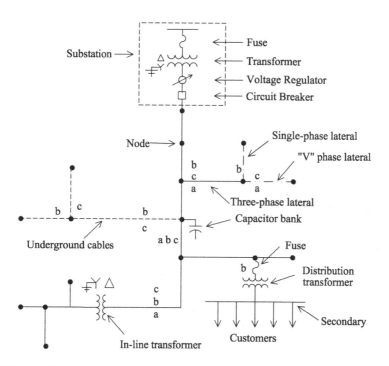

FIGURE 1.4 Simple distribution feeder.

1.4 DISTRIBUTION FEEDER MAP

The analysis of a distribution feeder is important to an engineer in order to determine the existing operating conditions of a feeder and to be able to play the "what if" scenarios of future changes to the feeder. Before the engineer can perform the analysis of a feeder, a detailed map of the feeder must be available. A sample of such a map is shown in Figure 1.5.

The map of Figure 1.5 contains most of the following information:

1. Lines (overhead and underground)
    a. Where
    b. Distances
    c. Details
        i. Conductor sizes (not on this map)
        ii. Phasing
2. Distribution transformers
    a. Location
    b. kVA rating
    c. Phase connection
3. In-line transformers
    a. Location
    b. kVA rating
    c. Connection
4. Shunt capacitors
    a. Location
    b. kvar rating
    c. Phase connection
5. Voltage regulators
    a. Location
    b. Phase connection
    c. Type (not shown on this map)
        i. Single phase
        ii. Three phase
6. Switches
    a. Location
    b. Normal open/close status

1.5 DISTRIBUTION FEEDER ELECTRICAL CHARACTERISTICS

Information from the map will define the physical location of the various devices. Electrical characteristics for each device will have to be determined before the analysis of the feeder can commence. In order to determine the electrical characteristics, the following data must be available:

1. Overhead and underground spacings
2. Conductor tables

Introduction to Distribution Systems

FIGURE 1.5 Institute of Electrical and Electronics Engineers (IEEE) 123 node test feeder.

 a. Geometric Mean Radius (GMR) (feet)
 b. Diameter (inches)
 c. Resistance (Ω/mile)

3. Voltage regulators
   a. Potential transformer ratios
   b. Current transformer ratios
   c. Compensator settings
      i. Voltage level
      ii. Bandwidth
      iii. R and X settings in volts
4. Transformers
   a. kVA rating
   b. Voltage ratings
   c. Impedance (R and X)
   d. No-load power loss

1.6 SUMMARY

As the smart grid [2] becomes a reality, it becomes increasingly more important to be able to accurately model and analyze each component of a distribution system. There are many different substation designs possible but, for the most part, the substation serves one or more radial feeders. Each component of a feeder must be modeled as accurately as possible for the analysis to have meaning. Sometimes the most difficult task for the engineer is to acquire all the necessary data. Feeder maps will contain most of the needed data. Additional data such as standard pole configurations, specific conductors used on each line segment, phasing, three-phase transformer connections, and voltage regulator settings must come from stored records. The remaining bits of information are the values of the loads. Chapter 2 will address the loads in a general sense. Again, as the smart grid, along with smart meters, become a reality, the load values will become much more accurate, which in turn will make the analysis more accurate. Once all the data has been acquired, the analysis can commence utilizing system models of the various devices that will be developed in later chapters.

REFERENCES

1. Carnovale, N., Fault Detection in Inverter-Based Microgrids Utilizing a Nonlinear Observer. Master's Thesis, University of Pittsburgh, Pittsburgh, PA, 2021.
2. Thomas, M.S. and McDonald, J. D., *Power System SCADA and Smart Grids*, CRC Press, Boca Raton, FL, 2015.

2 The Nature of Loads

The modeling and analysis of a power system depend upon the "load". What is load? The answer to that question depends upon what type of analysis is desired. For example, the steady-state analysis (power-flow study) of an interconnected transmission system will require a different definition of load than that used in the analysis of a secondary in a distribution feeder. The problem is that the "load" on a power system is constantly changing. The closer you are to the customer, the more pronounced will be the ever-changing load. There is no such thing as a "steady-state" load. To come to grips with load, it is first necessary to look at the "load" of an individual customer.

2.1 DEFINITIONS

The load that an individual customer or a group of customers presents to the distribution system is constantly changing. Every time a light bulb or an electrical appliance is switched on or off the load seen by the distribution feeder changes. To describe the changing load, the following terms are defined:

1. Demand
   - Load averaged over a specific period
   - Load can be kW, kvar, kVA, A
   - Must include the time interval
   - Example: The 15-minute kW demand is 100 kW
2. Maximum Demand
   - Greatest of all demands which occur over a specific period
   - Must include demand interval, period, and units
   - Example: The 15-minute maximum kW demand for the week was 150 kW
3. Average Demand
   - The average of the demands over a specified period (day, week, month, etc.)
   - Must include demand interval, period, and units
   - Example: The 15-minute average kW demand for the month was 350 kW
4. Diversified Demand
   - Sum of demands imposed by a group of loads over a particular period
   - Must include demand interval, period, and units
   - Example: The 15-minute diversified kW demand in the period ending at 9:30 was 200 kW

5. Maximum Diversified Demand
   - Maximum of the sum of the demands imposed by a group of loads over a particular period
   - Must include demand interval, period, and units
   - Example: The 15-minute maximum diversified kW demand for the week was 500 kW
6. Maximum Non-coincident Demand
   - For a group of loads, the sum of the individual maximum demands without any restriction that they occur at the same time
   - Must include demand interval, period, and units
   - Example: The maximum non-coincident 15-minute kW demand for the week was 700 kW
7. Demand Factor
   - Ratio of maximum demand to connected load
8. Utilization Factor
   - Ratio of the maximum demand to rated capacity
9. Load Factor
   - Ratio of the average demand of any individual load or a group of loads over a period to the maximum demand over the same period
10. Diversity Factor
    - Ratio of the "maximum non-coincident demand" to the "maximum diversified demand"
11. Load Diversity
    - Difference between "maximum non-coincident demand" and the "maximum diversified demand"

2.2 INDIVIDUAL CUSTOMER LOAD

Figure 2.1 illustrates how the instantaneous kW load of a customer changes for two 15-minute intervals.

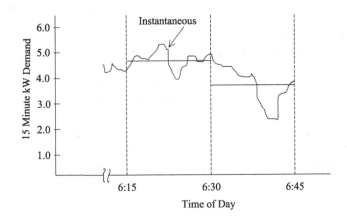

FIGURE 2.1 Customer demand curve.

The Nature of Loads

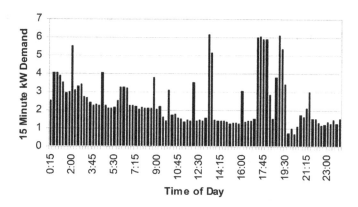

FIGURE 2.2 24-hour demand curve for Customer #1.

2.2.1 DEMAND

To define the load, the demand curve is broken into equal time intervals. In Figure 2.1, the selected time interval is 15 minutes. In each interval, the average value of the demand is determined. In Figure 2.1, the straight lines represent the average load in a time interval. The shorter the time interval, the more accurate will be the value of the load. This process is very similar to numerical integration. The average value of the load in an interval is defined as the **"15-minute kW demand"**.

The 24-hour, 15-minute kW demand curve for a customer is shown in Figure 2.2. This curve is developed from a spreadsheet that gives the 15-minute kW demand for a period of 24 hours.

2.2.2 MAXIMUM DEMAND

The demand curve shown in Figure 2.2 represents a typical residential customer. Each bar represents the "15-minute kW demand". Note that during the 24-hour period, there is a great variation in the demand. This customer has three periods in which the kW demand exceeds 6.0 kW. The greatest of these is the **"15-minute maximum kW demand"**. For this customer, the "15-minute maximum kW demand" occurs at 13:15 and has a value of 6.18 kW.

2.2.3 AVERAGE DEMAND

During the 24-hour period, energy (kWh) will be consumed. The energy in kWh used during each 15-minute time interval is computed by

$$kWh = (15 - \text{minute } kW \text{ demand}) \cdot \frac{1}{4} \cdot \text{hour} \qquad (2.1)$$

The total energy consumed during the day is then the summation of all of the 15-minute interval consumptions. From the spreadsheet, the total energy consumed during the period by Customer #1 is 58.96 kWh. The **"15-minute average kW demand"** is computed by

$$kW_{average} = \frac{Total\ Energy}{Hours} = \frac{58.96}{24} = 2.46\,kW \qquad (2.2)$$

2.2.4 LOAD FACTOR

"Load factor" is a term that is often referred to when describing a load. It is defined as the ratio of the average demand to the maximum demand. In many ways, load factor gives an indication of how well the utility's facilities are being utilized. From the utility's standpoint, the optimal load factor would be 1.00 since the system has to be designed to handle the maximum demand. Sometimes utility companies will encourage industrial customers to improve their load factor. One method of encouragement is to penalize the customer on the electric bill for having a low load factor.

For Customer #1 in Figure 2.2, the load factor is computed to be:

$$Load\ Factor = \frac{kW_{average}}{kW_{maximum}} = \frac{2.46}{6.18} = 0.40 \qquad (2.3)$$

2.3 DISTRIBUTION TRANSFORMER LOADING

A distribution transformer will provide service to one or more customers. Each customer will have a demand curve like that of Figure 2.2. However, the peaks and valleys and maximum demands will be different for each customer. Figures 2.3, 2.4, and 2.5 give the demand curves for the three additional customers connected to the same distribution transformer.

The load curves for the four customers show that each customer has its unique loading characteristic. The customers' individual maximum kW demand occurs at different times of the day. Customer #3 is the only customer who will have a high load factor. A summary of individual loads is given in Table 2.1.

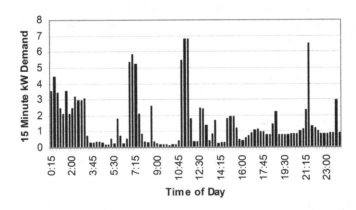

FIGURE 2.3 24-hour demand curve for Customer #2.

The Nature of Loads

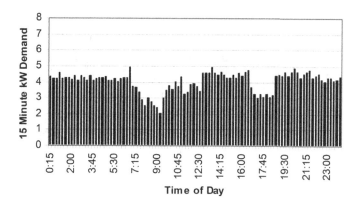

FIGURE 2.4 24-hour demand curve for Customer #3.

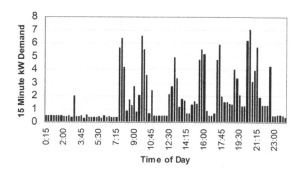

FIGURE 2.5 24-hour demand curve for Customer #4.

TABLE 2.1
Individual Customer Load Characteristics

	Cust. #1	Cust. #2	Cust. #3	Cust. #4
Energy Usage (kWh)	58.57	36.46	95.64	42.75
Maximum kW Demand	6.18	6.82	4.93	7.05
Time of Max. kW Demand	13:15	11:30	6:45	20:30
Average kW Demand	2.44	1.52	3.98	1.78
Load Factor	0.40	0.22	0.81	0.25

These four customers demonstrate that there is great diversity between their loads.

2.3.1 Diversified Demand

It is assumed that the same distribution transformer serves the four customers discussed previously. The sum of the four 15 kW demands for each time interval is the **"diversified demand"** for the group in that time interval, and in this case, the distribution transformer. The 15-minute diversified kW demand of the transformer for the

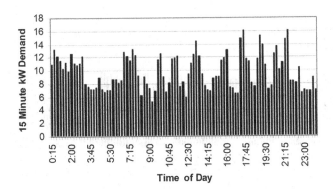

FIGURE 2.6 Transformer diversified demand curve.

day is shown in Figure 2.6. Note in this figure how the demand curve is beginning to smooth out. There are not as many significant changes as seen by some of the individual customer curves.

2.3.2 Maximum Diversified Demand

The transformer demand curve of Figure 2.6 demonstrates how the combined customer loads begin to smooth out the extreme changes of the individual loads. For the transformer, the 15-minute kW demand exceeds 16 kW twice. The greater of these is the **"15-minute maximum diversified kW demand"** of the transformer. It occurs at 17:30 and has a value of 16.16 kW. Note that this maximum demand does not occur at the same time as any one of the individual demands nor is this maximum demand the sum of the individual maximum demands.

2.3.3 Load Duration Curve

A **"load duration curve"** can be developed for the transformer serving the four customers. Sorting in descending order, the kW demand of the transformer develops the load duration curve shown in Figure 2.7.

The load duration curve plots the 15-minute kW demand versus the percent of time; the transformer operates at or above the specific kW demand. For example, the load duration curve shows the transformer operates with a 15-minute kW demand of 12 kW or greater than 22% of the time. This curve can be used to determine whether a transformer needs to be replaced due to an overloading condition.

2.3.4 Maximum Non-coincident Demand

The **"15-minute maximum non-coincident kW demand"** for the day is the sum of the individual customer's 15-minute maximum kW demands. For the transformer in question, the sum of the individual maximums is

$$kW_{maximum\ non-coincident\ demand} = 6.18 + 6.82 + 4.93 + 7.05 = 24.98\,\text{kW} \quad (2.4)$$

The Nature of Loads

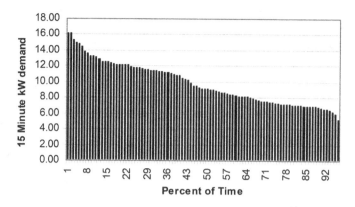

FIGURE 2.7 Transformer load duration curve.

2.3.5 DIVERSITY FACTOR

By definition, diversity factor is the ratio of the maximum non-coincident demand of a group of customers to the maximum diversified demand of the group. With reference to the transformer serving four customers, the diversity factor for the four customers would be

$$Diversity\ Factor = \frac{kW_{maximum\ non-coincident\ demand}}{kW_{maximum\ diversified\ demand}} = \frac{24.98}{16.15} = 1.5458 \quad (2.5)$$

The idea behind the diversity factor is that when the maximum demands of the customers are known, then the maximum diversified demand of a group of customers can be computed. There will be a different value of the diversity factor for different numbers of customers. The previously computed value would apply to four customers. If there are five customers, then a load survey would have to be set up to determine the diversity factor for five customers. This process would have to be repeated for all practical numbers of customers. Table 2.2 is an example of the diversity factors for the number of customers ranging from one up to 70. The table was developed from a different database than the four customers that have been discussed previously.

A graph of the diversity factors is shown in Figure 2.8.

Note in Table 2.2 and Figure 2.8 that the value of the diversity factor has basically leveled out when the number of customers has reached 70. This is an important observation because it means, at least for the system from which these diversity factors were determined, that the diversity factor will remain constant at 3.20 from 70 customers and up. In other words, as viewed from the substation, the maximum diversified demand of a feeder can be predicted by computing the total non-coincident maximum demand of all of the customers served by the feeder and dividing by 3.2.

TABLE 2.2
Diversity Factors

N	DF	N	DF	N	DF	N	DF	N	DF	N	DF	N	DF
1	1.0	11	2.67	21	2.90	31	3.05	41	3.13	51	3.15	61	3.18
2	1.60	12	2.70	22	2.92	32	3.06	42	3.13	52	3.15	62	3.18
3	1.80	13	2.74	23	2.94	33	3.08	43	3.14	53	3.16	63	3.18
4	2.10	14	2.78	24	2.96	34	3.09	44	3.14	54	3.16	64	3.19
5	2.20	15	2.80	25	2.98	35	3.10	45	3.14	55	3.16	65	3.19
6	2.30	16	2.82	26	3.00	36	3.10	46	3.14	56	3.17	66	3.19
7	2.40	17	2.84	27	3.01	37	3.11	47	3.15	57	3.17	67	3.19
8	2.55	18	2.86	28	3.02	38	3.12	48	3.15	58	3.17	68	3.19
9	2.60	19	2.88	29	3.04	39	3.12	49	3.15	59	3.18	69	3.20
10	2.65	20	2.90	30	3.05	40	3.13	50	3.15	60	3.18	70	3.20

Note: DF = diversity factors

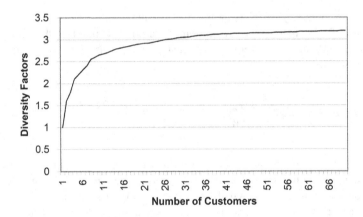

FIGURE 2.8 Diversity factors.

2.3.6 DEMAND FACTOR

The demand factor can be defined for an individual customer. For example, the 15-minute maximum kW demand of Customer #1 was found to be 6.18 kW. In order to determine the demand factor, the total connected load of the customer needs to be known. The total connected load will be the sum of the ratings of all of the electrical devices at the customer's location. Assume that this total comes to 35 kW, the demand factor is computed to be

$$\text{Demand Factor} = \frac{kW_{maximum\ demand}}{kW_{total\ connected\ load}} = \frac{6.18}{35} = 0.1766 \qquad (2.6)$$

The Nature of Loads

The demand factor gives an indication of the percentage of electrical devices that are on when the maximum demand occurs. The demand factor can be computed for an individual customer but not for a distribution transformer or the total feeder.

2.3.7 Utilization Factor

The utilization factor gives an indication of how well the capacity of an electrical device is being utilized. For example, the transformer serving the four loads is rated 15 kVA. Using the 16.16 kW maximum diversified demand and assuming a power factor of 0.9, the 15-minute maximum kVA demand on the transformer is computed by dividing the 16.16 kW maximum kW demand by the power factor and would be 17.96 kVA. The utilization factor is computed to be

$$Utilization\ Factor = \frac{kVA_{maximum\ demand}}{kVA_{transformer\ rating}} = \frac{17.96}{15} = 1.197 \quad (2.7)$$

2.3.8 Load Diversity

Load diversity is defined as the difference between the non-coincident maximum demand and the maximum diversified demand. For the transformer in question, the load diversity is computed to be

$$\begin{aligned} Load\ Diversity &= KW_{non-conincident\ demand} - KW_{maximum\ diversified\ demand} \\ &= 24.97 - 16.16 = 8.81\ KW \end{aligned} \quad (2.8)$$

2.4 FEEDER LOAD

The load that a feeder serves will display a smoothed-out demand curve as shown in Figure 2.9.

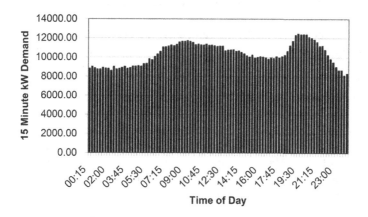

FIGURE 2.9 Feeder demand curve.

The feeder demand curve does not display any of the abrupt changes in demand of an individual customer demand curve or the semi-abrupt changes in the demand curve of a transformer. The simple explanation for this is that with several hundred customers served by the feeder, the odds are good that as one customer is turning off a light bulb another customer will be turning a light bulb on. The feeder load, therefore, does not experience a jump as would be seen in the individual customer's demand curve.

2.4.1 Load Allocation

In the analysis of a distribution feeder "load," data will have to be specified. The data provided will depend upon how detailed the feeder is to be modeled and the availability of customer load data. The most comprehensive model of a feeder will represent every distribution transformer. When this is the case, the load allocated to each transformer needs to be determined.

2.4.1.1 Application of Diversity Factors

The definition of the diversity factor is the ratio of the maximum non-coincident demand to the maximum diversified demand. A table of diversity factors is shown in Table 2.2. When such a table is available, then it is possible to determine the maximum diversified demand of a group of customers such as those served by a distribution transformer. That is, the maximum diversified demand can be computed by

$$kW_{maximum\ diversified\ demand} = \frac{kW_{maximum\ non-coincident\ demand}}{DF_{number\ of\ customers}} \quad (2.9)$$

This maximum diversified demand becomes the allocated "load" for the transformer.

2.4.1.2 Load Survey

Many times, the maximum demand of individual customers will be known either from metering or from a knowledge of the energy (kWh) consumed by the customer. Some utility companies will perform a load survey of similar customers in order to determine the relationship between the energy consumption in kWh and the maximum kW demand. Such a load survey requires the installation of a demand meter at each customer's location. The meter can be the same type as is used to develop the demand curves previously discussed, or it can be a simple meter that only records the maximum demand during the period. At the end of the survey period, the maximum demand vs. kWh for each customer can be plotted on a common graph. Linear regression is used to determine the equation of a straight line that gives the kW demand as a function of kWh. The plot of points for 15 customers, along with the resulting equation derived from a linear regression algorithm, is shown in Figure 2.10.

The straight-line equation derived is

$$kW_{maximum\ demand} = 0.1058 + 0.005014 \cdot kWh \quad (2.10)$$

The Nature of Loads

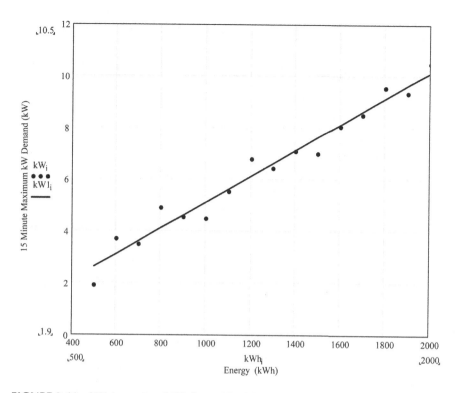

FIGURE 2.10 kW demand vs. kWh for residential customers.

Knowing the maximum demand for each customer is the first step in developing a table of diversity factors, as shown in Table 2.2. The next step is to perform a load survey where the maximum diversified demand of groups of customers is metered. This will involve selecting a series of locations where demand meters can be placed that will record the maximum demand for groups of customers ranging from at least two to 70. At each meter location, the maximum demand of all downstream customers must also be known. With that data, the diversity factor can be computed for the given number of downstream customers.

Example 2.1: A single-phase lateral provides service to three distribution transformers, as shown in Figure 2.11.

The energy in kWh consumed by each customer during a month is known. A load survey has been conducted for customers in this class, and it has been found that the customer 15-minute maximum kW demand is given by the equation

$$kW_{demand} = 0.2 + 0.008 \cdot kWh$$

The kWh consumed by Customer #1 is 1,523 kWh. The 15-minute maximum kW demand for Customer #1 is then computed as

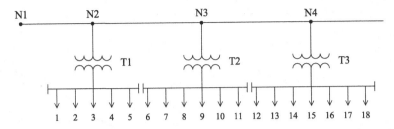

FIGURE 2.11 Single-phase lateral.

$$kW_1 = 0.2 + 0.008 \cdot 1523 = 12.4 \text{ kW}$$

The results of this calculation for the remainder of the customers are summarized next by transformer.

	Transformer T1				
Customer	#1	#2	#3	#4	#5
kWh	1,523	1,645	1,984	1,590	1,456
kW	12.4	13.4	16.1	12.9	11.9

	Transformer T2				
Customer	#6	#7	#8	#9	#10	#11
kWh	1,235	1,587	1,698	1,745	2,015	1,765
kW	10.1	12.9	13.8	14.2	16.3	14.3

	Transformer T3				
Customer	#12	#13	#14	#15	#16	#17	#18
kWh	2,098	1,856	2,058	2,265	2,135	1,985	2,103
kW	17.0	15.1	16.7	18.3	17.3	16.1	17.0

1. Determine for each transformer the 15-minute non-coincident maximum kW demand, and using the Table of Diversity Factors in Table 2.2, determine the 15-minute maximum diversified kW demand.

T1:
$$kW_{maximum\ non-coincident\ demand} = 12.4 + 13.4 + 16.1 + 12.9 + 11.8 = 66.6 \text{ kW}$$
$$kW_{maximum\ diversified\ demand} = \frac{kW_{maximum\ non-concident\ demand}}{DF_5} = 30.3 \text{ kW}$$

T2:
$$kW_{maximum\ non-coincident\ demand} = 10.1 + 12.9 + 13.8 + 14.2 + 16.3 + 14.3 = 81.6 \text{ kW}$$
$$kW_{maximum\ diversified\ demand} = \frac{kW_{maximum\ non-concident\ demand}}{DF_6} = 35.4 \text{ kW}$$

The Nature of Loads

$T3$:
$$kW_{maximum\ non-coincident\ demand} = 17.0 + 15.0 + 16.7 + 18.3 + 17.3 + 16.1 + 17.0$$
$$= 117.4\ kW$$
$$kW_{maximum\ diversified\ demand} = \frac{kW_{maximum\ non-concident\ demand}}{DF_7} = 48.9\ kW$$

Based upon the 15-minute maximum kW diversified demand on each transformer and an assumed power factor of 0.9, the 15-minute maximum kVA diversified demand on each transformer would be as follows:

$$kVA_{T1\ maximum\ diversified\ demand} = \frac{30.2}{.9} = 33.6\ kVA$$
$$kVA_{T2\ maximum\ diversified\ demand} = \frac{35.5}{.9} = 39.4\ kVA$$
$$kVA_{T3\ maximum\ diversified\ demand} = \frac{48.9}{.9} = 54.4\ kVA$$

The kVA ratings selected for the three transformers would be 25 kVA, 37.5 kV, and 50 kVA respectively. With those selections, only Transformer T1 would experience a significant maximum kVA demand greater than its rating (135%).

2. Determine the 15-minute non-coincident maximum kW demand and 15-minute maximum diversified kW demand for each of the line segments.

Segment N1 to N2: The maximum non-coincident kW demand is the sum of the maximum demands of all 18 customers.

$$kW_{maximum\ non-coincident\ demand} = 66.6 + 81.6 + 117.4 = 265.6\ kW$$

The maximum diversified kW demand is computed by using the diversity factor for 18 customers.

$$kW_{maximum\ diversified\ demand} = \frac{265.5}{2.86} = 92.8\ KW$$

Segment N2 to N3: This line segment "sees" 13 customers. The non-coincident maximum demand is the sum of Customers #6 through #18. The diversity factor for 13 (2.74) is used to compute the maximum diversified kW demand.

$$kW_{maximum\ non-coincident\ demand} = 81.6 + 117.4 = 199.0\ kW$$
$$kW_{maximum\ diversified\ demand} = \frac{199.0}{2.74} = 72.6\ kW$$

Segment N3 to N4: This line segment "sees" the same non-coincident demand and diversified demand as that of Transformer T3.

$$kW_{maximum\ non-coincident\ demand} = 117.4 = 117.4\ kW$$
$$kW_{maximum\ diversified\ demand} = \frac{199.0}{2.4} = 48.9\ kW$$

Example 2.1 demonstrates that Kirchhoff's Current Law (KCL) is not obeyed when the maximum diversified demands are used as the "load" flowing through the line segments and through the transformers. For example, at node N1, the maximum diversified demand flowing down the line segment N1–N2 is 92.8 kW and the maximum diversified demand flowing through Transformer T1 is 30.3 kW. KCL would then predict that the maximum diversified demand flowing down line segment N2–N3 would be the difference of these or 62.5 kW. However, the calculations for the maximum diversified demand in that segment were computed to be 72.6 kW. The explanation for this is that the maximum diversified demands for the line segments and transformers don't necessarily occur at the same time. At the time that the line segment N2–N3 is experiencing its maximum diversified demand, line segment N1–N2 and Transformer T1 are not at their maximum values. All that can be said is that at the time segment N2–N3 is experiencing its maximum diversified demand, the difference between the actual demand on the line segment N1–N2 and the demand of Transformer T1 will be 72.6 kW. There will be an infinite number of combinations of line flow down N1–N2 and through Transformer T1 that will produce the maximum diversified demand of 72.6 kW on line N2–N3.

2.4.1.3 Transformer Load Management

A transformer load management program is used by utilities to determine the loading on distribution transformers based upon a knowledge of the kWh supplied by the transformer during a peak loading month. The program is primarily used to determine when a distribution transformer needs to be changed out due to a projected overloading condition. The results of the program can also be used to allocate loads to transformers for feeder analysis purposes.

The transformer load management program relates the maximum diversified demand of a distribution transformer to the total kWh supplied by the transformer during a specific month. The usual relationship is the equation of a straight line. Such an equation is determined from a load survey. This type of load survey meters the maximum demand on the transformer in addition to the total energy in kWh of all of the customers connected to the transformer. With the information available from several sample transformers, a curve similar to that shown in Figure 2.10 can be developed, and the constants of the straight-line equation can then be computed. This method has the advantage because the utility will have in the billing database the kWh consumed by each customer every month. As long as the utility knows which customers are connected to each transformer, by using the developed equation, the maximum diversified demand (allocated load) on each transformer on a feeder can be determined for each billing period.

2.4.1.4 Metered Feeder Maximum Demand

The major disadvantage of allocating load using the diversity factors is that most utilities will not have a table of diversity factors. The process of developing such a table is generally not cost beneficial. The major disadvantage of the transformer load management method is that a database is required that specifies which transformers serve which customers. Again, this database is not always available.

Allocating load based upon the metered readings in the substation requires the least amount of data. Most feeders will have metering in the substation that will at

The Nature of Loads

minimum give either the total three-phase maximum diversified *kW* or *kVA* demand and/or the maximum current per phase during a month. The *kVA* ratings of all distribution transformers are always known as feeders. The metered readings can be allocated to each transformer based upon the transformer rating. An "allocation factor" (*AF*) can be determined based upon the metered three-phase *kW* or *kVA* demand and the total connected distribution transformer *kVA*.

$$AF = \frac{kVA_{metered\ demand}}{kVA_{total\ kVA\ rating}} \quad (2.11)$$

where $kVA_{total\ kVA\ rating}$ = Sum of the *kVA* ratings of all distribution transformers.
The allocated load per transformer is then determined by

$$kVA_{transformer\ demand} = AF \cdot kVA_{transformer\ rating} \quad (2.12)$$

The transformer demand will be either *kW* or *kVA* depending upon the metered quantity.

When the *kW* or *kVA* is metered by phase, then the load can be allocated by phase where it will be necessary to know the phasing of each distribution transformer.

When the maximum current per phase is metered, the load allocated to each distribution transformer can be done by assuming nominal voltage at the substation and then computing the resulting *kVA*. The load allocation will now follow the same procedure as outlined earlier.

If there is no metered information on the reactive power or power factor of the feeder, a power factor will have to be assumed for each transformer load.

Modern substations will have microprocessor-based metering that will provide kW, kvar, kVA, power factor, and current per phase. With this data, the reactive power can also be allocated. Since the metered data at the substation will include losses, an iterative process will have to be followed so that the allocated load plus losses will equal the metered readings.

Example 2.2: Assume that the metered maximum diversified kW demand for the system of Example 2.1 is 92.9 kW. Allocate this load according to the *kVA* ratings of the three transformers.

$$kVA_{total} = 25 + 37.5 + 50 = 112.5\ kVA$$

$$AF = \frac{92.9}{112.5} = 0.8258$$

The allocated *kW* for each transformer becomes

T1: $kW_1 = 0.8258 \cdot 25 = 20.64$ kW,
T2: $kW_2 = 0.8258 \cdot 37.5 = 30.97$ kW,
T3: $kW_3 = 0.8258 \cdot 50 = 41.29$ kW

2.4.1.5 What Method to Use?

Four different methods have been presented for allocating load to distribution transformers:

- Application of diversity factors
- Load survey
- Transformer load management
- Metered feeder maximum demand

Which method to use depends upon the purpose of the analysis. If the purpose of the analysis is to determine as closely as possible the maximum demand on a distribution transformer, then either the diversity factor or the transformer load management method can be used. Neither of these methods should be employed when the analysis of the total feeder is to be performed. The problem is that using either of those methods will result in a much larger maximum diversified demand at the substation than actually exists. When the total feeder is to be analyzed, the only method that gives good results is that of allocating load based upon the kVA ratings of the transformers.

2.4.2 Voltage Drop Calculations Using Allocated Loads

The voltage drops down line segments, and through distribution, transformers are of interest to the distribution engineer. Four different methods of allocating loads have been presented. The various voltage drops can be computed using the loads allocated by the three methods. For these studies, it is assumed that the allocated loads will be modeled as constant real power and reactive power.

2.4.2.1 Application of Diversity Factors

The loads allocated to a line segment or a distribution transformer using diversity factors are a function of the total number of customers "downstream" from the line segment or distribution transformer. The application of the diversity factors was demonstrated in Example 2.1. With a knowledge of the allocated loads flowing in the line segments and through the transformers and the impedances, the voltage drops can be computed. The assumption is that the allocated loads will be constant real power and reactive power. In order to avoid an iterative solution, the voltage at the source end is assumed and the voltage drops calculated from that point to the last transformer. Example 2.3 demonstrates how the method of load allocation using diversity factors is applied. The same system and allocated loads from Example 2.1 are used in Example 2.3.

2.4.2.2 Load Allocation Based Upon Transformer Ratings

When only the ratings of the distribution transformers are known, the feeder can be allocated based upon the metered demand and the transformer kVA ratings. This method was discussed in Section 2.3.3. Example 2.4 demonstrates this method.

The Nature of Loads

Example 2.3: For the system of Example 2.1, assume the voltage at N1 is 2,400 volts; compute the secondary voltages on the three transformers using the diversity factors.

The system of Example 2.1, including segment distances, is shown in Figure 2.12.

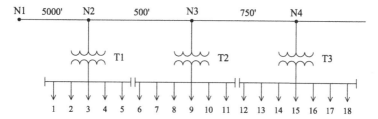

FIGURE 2.12 Single-phase lateral with distances.

Assume that the power factor of the loads is 0.9 lagging.
The impedance of the lines are $z = 0.3 + j0.6$ Ω/mile.
The ratings of the transformers are

T1: 25 kVA, 2400–240 volts, $Z = 1.8/\underline{40}\%$,
T2: 37.5 kVA, 2400–240 volts, $Z = 1.9/\underline{45}\%$,
T3: 50 kVA, 2400–240 volts, $Z = 2.0/\underline{50}\%$

From Example 2.1, the maximum diversified kW demands were computed. Using the 0.9 lagging power factor, the maximum diversified kW and kVA demands for the line segments and transformers are as follows:

Segment N1–N2:	$P_{12} = 92.9$ kW	$S_{12} = 92.9 + j45.0$ kVA
Segment N2–N3:	$P_{23} = 72.6$ kW	$S_{23} = 72.6 + j35.2$ kVA
Segment N3–N4:	$P_{34} = 49.0$ kW	$S_{34} = 49.0 + j23.7$ kVA
Transformer T1:	$P_{T1} = 30.3$ kW	$S_{T1} = 30.3 + j14.7$ kVA
Transformer T2:	$P_{T2} = 35.5$ kW	$S_{T2} = 35.5 + j17.2$ kVA
Transformer T3:	$P_{T3} = 49.0$ kW	$S_{T3} = 49.0 + j23.7$ kVA

Convert transformer impedances to ohms referred to the high voltage side.

T1: $$Z_{base1} = \frac{kV^2 \cdot 1000}{kVA_1} = \frac{2.4^2 \cdot 1000}{25} = 230.4 \; \Omega$$

$$Z_{T1} = (0.018/\underline{40}) \cdot 230.4 = 3.18 + j2.67 \; \Omega$$

T2: $$Z_{base2} = \frac{kV^2 \cdot 1000}{kVA_2} = \frac{2.4^2 \cdot 1000}{37.5} = 153.6 \, \Omega$$

$$Z_{T2} = (0.019 \underline{/45}) \cdot 153.6 = 2.06 + j2.06 \, \Omega$$

T3: $$Z_{base3} = \frac{kV^2 \cdot 1000}{kVA_3} = \frac{2.4^2 \cdot 1000}{50} = 115.2 \, \Omega$$

$$Z_{T3} = (0.02 \underline{/50}) \cdot 115.2 = 1.48 + j1.77 \, \Omega$$

Compute the line impedances.

N1 – N2: $$Z_{12}(0.3 + j0.6) \cdot \frac{5000}{5280} = 0.2841 + j0.5682 \, \Omega$$

N2 – N3: $$Z_{23}(0.3 + j0.6) \cdot \frac{500}{5280} = 0.0284 + j0.0568 \, \Omega$$

N3 – N4: $$Z_{34}(0.3 + j0.6) \cdot \frac{750}{5280} = 0.0426 + j0.0852 \, \Omega$$

Calculate the current flowing in segment N1 – N2.

$$I_{12} = \left(\frac{kW + jkvar}{kV}\right)^* = \left(\frac{92.9 + j45.0}{2.4 \underline{/0}}\right)^* = 43.0 \underline{/-25.84} \, A$$

Calculate the voltage at N2.

$$V_2 = V_1 - Z_{12} \cdot I_{12}$$

$$V_2 = 2400 \underline{/0} - (0.2841 + j0.5682) \cdot 43.0 \underline{/-25.84} = 2378.4 \underline{/-0.4} \, V$$

Calculate the current flowing into T1.

$$I_{T1} = \left(\frac{kW + jkvar}{kV}\right)^* = \left(\frac{30.3 + j14.7}{2.378 \underline{/-0.4}}\right)^* = 14.16 \underline{/-26.84} \, A$$

Calculate the secondary voltage referred to the high side.

$$V_{T1} = V_2 - Z_{T1} \cdot I_{T1}$$

$$V_{T1} = 2378.4 \underline{/-0.4} - (3.18 + j2.67) \cdot 14.16 \underline{/-26.24} = 2321.5 \underline{/-0.8} \, V$$

Compute the secondary voltage by dividing by the turns ratio of 10.

$$Vlow_{T1} = \frac{2321.5 / -0.8}{10} = 232.15 / -0.8 \text{ V}$$

Calculate the current flowing in line section N2–N3.

$$I_{23} = \left(\frac{kW + jkvar}{kV}\right)^* = \left(\frac{72.6 + j35.2}{2.378 / -0.4}\right)^* = 33.9 / -25.24 \text{ A}$$

Calculate the voltage at N3.

$$V_3 = V_2 - Z_{23} \cdot I_{23}$$

$$V_2 = 2378.4 / -0.4 - (0.0284 + j0.0568) \cdot 33.9 / -26.24 = 2376.7 / -0.4 \text{ V}$$

Calculate the current flowing into T2.

$$I_{T2} = \left(\frac{kW + jkvar}{kV}\right)^* = \left(\frac{35.5 + j17.2}{2.3767 / -0.4}\right)^* = 16.58 / -26.27 \text{ A}$$

Calculate the secondary voltage referred to the high side.

$$V_{T2} = V_3 - Z_{T2} \cdot I_{T2}$$

$$V_{T2} = 2376.7 / -0.4 - (2.06 + j2.06) \cdot 16.58 / -26.27 = 2331.1 / -0.8 \text{ V}$$

Compute the secondary voltage by dividing by the turns ratio of 10.

$$Vlow_{T2} = \frac{2331.1 / -0.8}{10} = 233.1 / -0.8 \text{ V}$$

Calculate the current flowing in line section N3–N4.

$$I_{34} = \left(\frac{kW + jkvar}{kV}\right)^* = \left(\frac{49.0 + j23.7}{2.3767 / -0.4}\right)^* = 22.9 / -25.27 \text{ A}$$

Calculate the voltage at N4.

$$V_4 = V_3 - Z_{34} \cdot I_{34}$$

$$V_4 = 2376.7 / -0.4 - (0.0426 + j0.0852) \cdot 22.9 / -26.27 = 2375.0 / -0.5 \text{ V}$$

The current flowing into T3 is the same as the current from N3 to N4.

$$I_{T3} = \left(\frac{kW + jkvar}{kV}\right)^* = \left(\frac{51.0 + j24.7}{2.375.0/-0.5}\right)^* = 23.91/\underline{-26.30}\,\text{A}$$

Calculate the secondary voltage referred to the high side.

$$V_{T3} = V_4 - Z_{T3} \cdot I_{T3}$$

$$V_{T3} = 2375.0/\underline{-0.5} - (1.48 + j1.77) \cdot 23.91/\underline{-26.30} = 2326.1/\underline{-1.0}\,\text{V}$$

Compute the secondary voltage by dividing by the turns ratio of 10.

$$Vlow_{T3} = \frac{2326.1/\underline{-1.0}}{10} = 232.6/\underline{-1.0}\,\text{V}$$

Calculate the percent voltage drop to the secondary of Transformer T3. Use the secondary voltage referred to the high side.

$$V_{drop} = \frac{|V_1| - |V_{T3}|}{|V_1|} \cdot 100 = \frac{2400 - 2326.1}{2400} \cdot 100 = 3.0789\%$$

Example 2.4: For the system of Example 2.1, assume the voltage at N1 is 2,400 volts, compute the secondary voltages on the three transformers allocating the loads based upon the transformer ratings. Assume that the metered kW demand at N1 is 92.9 kW.

The impedances of the line segments and transformers are the same as in Example 2.3.

Assume the load power factor is 0.9 lagging and compute the kVA demand at N1 from the metered demand.

$$S_{12} = \frac{92.9}{0.9}/\cos^{-1}(0.9) = 92.9 + j45.0 = 103.2/\underline{25.84}\,\text{kVA}$$

Calculate the AF.

$$AF = \frac{103.2/\underline{25.84}}{25 + 37.5 + 50} = 0.9175/\underline{25.84}$$

Allocate the loads to each transformer.

$$S_{T1} = AF \cdot kVA_{T1} = (0.9175/\underline{25.84}) \cdot 25 = 20.6 + j10.0\,\text{kVA}$$

$$S_{T2} = AF \cdot kVA_{T2} = (0.9175/\underline{25.84}) \cdot 37.5 = 31.0 + j15.0\,\text{kVA}$$

$$S_{T3} = AF \cdot kVA_{T1} = (0.9175 / \underline{25.84}) \cdot 50 = 41.3 + j20.0 \text{ kVA}$$

Calculate the line flows.

$$S_{12} = S_{T1} + S_{T2} + S_{T3} = 92.9 + j45.0 \text{ kVA}$$

$$S_{23} = S_{T2} + S_{T3} = 72.3 + j35.0 \text{ kVA}$$

$$S_{34} = S_{T3} = 41.3 + j20.0 \text{ kVA}$$

Using these values of line flows and flows into transformers, the procedure for computing the transformer secondary voltages is exactly the same as in Example 2.3. When this procedure is followed, the node and secondary transformer voltages are as follows:

$V_2 = 2378.1 / \underline{-0.4}$ V $Vlow_{T1} = 234.0 / \underline{-0.6}$ V

$V_3 = 2376.4 / \underline{-0.4}$ V $Vlow_{T2} = 233.7 / \underline{-0.6}$ V

$V_4 = 2374.9 / \underline{-0.5}$ V $Vlow_{T3} = 233.5 / \underline{-0.9}$ V

The percent voltage drop for this case is

$$V_{drop} = \frac{|V_1| - |V_{T3}|}{|V_1|} \cdot 100 = \frac{2400 - 2334.8}{2400} \cdot 100 = 2.7179\%$$

2.5 INDIVIDUAL CUSTOMER LOADS WITH ELECTRIC VEHICLES

Electric vehicles (EVs) are quickly becoming popular and are a considerable load on the grid compared with the remaining connected load for residential and commercial customers. Figure 2.2 showed a 24-hour demand curve for Customer #1. This section will show the effect of adding a Level 2 EV charger to this customer.

Chapter 9 explains EV chargers in detail and how they behave as a load from a circuit perspective, but in this section, they will be covered at a high level. Figure 2.13 shows the 24-hour demand curve for Customer #1 with the addition of a Level 2 EV charger.

Level 2 chargers commonly consume 9.6 kW of real power while charging. For Customer #1, the EV charger begins charging at 18:15 and charges for one hour until 19:15. To show the drastic difference of this load with and without the EV charger, Figure 2.2 is shown again in Figure 2.14 using the same scale as Figure 2.13.

This shifts the occurrence of the maximum kW demand from 13:15 to 19:00 and, in addition, changes the maximum kW demand from 6.18 kW to 15.73 kW. This is more than double the original maximum kW demand. In addition, this short charging period increased the energy consumed by 9.6 kWh and the average kW demand by 0.44 kW. The most significant impact, however, is on the load factor. The load factor

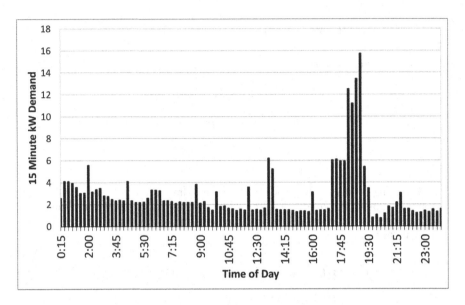

FIGURE 2.13 24-Hour demand curve for Customer #1 with EV charger.

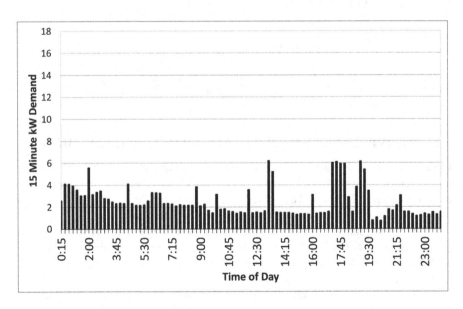

FIGURE 2.14 24-hour demand curve for Customer #1 without EV charger.

was severely reduced with the addition of the EV charger from 0.4 to 0.18. Lastly, the loading on the distribution transformer is a concern. It was previously calculated that the maximum diversified demand for Customers #1 through #4 was 16.16 kW, and those four customers were connected to a 15 kVA transformer. During the period of maximum diversified demand, the kVA load on the transformer was calculated to be

The Nature of Loads

17.96, which produced a transformer utilization factor of 1.197. This is slightly overloaded but tolerable for short periods of time. EV chargers that use smart charging assume a power factor of 0.95 lagging, so the total kVA load on the transformer for the new maximum diversified demand is

$$kVA_{maximum\ demand} = \frac{16.16}{0.9} + \frac{9.6}{0.95} = 28.06\ kVA.$$

This gives a new transformer utilization factor of

$$Utilization\ Factor = \frac{kVA_{maximum\ demand}}{kVA_{transformer\ rating}} = \frac{28.06}{15} = 1.871,$$

which is an unacceptable kVA load on the transformer, even for short periods of time. This would require the distribution engineer to intervene by replacing this transformer. In addition, it will be shown in Chapter 9 that the high current drawn by these EV chargers has a significant impact on the customer's service voltage, which also must be considered.

The impact that EV chargers have on the grid is significant; however, the cost of an "electrical gallon" or an eGallon as defined by the Department of Energy, is still much cheaper than a gallon of petroleum-based gasoline [1]. Due to the differential in cost of petroleum-based energy and electrical energy, and the continued electrification of traditionally non-electrified devices, the trend of EVs is likely to continue to grow. This will have to be addressed by distribution engineers to keep the grid stable and reliable and to keep voltages within limits.

2.6 SUMMARY

This chapter has demonstrated the nature of the loads on a distribution feeder. There is a great diversity between individual customer demands, but as the demand is monitored on line segments working back toward the substation, the effect of the diversity between demands becomes very slight. It was shown that the effect of diversity between customer demands must be taken into account when the demand on a distribution transformer is computed. The effect of diversity for short laterals can be taken into account in determining the maximum flow on the lateral. For the diversity factors of Table 2.2, it was shown that when the number of customers exceeds 70, the effect of diversity has pretty much disappeared. This is evidenced by the fact that the diversity factor has become almost constant as the number of customers approached 70. It must be understood that the number 70 will apply only to the diversity factors of Table 2.2. If a utility is going to use diversity factors, then that utility must perform a comprehensive load survey to develop the table of diversity factors that apply to that particular system.

Examples 2.3 and 2.4 show that the final node and transformer voltages are approximately the same. There is very little difference between the voltages when the loads were allocated using the diversity factors and when the loads were allocated based upon the transformer kVA ratings.

Section 2.5 showed that EV chargers add a significant load to the grid. The load due to just one EV charger is substantial enough to cause overloading of transformers, which could require replacement of the device.

PROBLEMS

2.1 The following are the 15-minute kW demands for four customers between the hours of 17:00 and 21:00. A 25 kVA single-phase transformer serves the four customers.

Time	Cust #1 KW	Cust #2 KW	Cust #3 KW	Cust #4 kW
17:00	8.81	4.96	11.04	1.44
17:15	2.12	3.16	7.04	1.62
17:30	9.48	7.08	7.68	2.46
17:45	7.16	5.08	6.08	0.84
18:00	6.04	3.12	4.32	1.12
18:15	9.88	6.56	5.12	2.24
18:30	4.68	6.88	6.56	1.12
18:45	5.12	3.84	8.48	2.24
19:00	10.44	4.44	4.12	1.12
19:15	3.72	8.52	3.68	0.96
19:30	8.72	4.52	0.32	2.56
19:45	10.84	2.92	3.04	1.28
20:00	6.96	2.08	2.72	1.92
20:15	6.62	1.48	3.24	1.12
20:30	7.04	2.33	4.16	1.76
20:45	6.69	1.89	4.96	2.72
21:00	1.88	1.64	4.32	2.41

(a) For each of the customers, determine
  (1) maximum 15-minute kW demand,
  (2) average 15-minute kW demand,
  (3) total kWh usage in the time period, and
  (4) load factor.
(b) For the 25 kVA transformer, determine
  (1) maximum 15-minute diversified demand,
  (2) maximum 15-minute non-coincident demand,
  (3) utilization factor (assume unity power factor),
  (4) diversity factor, and
  (5) load diversity.
(c) Plot the load duration curve for the transformer.

2.2 Two transformers each serving four customers are shown in Figure 2.15: The following table gives the time interval and kVA demand of the four customer demands during the peak load period of the year. Assume a power factor of 0.9 lagging.

The Nature of Loads

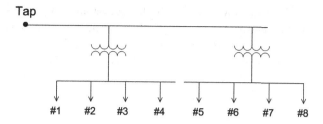

FIGURE 2.15 System for Problem 2.2.

Time	#1	#2	#3	#4	#5	#6	#7	#8
3:00–3:30	10	0	10	5	15	10	50	30
3:30–4:00	20	25	15	20	25	20	30	40
4:00–4:30	5	30	30	15	10	30	10	10
4:30–5:00	0	10	20	10	13	40	25	50
5:00–5:30	15	5	5	25	30	30	15	5
5:30–6:00	15	15	10	10	5	20	30	25
6:00–6:30	5	25	25	15	10	10	30	25
6:30–7:00	10	50	15	30	15	5	10	30

(a) For each transformer, determine the following:
  (1) 30-minute maximum *kVA* demand,
  (2) non-coincident maximum *kVA* demand,
  (3) load factor,
  (4) diversity factor,
  (5) suggested transformer rating (50, 75, 100, 167),
  (6) utilization factor, and
  (7) energy (kWh) during the four-hour period.
(b) Determine the maximum diversified 30-minute kVA demand at the "tap"

2.3 Two single-phase transformers serving 12 customers are shown in Figure 2.16.

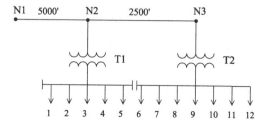

FIGURE 2.16 Circuit for Problem 2.3.

The 15-minute kW demands for the 12 customers between the hours of 5:00 p.m. and 9:00 p.m. are given on the next page. Assume a load power factor of 0.95 lagging. The impedance of the lines are $z = 0.306 + j0.6272$ Ω/mile. The voltage at node N1 is 2,500/0 volts.

Transformer ratings:

T1:	25 kVA	2400–240 volts	$Z_{pu} = 0.018/40$
T2:	37.5 kVA	2400–240 volts	$Z_{pu} = 0.020/50$

(a) Determine the maximum kW demand for each customer
(b) Determine the average kW demand for each customer
(c) Determine the kWH consumed by each customer in this time period
(d) Determine the load factor for each customer
(e) Determine the maximum diversified demand for each transformer
(f) Determine the maximum non-coincident demand for each transformer
(g) Determine the utilization factor (assume 1.0 power factor) for each transformer
(h) Determine the diversity factor of the load for each transformer
(i) Determine the maximum diversified demand at Node N1
(j) Compute the secondary voltage for each transformer taking diversity into account

Transformer #1– 25 kVA

Time	#1 KW	#2 KW	#3 KW	#4 KW	#5 kW
05:00	2.13	0.19	4.11	8.68	0.39
05:15	2.09	0.52	4.11	9.26	0.36
05:30	2.15	0.24	4.24	8.55	0.43
05:45	2.52	1.80	4.04	9.09	0.33
06:00	3.25	0.69	4.22	9.34	0.46
06:15	3.26	0.24	4.27	8.22	0.34
06:30	3.22	0.54	4.29	9.57	0.44
06:45	2.27	5.34	4.93	8.45	0.36
07:00	2.24	5.81	3.72	10.29	0.38
07:15	2.20	5.22	3.64	11.26	0.39
07:30	2.08	2.12	3.35	9.25	5.66
07:45	2.13	0.86	2.89	10.21	6.37
08:00	2.12	0.39	2.55	10.41	4.17
08:15	2.08	0.29	3.00	8.31	0.85
08:30	2.10	2.57	2.76	9.09	1.67
08:45	3.81	0.37	2.53	9.58	1.30
09:00	2.04	0.21	2.40	7.88	2.70

Transformer #2 – 37.5 kVA

Time	#6 KW	#7 KW	#8 KW	#9 KW	#10 KW	#11 KW	#12 kW
05:00	0.87	2.75	0.63	8.73	0.48	9.62	2.55
05:15	0.91	5.35	1.62	0.19	0.40	7.98	1.72
05:30	1.56	13.39	0.19	5.72	0.70	8.72	2.25
05:45	0.97	13.38	0.05	3.28	0.42	8.82	2.38
06:00	0.76	13.23	1.51	1.26	3.01	7.47	1.73
06:15	1.10	13.48	0.05	7.99	4.92	11.60	2.42
06:30	0.79	2.94	0.66	0.22	3.58	11.78	2.24
06:45	0.60	2.78	0.52	8.97	6.58	8.83	1.74
07:00	0.60	2.89	1.80	0.11	7.96	9.21	2.18
07:15	0.87	2.75	0.07	7.93	6.80	7.65	1.98
07:30	0.47	2.60	0.16	1.07	7.42	7.78	2.19
07:45	0.72	2.71	0.12	1.35	8.99	6.27	2.63
08:00	1.00	3.04	1.39	6.51	8.98	10.92	1.59
08:15	0.47	1.65	0.46	0.18	7.99	5.60	1.81
08:30	0.44	2.16	0.53	2.24	8.01	7.74	2.13
08:45	0.95	0.88	0.56	0.11	7.75	11.72	1.63
09:00	0.79	1.58	1.36	0.95	8.19	12.23	1.68

2.4 On a different day, the metered 15-minute kW demand at node N1 for the system of Problem 2.3 is 72.43 kW. Assume a power factor of 0.95 lagging. Allocate the metered demand to each transformer based upon the transformer kVA rating. Assume the loads are constant current and compute the secondary voltage for each transformer.

2.5 A single-phase lateral serves four transformers as shown in Figure 2.17.

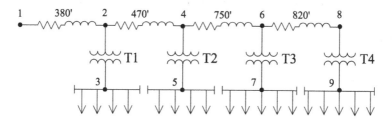

FIGURE 2.17 System for Problem 2.5.

Assume that each customer's maximum demand is 15.5 kW + $j7.5$ kvar. The impedance of the single-phase lateral is $z = 0.4421 + j0.3213$ Ω/1000 ft. The four transformers are rated as follows:

$T1$ and $T2$: 37.5 kVA, 2,400–240 volts, $Z = 0.01 + j0.03$ per unit
$T3$ and $T4$: 50 kVA, 2,400–240 volts, $Z = 0.015 + j0.035$ per unit

Use the diversity factors found in Table 2.2 to complete the following:

(a) Determine the 15-minute maximum diversified kW and kvar demands on each transformer.

(a) Determine the 15-minute maximum diversified kW and kvar demands for each line section.

(b) If the voltage at Node 1 is 2,600/$\underline{0}$ volts, determine the voltage at Nodes 2, 3, 4, 5, 6, 7, 8, 9. In calculating the voltages, take into account diversity using the answers from (1) and (b) above.

(c) Use the 15-minute maximum diversified demands at the lateral tap (section 1–2) from part (b) above. Divide these maximum demands by 18 (number of customers) and assign that as the "instantaneous load" for each customer. Now calculate the voltages at all of the nodes listed in part (c) using the instantaneous loads.

(d) Repeat part (d) above except assume the loads are "constant current". To do this, take the current flowing from Node 1 to Node 2 from part (d) above, divide by 18 (number of customers), and assign that as the "instantaneous constant current load" for each customer. Again, calculate all of the voltages.

(e) Take the maximum diversified demand from Node 1 to Node 2 and "allocate" that out to each of the four transformers based upon their kVA ratings. To do this, take the maximum diversified demand and divide it by 175 (total kVA of the four transformers). Now multiply each transformer kVA rating by that number to give how much of the total diversified demand is being served by each transformer. Again, calculate all of the voltages.

(f) Compute the percent differences in the voltages for parts (d), (e), and (f) at each of the nodes using the part (c) answer as the base.

REFERENCES

1. "eGallon". Accessed on Dec. 8, 2021. [Online]. Available: https://www.energy.gov/maps/egallon.

3 Balanced System Method of Analysis

A distribution feeder provides service to unbalanced three-phase, two-phase, and single-phase loads over untransposed three-phase, two-phase and single-phase line segments. This combination leads to the three-phase line currents and the line voltages being unbalanced. To analyze these conditions as precisely as possible, it will be necessary to model all three phases of the feeder as accurately as possible. However, many times only a "ballpark" answer is needed. When this is the case, assuming balanced three-phase loads and transposed lines, methods of modeling and analysis can be employed. It is the purpose of this chapter to develop methods of analysis assuming balanced loads and transposed lines. In later chapters, the exact models and analysis of a distribution system will be developed. The approximate methods of modeling and analysis will assume perfectly balanced three-phase loads and transposed three-phase lines. With these assumptions, a single line-to-neutral equivalent circuit for the feeder can be used.

3.1 LINE IMPEDANCE

For the approximate modeling of a three-phase line segment, it will be assumed that the line segment is transposed even though that is not the case in distribution systems. However, with line transposed assumption, only the equivalent impedance of the line segment needs to be determined. A typical three-phase line configuration is shown in Figure 3.1.

The equation for the equivalent line impedance for the configuration shown in Figure 3.1 is given by [3]

$$z_{line} = r + j0.12134 \cdot \ln\left(\frac{D_{eq}}{GMR}\right) \Omega/\text{mile}, \tag{3.1}$$

where r = conductor resistance (from tables) in Ω/mile,

$$D_{eq} = \sqrt[3]{D_{ab} \cdot D_{bc} \cdot D_{ca}} \text{ ft}, \tag{3.2}$$

GMR = conductor geometric mean radius in feet from tables.

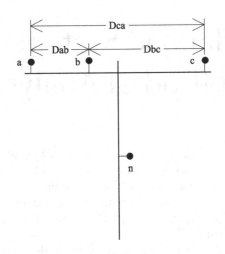

FIGURE 3.1 Three-phase line configuration.

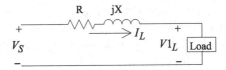

FIGURE 3.2 Line-to-neutral equivalent circuit.

3.2 VOLTAGE DROP

A line-to-neutral equivalent circuit of a three-phase line segment serving a balanced three-phase load is shown in Figure 3.2. In this figure, the "load" is equal to the three-phase load divided by three.

Kirchhoff's Voltage Law (KVL) applied to the circuit of Figure 3.2 gives

$$V_S = V_L + (R + jX) \cdot I_L = V_L + R \cdot I_L + jX \cdot I_L \tag{3.3}$$

The phasor diagram for Equation 3.3 is shown in Figure 3.3 where the voltages are line to neutral.

In Figure 3.3, the phasor for the voltage drop through the line resistance (RI_L) is shown in the phase with the current phasor, and the phasor for the voltage drop through the reactance (jXI_L) is shown leading the current phasor by 90 degrees. The

FIGURE 3.3 Phasor diagram.

Balanced System Method of Analysis

voltage drop is defined as the difference between the magnitudes of the source and the load voltages.

$$V_{drop} = \frac{\|V_S| - |V1_L\|}{|V_S|} \cdot 100\% \tag{3.4}$$

The angle between the source voltage and the load voltage (δ) is very small.

Example 3.1: A three-phase line segment has the configuration, as shown in Figure 3.1. The equation for the line impedance for the configuration shown in Figure 3.1 is given in Equations 3.3 and 3.4.

The spacings between conductors are

$D_{ab} = 2.5$ ft., $D_{bc} = 4.5$ ft., $D_{ca} = 7.0$ ft

The conductors of the line are 336,400 26/7 ACSR (Aluminum Conductor Steel Reinforced).
Determine the impedance of the line in ohms/mile:

Solution:

From the table of conductor data in Appendix A.

$r = 0.306$ Ω/mile
$GMR = 0.0244$ ft

Compute the equivalent spacing.

$$D_{eq} = \sqrt[3]{2.5 \cdot 4.5 \cdot 7.0} = 4.2863 \, \text{ft}$$

Using Equation 3.4,

$$z_{line} = 0.306 + j0.12134 \cdot \ln\left(\frac{4.2863}{0.0244}\right) = 0.306 + j0.6272 \, \Omega/\text{mile}$$

Remember that the line impedance computed in Example 3.1 is for a transposed three-phase line.
The impedance of the line will be used for development of the approximate results. The assumption of transposition will be made, and the calculation of the line impedances of a typical three-phase distribution line will be developed in Chapter 4.

Example 3.2: The length of the line in Figure 3.2 is 5,000 feet. The transposed impedance of the line was computed in Example 3.1. The line-to-neutral voltage source is 7,200 volts, and it is serving a load impedance of 8 + j6 Ohms per phase at the end of the line. Compute the load line-to-neutral voltage and percent voltage drop.

Length = 5000 ft E = 7200 V Z_{load} = 8 + j6 Ω

Calculate the total line impedance.

$$Z_{line} = z \cdot \frac{5000}{5280} = 0.2893 + j0.5939 \ \Omega$$

Compute the total current.

$$I_L = \frac{E}{Z_{line} + Z_{load}} = 679.73 \underline{/-38.5} \ A$$

Compute the load voltage.

$$V_L = E - Z_{line} \cdot I_L = 6797.3 \underline{/-1.63} \ V$$

Compute the percent voltage drop.

$$Drop_{percent} = \frac{|E| - |V_L|}{|E|} \cdot 100 = 5.59\%$$

The system of Example 3.2 is linear since there is a direct relationship between the load voltage and the load current, as shown in Equation 3.3.

$$V_L = Z_{load} \cdot I_L \tag{3.5}$$

Suppose now that the load is changed to a fixed value of constant three-phase kVA and power factor ($S_L = P_L + jQ_L$). The relationship between the load voltage and load current is now as follows:

$$S_{3-phase} = 15,000 \ kVA$$

$$S_L = \frac{S_{3-phase}}{3} = 5000 \ kVA$$

$$I_L = \left(\frac{S_L \cdot 1000}{V_L}\right)^* \tag{3.6}$$

No longer is the load current directly proportional to the load voltage, thus leading to a non-linear system.

Balanced System Method of Analysis

Example 3.3: The load per phase of Example 3.2 is changed to the following:

$$S_L = 5000 \text{ kVA at } 0.85 \text{ lagging power factor}$$

$$S_L = 4250 + j2633.9 \text{ kVA}$$

The problem is the load current must follow Equation 3.6, so it is impossible to compute the load current without knowing the load voltage. To start the analysis, it is assumed that the load current is zero. By setting the current to zero, the voltages at all nodes are computed using the forward sweep. At the start, the load voltage at Node 4 is as follows:

$$I_S = I_L$$

$$I_L = 0$$

$$V_{Lold} = E - Z_{line} \cdot I_S = 7200\underline{/0} \text{ V}$$

With the computed voltage, it is possible to compute the load current.

$$I_L = \left(\frac{S_L \cdot 1000}{V_L}\right)^* = 694.4\underline{/-31.8} \text{ A}$$

Work the problem backward (backward sweep) to compute the new source current.

$$I_S = I_L = 694.4\underline{/-31.8} \text{ A}$$

The specified source voltage and the new current is used to go forward and again calculate a new value of load voltage. This is called the "forward sweep".

$$V_{Lnew} = E - Z_{line} \cdot I_L = 6816.1\underline{/-2.06} \text{ V}$$

This voltage is compared to the previous computed voltage.

$$Error = \frac{\||V_{Lnew}| - |V_{Lold}|\|}{|E|} = 0.0533,$$

where V_{Lold} = load voltage from a previous iteration.

If the error is greater than a specified tolerance (0.0001), then the new load current is the current provided by the source. The forward sweep using the new load current and given source voltage computes a new load voltage that leads to a new load current. The process of forward and backward sweeps continue until the error is less than the specified tolerance of 0.0001. After five iterations, the load voltage is computed to be

$$V_L = 6783.2 /\!\!-\!2.07 \text{ V}$$

The error is computed to be

$$\text{Error} = 0.00002$$

Since the error is less than the tolerance, another backward sweep is not needed.

The final load current is

$$I_L = 737.1 /\!\!-\!33.9 \text{ A}$$

With these values of load voltage and currents computed, the equivalent single-phase load is

$$5,000 \text{ at } 0.85 \text{ power factor}$$

The problem with the non-linear system is the process of manually doing the forward and backward sweeps. A method known as the ladder iterative technique (LIT) can be applied to a non-linear system.

3.3 THE LIT

Because a distribution feeder is radial, iterative techniques commonly used in transmission network power-flow studies are not used because of poor convergence characteristics [1]. The "LIT" specifically designed for a radial system can be used [2].

3.3.1 Linear Network

A distribution feeder is non-linear because most loads are assumed to be constant kW and kvar. However, the approach taken for a linear system can be modified to account for the non-linear characteristics of the distribution feeder. Figure 3.4 shows a linear ladder network.

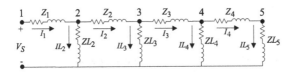

FIGURE 3.4 Linear ladder network.

Balanced System Method of Analysis

The linear ladder network assumes that the line impedances and load impedances are known along with the voltage (V_S) at the source. The solution for this network is to calculate the voltage at every node under a no-load condition. With no-load currents, the computed voltage at Node 5 will equal that of the specified voltage at the source. The "backward sweep" commences by computing the load current at Node 5. The load current I_5 is

$$IL_5 = \frac{V_5}{ZL_5} \tag{3.7}$$

For this "end node" case, the line current I_4 is equal to the load current IL_5. KCL calculates the upstream node voltage. At each node, the load current is computed using the computed node voltage. The load current at Node 4 is then

$$IL_4 = \frac{V_4}{ZL_4} \tag{3.8}$$

With the load current (IL_4) computed, KCL is used to determine the line current.

$$I_3 = IL_4 + I_4 \tag{3.9}$$

This continues until a voltage V_1 has been computed at the source. The computed voltage V_1 is compared to the specified voltage V_S. There will be a difference between these two voltages. The ratio of the specified voltage to the computed voltage can be determined as

$$Ratio = \frac{V_S}{V_1} \tag{3.10}$$

Since the network is linear, the line and load currents and node voltages in the network can be multiplied by the ratio for the final solution to the network.

Example 3.4: The system of Figure 3.4 has the following line and load impedances:

$$[Z_{line}] = \begin{bmatrix} 0.1739 + j0.3564 \\ 0.1449 + j0.3970 \\ 0.1159 + j0.2376 \\ 0.2028 + j0.4158 \end{bmatrix} \Omega \quad [ZL] = \begin{bmatrix} 99999 \\ 75 + j45 \\ 50 + j22.5 \\ 45 + j25 \\ 80 + j42.5 \end{bmatrix} \Omega$$

The source voltage for the system is $V_1 = V_S = 7200/\underline{0}$ volts.
Compute:

- Node voltages
- Load currents
- Line currents

Initially, it is assumed that the line currents are zero, so the voltages at all nodes are equal to the source voltage. A simple computer routine (Figure 3.5) is developed for computing the required voltages and currents:

$$X := \begin{vmatrix} I \leftarrow \text{start} \\ V_5 \leftarrow 7200 \\ \text{for } i \in 5..2 \\ \quad \begin{vmatrix} IL_i \leftarrow \dfrac{V_i}{ZL_i} \\ I_{i-1} \leftarrow I_i + IL_i \\ V_{i-1} \leftarrow V_i + Z_{i-1} \cdot I_{i-1} \end{vmatrix} \\ Out_1 \leftarrow V \\ Out_2 \leftarrow IL \\ Out_3 \leftarrow I \\ Out \end{vmatrix}$$

FIGURE 3.5 Computer routine.

When the computer program is run, the computed voltage at the source is

$$V_1 = 7512.2326/\underline{1.75}\ \text{V} \qquad (3.11)$$

Since the system is linear, the ratio of the given source voltage to the computed source voltage is

$$Ratio = \dfrac{V_S}{V_1} = 0.9580 - j0.0292 \qquad (3.12)$$

All voltages and currents are multiplied by the ratio to give the final linear values:

Balanced System Method of Analysis

$$V_{node} = \begin{bmatrix} 7200.00/0 \\ 7064.51/-0.77 \\ 6974.84/-1.31 \\ 6929.27/-1.57 \\ 6900.73/-1.75 \end{bmatrix} \quad I_{line} = \begin{bmatrix} 418.38/-29.13 \\ 337.71/-28.51 \\ 210.77/-30.30 \\ 76.18/-29.73 \\ 0 \end{bmatrix} \quad IL = \begin{bmatrix} 0 \\ 80.77/-31.73 \\ 127.21/-25.54 \\ 134.61/-30.63 \\ 76.18/-29.73 \end{bmatrix} \quad (3.13)$$

The results can be computed again using the voltage at Node 5, as shown in Equation 3.13. The new computer routine is shown in Figure 3.6 with the only change being the initial value of the Node 5 voltage.

$$X := \begin{vmatrix} I \leftarrow \text{start} \\ V_5 \leftarrow 6900.73 \cdot e^{-j \cdot 1.75 \cdot \text{deg}} \\ \text{for } i \in 5..2 \\ \quad \begin{vmatrix} IL_i \leftarrow \dfrac{V_i}{ZL_i} \\ I_{i-1} \leftarrow I_i + IL_i \\ V_{i-1} \leftarrow V_i + Z_{i-1} \cdot I_{i-1} \end{vmatrix} \\ Out_1 \leftarrow V \\ Out_2 \leftarrow IL \\ Out_3 \leftarrow I \\ Out \end{vmatrix}$$

FIGURE 3.6 Revised computer routine.

When this is done, the same results as shown in Equation 3.13 are computed.
The percent voltage drop to the last node is of interest. The percent voltage drop computed is

$$V_{drop} = \frac{\||V_1| - |V_5|\|}{|V_1|} \cdot 100 = 5.11\% \quad (3.14)$$

Most distribution systems limit the percentage voltage drop to 3%, so this case needs to be worked on to keep the voltage drop less than 3%. The application of shunt capacitors used for the purpose of limiting voltage drop will be addressed after the non-linear example.

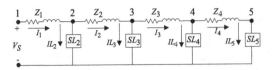

FIGURE 3.7 Non-linear ladder network.

3.3.2 NON-LINEAR NETWORK

The linear network of Figure 3.4 is modified to a non-linear network by replacing the constant load impedances by constant complex power loads, as shown in Figure 3.7.

As with the linear network, the initial "forward sweep" computes the voltage at all nodes, assuming the line currents are zero. Since the initial currents are zero, the Node 5 (end node) voltage will equal that of the specified source voltage. After the forward sweep, the load current at each node is computed using the initial node voltages.

$$I_n = \left(\frac{S_n}{V_n}\right)^* \qquad (3.15)$$

The "backward sweep" will determine the load current at each node and the resulting line current feeding each node. As in the linear case, this first "iteration" will produce a voltage that is not equal to the specified source voltage (V_S). Because the network is non-linear, multiplying currents and voltages by the ratio of the specified voltage to the computed voltage will not give the solution. The most direct modification using the LIT is to perform a "forward sweep". The "forward sweep" commences by using the specified source voltage and the line currents from the previous "backward" sweep. For example, KVL is used to compute the voltage at Node 2 by

$$V_2 = V_S - Z_1 \cdot I_1 \qquad (3.16)$$

An "error" between the present V_{new} and old V_{old} at all nodes is computed. If the maximum node error is greater than a specified tolerance, a backward sweep is started that will lead to "new" line currents. These currents and the original source voltage are used for the new "forward sweep". Convergence is achieved when the phase voltages at all nodes satisfy:

$$Error_n = \frac{\left||V_n| - |V_{n-1}|\right|}{V_{nominal}} \leq \text{specified tolerance,} \qquad (3.17)$$

where n = iteration.

The system of Example 3.3 is used, and a program using the ladder method is shown in Figure 3.5.

Balanced System Method of Analysis

Example 3.5: The non-linear system of Figure 3.5 is the same as in Example 3.4, with the only difference being the loads are modeled as constant power (P) and reactive power (Q). The non-lineal circuit is shown in Figure 3.8.

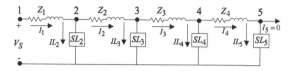

FIGURE 3.8 Non-linear circuit.

The node equivalent single-phase loads are

$$[SL] = \begin{bmatrix} SL_1 \\ SL_2 \\ SL_3 \\ SL_4 \\ SL_5 \end{bmatrix} = \begin{bmatrix} 0 \\ 500 + j300 \\ 825 + j400 \\ 800 + j500 \\ 475 + j275 \end{bmatrix} \text{kVA.} \quad (3.18)$$

To compute the node voltages for a non-linear system, the LIT is used. The forward sweep computes voltages at each node using the most recent values of line currents. The backward sweep uses the new voltages to calculate the load currents, and then KCL is used to compute each of the line currents. A computer routine listing is shown in Figure 3.9.

$$Tol = .0001 \quad V_S = 7200 \text{ volts } start = \begin{bmatrix} 0 \\ 0 \\ 0 \\ 0 \\ 0 \end{bmatrix}$$

The computer routine shown in Figure 3.9 stops when the maximum error is less than the specified tolerance. In the routine, the word "break" really means to stop the program.

The results are as follows:
Iterations = 4

$$[V_{node}] = \begin{bmatrix} 7200/\underline{0} \\ 7057.8/\underline{-0.8} \\ 6962.9/\underline{-1.3} \\ 6914.8/\underline{-1.5} \\ 6884.1/\underline{-1.7} \end{bmatrix} [IL] = \begin{bmatrix} 0 \\ 82.6/\underline{-31.7} \\ 131.7/\underline{-27.1} \\ 136.4/\underline{-33.5} \\ 79.7/\underline{-31.8} \end{bmatrix} [I] = \begin{bmatrix} 430.0/\underline{-30.9} \\ 347.4/\underline{-30.7} \\ 216.1/\underline{-32.9} \\ 79.7/\underline{-31.8} \\ 0 \end{bmatrix} \quad (3.19)$$

$$\text{Tol} = .0001 \quad V_S = 7200 \text{ volts} \quad \text{start} = \begin{bmatrix} 0 \\ 0 \\ 0 \\ 0 \\ 0 \end{bmatrix}$$

$$Y := \begin{vmatrix} I \leftarrow \text{start} \\ V_{old} \leftarrow \text{start} \\ V_1 \leftarrow V_S \\ \text{for } n \in 1..20 \\ \quad \begin{vmatrix} \text{for } j \in 2..5 \\ \quad \begin{vmatrix} V_j \leftarrow V_{j-1} - Z_{j-1} \cdot I_{j-1} \\ \text{Error}_j \leftarrow \dfrac{\left| |V_j| - |V_{old_j}| \right|}{|V_S|} \\ V_{old_j} \leftarrow V_j \end{vmatrix} \\ \text{Emax} \leftarrow \max(\text{Error}) \\ \text{break if Emax} < \text{Tol} \\ \text{for } i \in 5..2 \\ \quad \begin{vmatrix} IL_i \leftarrow \dfrac{\overline{SL_i \cdot 1000}}{V_i} \\ I_{i-1} \leftarrow IL_i + I_i \end{vmatrix} \end{vmatrix} \\ \text{Out}_1 \leftarrow V \\ \text{Out}_2 \leftarrow IL \\ \text{Out}_3 \leftarrow I \\ \text{Out}_4 \leftarrow n \\ \text{Out} \end{vmatrix}$$

FIGURE 3.9 Non-linear system.

In Example 3.3, it was shown that when the load is specified as constant real and reactive power (PQ), the forward/backward sweeps are required to compute the results. In the example, the LIT was recommended for performing the forward/backward sweeps. The load was initially a constant Z per phase but then was changed to a constant phase PQ load, which made the system non-linear. With the ladder technique presented, Example 3.5 with the constant PQ load is modeled in Example 3.6.

Balanced System Method of Analysis

Example 3.6: Use the LIT to solve the problem from Example 3.3 with the load converted to a constant PQ. The computer routine applying the LIT is shown in Figure 3.10.

$$\text{Tol} := .0001 \quad E := 7200 \quad S_L := 5000 \cdot e^{j \cdot \text{acos}(.85)} \quad S_L = 4250 + 2633.9134j$$

$$X := \begin{vmatrix} I_L \leftarrow 0 \\ V_{old} \leftarrow 0 \\ \text{for } n \in 1..10 \\ \quad \begin{vmatrix} V_L \leftarrow E - Z_{line} \cdot I_L \\ \text{Error} \leftarrow \dfrac{||V_L| - |V_{old}||}{E} \\ \text{break if Error} < \text{Tol} \\ I_L \leftarrow \dfrac{\overline{S_L \cdot 1000}}{\overline{V_L}} \\ V_{old} \leftarrow V_L \end{vmatrix} \\ \text{Out}_1 \leftarrow n \\ \text{Out}_2 \leftarrow V_L \\ \text{Out}_3 \leftarrow I_L \\ \text{Out}_4 \leftarrow \text{Error} \\ \text{Out} \end{vmatrix}$$

FIGURE 3.10 Example 3.6 ladder program.

When the program has been run, the results are as follows:

$$V_l = 6783.9/-2.1 \text{ V}$$

$$I_l = 737.0/-33.8 \text{ A}$$

$$\text{Error} = 0.00002$$

$$V_{drop} = 5.79\%$$

$$S_{source} = 4407 + j2956 \text{ kVA} \quad (3.20)$$

The results using the ladder routine match those in the original Example 3.5 where the forward/backward sweeps were computed one at a time. The voltage drop is greater than 3%, which is the maximum accepted voltage drop. A common method of reducing the voltage drop is the application of a shunt capacitor. An approximate method for the selection of the capacitor bank size is to select a capacitor bank that will provide approximately 80.5% of the total reactive power of the system. The circuit with the capacitor bank is shown in Figure 3.11.

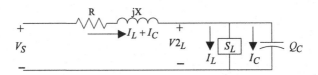

FIGURE 3.11 Line with load and capacitor bank.

The phasor diagram showing the load voltages with and without the capacitor bank are in Figure 3.12. In the phasor diagram, $V1_L$ is without the capacitor band and $V2_L$ with the capacitor bank. The phasor diagram shows that with the capacitor bank installed, the load voltage is larger than without the capacitor bank.

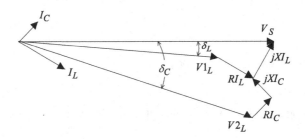

FIGURE 3.12 Phasor diagram with capacitor bank.

With the capacitor bank added to the system, the voltage at the load is given by

$$V2_L = E - Z_{line} \cdot (I_L + I_C) \qquad (3.21)$$

The load voltage will increase because of the shunt capacitor. For this case, the reactive part of the source is $j2956$ kvar. A capacitor bank rated at $-j2400$ kvar is used in Example 3.7. The value was selected to be approximately 80% of the inductive reactive part of the source.

Balanced System Method of Analysis

Example 3.7: With the capacitor bank added, the computer routine using the ladder method is modified to that shown in Figure 3.13.

Example 3.7 Tol := .0001 E := 7200 Z_{line} := 0.2893 + 0.5929j

S_{load} := 5000·$e^{j \cdot acos(.85)}$ S_{load} = 4250 + 2633.9134j S_C := −j·2400

$$X := \begin{vmatrix} I_L \leftarrow 0 \\ I_C \leftarrow 0 \\ V_{old} \leftarrow 0 \\ \text{for } n \in 1..10 \\ \quad \begin{vmatrix} V_L \leftarrow E - Z_{line} \cdot (I_L + I_C) \\ \text{Error} \leftarrow \dfrac{||V_L| - |V_{old}||}{E} \\ \text{break if Error} < \text{Tol} \\ I_L \leftarrow \dfrac{\overline{S_{load}} \cdot 1000}{\overline{V_L}} \\ I_C \leftarrow \dfrac{\overline{S_C} \cdot 1000}{\overline{V_L}} \\ V_{old} \leftarrow V_L \end{vmatrix} \\ I_{line} \leftarrow I_L + I_C \\ Out_1 \leftarrow n \\ Out_2 \leftarrow V_L \\ Out_3 \leftarrow I_L \\ Out_4 \leftarrow I_C \\ Out_5 \leftarrow I_{line} \\ Out_6 \leftarrow \text{Error} \\ Out \end{vmatrix}$$

FIGURE 3.13 Capacitor bank added.

The results with the capacitor bank are as follows:

$$V_L = 6996.0/\underline{-2.8} \text{ V}$$

$$I_L = 714.6/\underline{-34.7} \text{ A}$$

$$I_C = 343.0/\underline{87.2} \text{ A}$$

$$I_{line} = 608.4/\underline{-5.9} \text{ A}$$

$$\text{Error} = 0.00008$$

$$V_{drop} = 2.83\% \tag{3.22}$$

3.4 SUMMARY

This chapter has presented methods of analysis for balanced three-phase systems with transposed lines. The purpose is to give an idea of what the approximate solutions will be when a three-phase distribution system without transposed lines is analyzed. Examples demonstrated how the balanced system is analyzed. The most important feature of the chapter has been the introduction of the LIT. This technique will be used throughout the remainder of the text.

PROBLEMS

3.1 Shown in Figure 3.14 is the pole configuration for a three-phase primary feeder. The conductors are 266,800 26/7 ACSR. Assume the line is transposed. Alter the MATLAB script M0301.m to perform the following calculations.

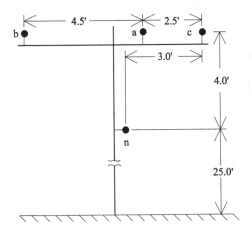

FIGURE 3.14 Transposed pole configuration.

(a) Determine the equivalent phase spacing (D_{eq})
(b) Compute the series line impedance in ohms/mile

3.2 The line in Problem 3.1 serves the primary feeder shown in Figure 3.15.

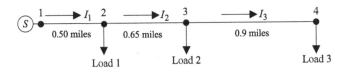

FIGURE 3.15 Primary feeder.

The source line-to-line voltage is 12.47 kV.
The loads are constant impedances of

Load 1: $ZL1 = 93.3 + j45.2$,
Load 2: $ZL2 = 36.7 + j22.8$,
Load 3: $ZL3 = 54.7 + j18.0$

Alter the MATLAB script M0304.m to compute
(a) node voltages
(b) load currents
(c) line currents
(d) voltage drop

3.3 The loads in Figure 3.15 are changed to the following constant PQ.

$SL1 = 0$, $SL2 = 450 + j218$, $SL3 = 1020 + j632$, $SL4 = 855 + j281$

Alter the MATLAB script M0305.m to compute the node voltages, load currents, line currents, and percent voltage drop to from Node 1 to Node 4.

3.4 In Problem 3.3, the voltage drop exceeds 3%. Determine the kvar rating of a three-phase capacitor bank to be installed at Node 3 to limit the total voltage drop to approximately 3%. The size of the capacitor bank must be composed of 100 kvar units.

REFERENCES

1. Trevino, C., Cases of difficult convergence in load-flow problems, *IEEE Paper n.71-62-PWR*, Presented at the IEEE Summer Power Meeting, Los Angeles, 1970.
2. Kersting, W. H. and Mendive, D. L., An application of ladder network theory to the solution of three-phase radial load-flow problems, *IEEE Conference Paper*, Presented at the IEEE Winter Power Meeting, New York, January 1976.
3. Glover, J.D., Overbye T, and Sarma, M., *Power System Analysis and Design*, PWS Publishing Co., Boston, MA, 5th Edition, 2011.

4 Series Impedance of Overhead and Underground lines

The determination of the series impedance for overhead and underground lines is a critical step before the analysis of a distribution feeder can begin. The series impedance of a single-phase, two-phase (V-phase), or three-phase distribution line consists of the resistance of the conductors and the self and mutual inductive reactances resulting from the magnetic fields surrounding the conductors. The resistance component for the conductors will typically come from a table of conductor data such as found in Appendix A.

4.1 SERIES IMPEDANCE OF OVERHEAD LINES

The inductive reactance (self and mutual) component of the impedance is a function of the total magnetic fields surrounding a conductor. Figure 4.1 shows conductors 1 through n with the magnetic flux lines created by currents flowing in each of the conductors.

The currents in all conductors are assumed to be flowing out of the page. It is further assumed that the sum of the currents will add to zero. That is,

$$I_1 + I_2 + \cdots I_i + \cdots I_n = 0 \tag{4.1}$$

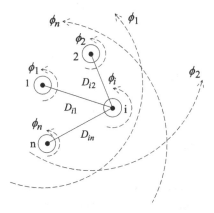

FIGURE 4.1 Magnetic fields.

The total flux linking conductor i is given by

$$\lambda_i = 2 \cdot 10^{-7} \cdot \left(I_1 \cdot \ln\frac{1}{D_{i1}} + I_2 \cdot \ln\frac{1}{D_{i2}} + \ldots + I_i \cdot \ln\frac{1}{GMR_i} + \ldots + I_n \cdot \ln\frac{1}{D_{in}} \right) \text{W} - \text{T/m}, \tag{4.2}$$

where
D_{in} = Distance between conductor i and conductor n (ft),
GMR_i = Geometric mean radius of conductor i (ft).

The inductance of conductor i consists of the "self-inductance" of conductor i and the "mutual inductance" between conductor i and all of the other $n - 1$ conductors. By definition,

$$Self - Inductance: L_{ii} = \frac{\lambda_{ii}}{I_i} = 2 \cdot 10^{-7} \cdot \ln\frac{1}{GMR_i} \text{ H/m}, \tag{4.3}$$

$$Mutual\ Inductance: L_{in} = \frac{\lambda_{in}}{I_n} = 2 \cdot 10^{-7} \cdot \ln\frac{1}{D_{in}} \text{ H/m} \tag{4.4}$$

4.1.1 Transposed and Balanced Three-Phase Lines

Chapter 3 was devoted to the analysis of transposed lines and balanced three-phase loads. In that chapter, the equation to calculate the impedance of the transposed line was given. It is assumed that a three-phase transposed line has each phase occupying each physical position for one-third the length of the line. Assuming a frequency of 60 Hz, the equation for calculating the impedance of the transposed line was given as

$$z_{line} = r + j0.12134 \cdot \ln\left(\frac{D_{eq}}{GMR}\right) \Omega/\text{mile}, \tag{4.5}$$

where r = conductor resistance (from tables) in Ω/mile,

$$D_{eq} = \sqrt[3]{D_{ab} \cdot D_{bc} \cdot D_{ca}}\ (\text{ft}), \tag{4.6}$$

GMR = conductor geometric mean radius (from tables) in (ft.).

4.1.2 Untransposed Distribution Lines

Because distribution systems consist of single-phase, two-phase, and untransposed three-phase lines serving unbalanced loads, it is necessary to retain the identity of the self and mutual impedance terms of the conductors in addition to including the ground return path for the unbalanced currents. The resistance of the conductors is taken directly from a table of conductor data. Equations 4.3 and 4.4 are used to

Series Impedance of Overhead and Underground lines

compute the self and mutual inductive reactances of the conductors. The inductive reactance is assumed to be at a frequency of 60 Hz, and the length of the conductor will be assumed to be one mile. With those assumptions, the self and mutual impedances are given by

$$\bar{z}_{ii} = r_i + j0.12134 \cdot \ln\frac{1}{GMR_i} \; \Omega/\text{mile}, \quad (4.7)$$

$$\bar{z}_{ij} = j0.12134 \cdot \ln\frac{1}{D_{ij}} \; \Omega/\text{mile} \quad (4.8)$$

In 1926, John Carson published a paper in which he developed a set of equations for computing the self and mutual impedances of lines, including the return path of current through the ground [2]. Carson's approach was to represent a line with the conductors connected to a source at one end and grounded at the remote end. Figure 4.2 illustrates a line consisting of two conductors (i and j) carrying currents (I_i and I_j), with the remote ends of the conductors tied to the ground. A fictitious "dirt" conductor carrying current (I_d) is used to represent the return path for the currents.

In Figure 4.2, Kirchhoff's Voltage Law (KVL) is used to write the equation for the voltage between conductor i and ground.

$$V_{ig} = \bar{z}_{ii} \cdot I_i + \bar{z}_{ij} \cdot I_j + \bar{z}_{id} \cdot I_d - \left(\bar{z}_{dd} \cdot I_d + \bar{z}_{di} \cdot I_i + \bar{z}_{dj} \cdot I_j\right) \quad (4.9)$$

Collecting terms in Equation 4.9 yields:

$$V_{ig} = \left(\bar{z}_{ii} - \bar{z}_{di}\right) \cdot I_i + \left(\bar{z}_{ij} - \bar{z}_{dj}\right) \cdot I_j + \left(\bar{z}_{id} - \bar{z}_{dd}\right) \cdot I_d \quad (4.10)$$

From KCL:

$$I_i + I_j + I_d = 0$$

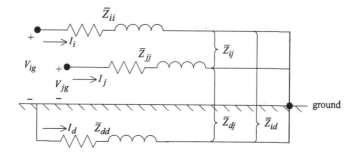

FIGURE 4.2 Two conductors including the dirt return path.

$$I_d = -I_i - I_j \tag{4.11}$$

Substitute Equation 4.11 into Equation 4.10 and collect terms.

$$V_{ig} = \left(\bar{z}_{ii} + \bar{z}_{dd} - \bar{z}_{di} - \bar{z}_{id}\right) \cdot I_i + \left(\bar{z}_{ij} + \bar{z}_{dd} - \bar{z}_{dj} - \bar{z}_{id}\right) \cdot I_j \tag{4.12}$$

Equation 4.12 is of the general form

$$V_{ig} = \hat{z}_{ii} \cdot I_i + \hat{z}_{ij} \cdot I_j, \tag{4.13}$$

where

$$\hat{z}_{ii} = \bar{z}_{ii} + \bar{z}_{dd} - \bar{z}_{di} - \bar{z}_{id}, \tag{4.14}$$

$$\hat{z}_{ij} = \bar{z}_{ij} + \bar{z}_{dd} - \bar{z}_{dj} - \bar{z}_{id} \tag{4.15}$$

In Equations 4.14 and 4.15, the "bar" impedances are given by Equations 4.7 and 4.8. Note that in these two equations, the effect of the ground return path is being "folded" into what will now be referred to as the "primitive" self and mutual impedances of the line. The "equivalent primitive circuit" of Figure 4.2 is shown in Figure 4.3.

Substituting Equations 4.7 and 4.8 of the "bar" impedances into Equations 4.14 and 4.15, the primitive self-impedance is given by

$$\hat{z}_{ii} = r_i + jx_{ii} + r_d + jx_{dd} - jx_{id} - jx_{di},$$

$$\hat{z}_{ii} = r_i + r_d + j0.12134 \cdot \left(\ln\frac{1}{GMR_i} + \ln\frac{1}{GMR_d} - \ln\frac{1}{D_{id}} - \ln\frac{1}{D_{di}}\right),$$

$$\hat{z}_{ii} = r_i + r_d + j0.12134 \cdot \left(\ln\frac{1}{GMR_i} + \ln\frac{D_{id} \cdot D_{di}}{GMR_d}\right) \tag{4.16}$$

In a similar manner, the primitive mutual impedance can be expanded:

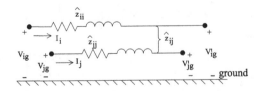

FIGURE 4.3 Equivalent primitive circuit.

Series Impedance of Overhead and Underground lines

$$\hat{z}_{ij} = jx_{ij} + r_d + jx_{dd} - jx_{dj} - jx_{id},$$

$$\hat{z}_{ij} = r_d + j0.12134 \cdot \left(\ln\frac{1}{D_{ij}} + \ln\frac{1}{GMR_d} - \ln\frac{1}{D_{dj}} - \ln\frac{1}{D_{id}} \right),$$

$$\hat{z}_{ij} = r_d + j0.12134 \left(\ln\frac{1}{D_{ij}} + \ln\frac{D_{dj} \cdot D_{id}}{GMR_d} \right) \quad (4.17)$$

The obvious problem in using Equations 4.16 and 4.17 is the fact that we do not know the values of the resistance of dirt (r_d), the geometric mean radius of dirt (GMR_d), and the distances from the conductors to dirt (D_{di} and D_{dj}). This is where John Carson's work bails us out.

4.1.3 CARSON'S EQUATIONS

Since a distribution feeder is inherently unbalanced, the most accurate analysis should not make any assumptions regarding the spacing between conductors, conductor sizes, and transposition. In Carson's 1926 paper, he developed a technique whereby the self and mutual impedances for the number of overhead conductors (**ncond**) can be determined. The equations can also be applied to underground cables. In 1926, this technique was not met with a lot of enthusiasm because of the tedious calculations that would have to be done on the slide rule and by hand. With the advent of the digital computer, Carson's equations have now become widely used.

In his paper, Carson assumes the earth is an infinite, uniform solid with a flat uniform upper surface and a constant resistivity. Any "end effects" introduced at the neutral grounding points are not large at power frequencies and therefore are neglected.

Carson made use of conductor images; that is, every conductor at a given distance above ground has an image conductor the same distance below ground. This is illustrated in Figure 4.4.

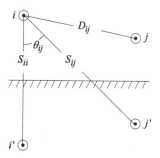

FIGURE 4.4 Conductors and images.

Referring to Figure 4.4, the original Carson equations are given in Equations 4.18 and 4.19.

Self-Impedance:

$$\hat{z}_{ii} = r_i + 4\omega P_{ii}G + j\left(X_i + 2\omega G \cdot \ln\frac{S_{ii}}{RD_i} + 4\omega Q_{ii}G\right) \Omega/\text{mile} \quad (4.18)$$

Mutual Impedance:

$$\hat{z}_{ij} = 4\omega P_{ij}G + j\left(2\omega G \cdot \ln\frac{S_{ij}}{D_{ij}} + 4\omega Q_{ij}G\right) \Omega/\text{mile}, \quad (4.19)$$

where
$\hat{z}_{ii}$ = self-impedance of conductor i in Ω/mile,
$\hat{z}_{ij}$ = mutual impedance between conductors i and j in Ω/mile,
r_i = resistance of conductor i in Ω/mile,
$\omega = 2\pi f$ = system angular frequency in radians per second,
$G = 0.1609347 \times 10^{-3}$ Ω/mile,
RD_i = radius of conductor i in feet,
GMR_i = geometric mean radius of conductor i in feet,
f = system frequency in Hertz,
ρ = resistivity of the earth in Ω-meters,
D_{ij} = distance between conductors i and j in feet (see Figure 4.4),
S_{ij} = distance between conductor i and image j in feet (see Figure 4.4),
θ_{ij} = angle between a pair of lines drawn from conductor i to its own image and to the image of conductor j (see Figure 4.4).

$$X_i = 2\omega G \cdot \ln\frac{RD_i}{GMR_i} \Omega/\text{mile} \quad (4.20)$$

$$P_{ij} = \frac{\pi}{8} - \frac{1}{3\sqrt{2}}k_{ij}\cos(\theta_{ij}) + \frac{k_{ij}^2}{16}\cos(2\theta_{ij}) \cdot \left(0.6728 + \ln\frac{2}{k_{ij}}\right) \quad (4.21)$$

$$Q_{ij} = -0.0386 + \frac{1}{2}\cdot\ln\frac{2}{k_{ij}} + \frac{1}{3\sqrt{2}}k_{ij}\cos(\theta_{ij}) \quad (4.22)$$

Series Impedance of Overhead and Underground lines

$$k_{ij} = 8.565 \times 10^{-4} \cdot S_{ij} \cdot \sqrt{\frac{f}{\rho}} \qquad (4.23)$$

4.1.4 Modified Carson's Equations

Only two approximations are made in deriving the "modified Carson equations". These approximations involve the terms associated with P_{ij} and Q_{ij}. The approximations use only the first term of the variable P_{ij} and the first two terms of Q_{ij}.

$$P_{ij} = \frac{\pi}{8} \qquad (4.24)$$

$$Q_{ij} = -0.0386 + \frac{1}{2} \ln \frac{2}{k_{ij}} \qquad (4.25)$$

Substitute X_i (Equation 4.20) into Equation 4.18.

$$\hat{z}_{ii} = r_i + 4\omega P_{ii} G + j\left(2\omega G \cdot \ln \frac{RD_i}{GMR_i} + 2\omega G \cdot \ln \frac{S_{ii}}{RD_i} + 4\omega Q_{ii} G \right) \qquad (4.26)$$

Combine terms and simplify.

$$\hat{z}_{ii} = r_i + 4\omega P_{ii} G + j2\omega G\left(\ln \frac{S_{ii}}{GMR_i} + \ln \frac{RD_i}{RD_i} + 2Q_{ii} \right) \qquad (4.27)$$

Simplify Equation 4.19.:

$$\hat{z}_{ij} = 4\omega P_{ij} G + j2\omega G\left(\ln \frac{S_{ij}}{D_{ij}} + 2Q_{ij} \right) \qquad (4.28)$$

Substitute expressions for P (Equation 4.24) and ω ($2 \cdot \pi \cdot f$) into Equations 4.27 and 4.28.

$$\hat{z}_{ii} = r_i + \pi^2 fG + j4\pi fG\left(\ln \frac{S_{ii}}{GMR_i} + 2Q_{ii} \right) \qquad (4.29)$$

$$\hat{z}_{ij} = \pi^2 fG + j4\pi fG\left(\ln \frac{S_{ij}}{D_{ij}} + 2Q_{ij} \right) \qquad (4.30)$$

Substitute the expression for k_{ij} (Equation 4.23) into the approximate expression for Q_{ij} (Equation 4.25).

$$Q_{ij} = -0.0386 + \frac{1}{2}\ln\left(\frac{2}{8.565 \times 10^{-4} \cdot S_{ij} \cdot \sqrt{\frac{f}{\rho}}}\right) \quad (4.31)$$

Expand:

$$Q_{ij} = -0.0386 + \frac{1}{2}\ln\left(\frac{2}{8.565 \times 10^{-4}}\right) + \frac{1}{2}\ln\frac{1}{S_{ij}} + \frac{1}{2}\ln\sqrt{\frac{\rho}{f}} \quad (4.32)$$

Equation 4.32 can be reduced to

$$Q_{ij} = 3.8393 - \frac{1}{2}\ln S_{ij} + \frac{1}{4}\ln\frac{\rho}{f}, \quad (4.33)$$

or

$$2Q_{ij} = 7.6786 - \ln S_{ij} + \frac{1}{2}\ln\frac{\rho}{f}, \quad (4.34)$$

Substitute Equation 4.34 into Equation 4.29 and simplify.

$$\hat{z}_{ii} = r_i + \pi^2 fG + j4\pi fG\left(\ln\frac{S_{ii}}{GMR_i} + 7.6786 - \ln S_{ii} + \frac{1}{2}\ln\frac{\rho}{f}\right)$$

$$\hat{z}_{ii} = r_i + \pi^2 fG + j4\pi fG\left(\ln\frac{1}{GMR_i} + 7.6786 + \frac{1}{2}\ln\frac{\rho}{f}\right) \quad (4.35)$$

Substitute Equation 4.34 into Equation 4.30 and simplify.

$$\hat{z}_{ij} = \pi^2 fG + j4\pi fG\left(\ln\frac{S_{ij}}{D_{ij}} + 7.6786 - \ln S_{ij} + \frac{1}{2}\ln\frac{\rho}{f}\right)$$

$$\hat{z}_{ij} = \pi^2 fG + j4\pi fG\left(\ln\frac{1}{D_{ij}} + 7.6786 + \frac{1}{2}\ln\frac{\rho}{f}\right) \quad (4.36)$$

Substitute in the values of π and G.

$$\hat{z}_{ii} = r_i + 0.00158836 \cdot f + j0.00202237 \cdot f \left(\ln \frac{1}{GMR_i} + 7.6786 + \frac{1}{2} \ln \frac{\rho}{f} \right) \quad (4.37)$$

$$\hat{z}_{ij} = 0.00158836 \cdot f + j0.00202237 \cdot f \left(\ln \frac{1}{D_{ij}} + 7.6786 + \frac{1}{2} \ln \frac{\rho}{f} \right) \quad (4.38)$$

It is now assumed
f = frequency = 60 Hertz,
ρ = earth resistivity = 100 Ω-meters.

Using these approximations and assumptions, the "modified Carson's equations" are

$$\hat{z}_{ii} = r_i + 0.09530 + j0.12134 \left(\ln \frac{1}{GMR_i} + 7.93402 \right) \Omega/\text{mile}, \quad (4.39)$$

$$\hat{z}_{ij} = 0.09530 + j0.12134 \left(\ln \frac{1}{D_{ij}} + 7.93402 \right) \Omega/\text{mile} \quad (4.40)$$

It will be recalled that Equations 4.16 and 4.17 could not be used because the resistance of dirt, the *GMR* of dirt, and the various distances from conductors to dirt were not known. A comparison of Equations 4.16 and 4.17 to Equations 4.41 and 4.42 demonstrates that the modified Carson's equations have defined the missing parameters. A comparison of the two sets of equations shows that

$$r_d = 0.09530 \; \Omega/\text{mile}, \quad (4.41)$$

$$\text{Ln} \frac{D_{id} \cdot D_{di}}{GMR_d} = \ln \frac{D_{dj} \cdot D_{id}}{GMR_d} = 7.93402 \quad (4.42)$$

The "modified Carson's equations" will be used to compute the primitive self and mutual impedances of overhead and underground lines.

4.1.5 PRIMITIVE IMPEDANCE MATRIX FOR OVERHEAD LINES

Equations 4.39 and 4.40 are used to compute the elements of an ***ncond*** × ***ncond*** "primitive impedance matrix". An overhead four-wire, grounded wye distribution line segment will result in a 4×4 matrix. For an underground grounded wye line segment consisting of three concentric neutral cables, the resulting matrix will be 6×6.

The primitive impedance matrix for a three-phase line consisting of *m* neutrals will be of the form

$$[\hat{z}_{primitive}] = \begin{bmatrix} \hat{z}_{aa} & \hat{z}_{ab} & \hat{z}_{ac} & | & \hat{z}_{an1} & \hat{z}_{an2} & \hat{z}_{anm} \\ \hat{z}_{ba} & \hat{z}_{bb} & \hat{z}_{bc} & | & \hat{z}_{bn1} & \hat{z}_{bn2} & \hat{z}_{bnm} \\ \hat{z}_{ca} & \hat{z}_{cb} & \hat{z}_{cc} & | & \hat{z}_{cn1} & \hat{z}_{cn2} & \hat{z}_{cnm} \\ --- & --- & --- & --- & --- & --- & --- \\ \hat{z}_{n1a} & \hat{z}_{n1b} & \hat{z}_{n1c} & | & \hat{z}_{n1n1} & \hat{z}_{n1n2} & \hat{z}_{n1nm} \\ \hat{z}_{n2a} & \hat{z}_{n2b} & \hat{z}_{n2c} & | & \hat{z}_{n2n1} & \hat{z}_{n2n2} & \hat{z}_{n2nm} \\ \hat{z}_{nma} & \hat{z}_{nmb} & \hat{z}_{nmc} & | & \hat{z}_{nmn1} & \hat{z}_{nmn2} & \hat{z}_{nmnm} \end{bmatrix} \quad (4.43)$$

In partitioned form, Equation 4.43 becomes

$$[\hat{z}_{primitive}] = \begin{bmatrix} [\hat{z}_{ij}] & [\hat{z}_{in}] \\ [\hat{z}_{nj}] & [\hat{z}_{nn}] \end{bmatrix} \quad (4.44)$$

4.1.6 Phase Impedance of Matrix for Overhead Lines

Except for a three-wire line, the primitive impedance matrix must be reduced to a 3×3 "phase impedance" matrix consisting of the self and mutual equivalent impedances for the three phases where the effect of the neutral impedance has been "folded" into the phase impedances.

A four-wire, grounded neutral line segment is shown in Figure 4.5.

A standard method of the reduction is the "Kron" method [3]. The assumption is made that the line has a multi-grounded neutral, as shown in Figure 4.5. The Kron reduction method applies KVL to the circuit.

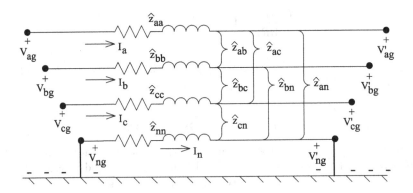

FIGURE 4.5 Four-wire, grounded wye line segment.

Series Impedance of Overhead and Underground lines

$$\begin{bmatrix} V_{ag} \\ V_{bg} \\ V_{cg} \\ V_{ng} \end{bmatrix} = \begin{bmatrix} V'_{ag} \\ V'_{bg} \\ V'_{cg} \\ V'_{ng} \end{bmatrix} + \begin{bmatrix} \hat{z}_{aa} & \hat{z}_{ab} & \hat{z}_{ac} & \hat{z}_{an} \\ \hat{z}_{ba} & \hat{z}_{bb} & \hat{z}_{bc} & \hat{z}_{bn} \\ \hat{z}_{ca} & \hat{z}_{cb} & \hat{z}_{cc} & \hat{z}_{cn} \\ \hat{z}_{na} & \hat{z}_{nb} & \hat{z}_{nc} & \hat{z}_{nn} \end{bmatrix} \cdot \begin{bmatrix} I_a \\ I_b \\ I_c \\ I_n \end{bmatrix} \quad (4.45)$$

In partitioned form, Equation 4.45 becomes

$$\begin{bmatrix} [V_{abc}] \\ [V_{ng}] \end{bmatrix} = \begin{bmatrix} [V'_{abc}] \\ [V'_{ng}] \end{bmatrix} + \begin{bmatrix} [\hat{z}_{ij}] & [\hat{z}_{in}] \\ [\hat{z}_{nj}] & [\hat{z}_{nn}] \end{bmatrix} \cdot \begin{bmatrix} [I_{abc}] \\ [I_n] \end{bmatrix} \quad (4.46)$$

Because the neutral is grounded, the voltages V_{ng} and V'_{ng} are equal to zero. Substitute those values into Equation 4.46 and expand results in

$$[V_{abc}] = [V'_{abc}] + [\hat{z}_{ij}] \cdot [I_{abc}] + [\hat{z}_{in}] \cdot [I_n], \quad (4.47)$$

$$[0] = [0] + [\hat{z}_{nj}] \cdot [I_{abc}] + [\hat{z}_{nn}] \cdot [I_n] \quad (4.48)$$

Solve Equation 4.48 for $[I_n]$:

$$[I_n] = -[\hat{z}_{nn}]^{-1} \cdot [\hat{z}_{nj}] \cdot [I_{abc}] \quad (4.49)$$

Note in Equation 4.49 that once the line currents have been computed, it is possible to determine the current flowing in the neutral conductor. Because this will be a useful concept, the "neutral transformation matrix" is defined as

$$[t_n] = -[\hat{z}_{nn}]^{-1} \cdot [\hat{z}_{nj}] \quad (4.50)$$

such that

$$[I_n] = [t_n] \cdot [I_{abc}] \quad (4.51)$$

Substitute Equation 4.49 into Equation. 4.47:

$$[V_{abc}] = [V'_{abc}] + \left([\hat{z}_{ij}] - [\hat{z}_{in}] \cdot [\hat{z}_{nn}]^{-1} \cdot [\hat{z}_{nj}] \right) \cdot [I_{abc}],$$

$$[V_{abc}] = [V'_{abc}] + [z_{abc}] \cdot [I_{abc}], \quad (4.52)$$

where

$$[z_{abc}] = [\hat{z}_{ij}] - [\hat{z}_{in}] \cdot [\hat{z}_{nn}]^{-1} \cdot [\hat{z}_{nj}] \quad (4.53)$$

Equation 4.53 is the final form of the "Kron" reduction method. The final phase impedance matrix becomes

$$[z_{abc}] = \begin{bmatrix} z_{aa} & z_{ab} & z_{ac} \\ z_{ba} & z_{bb} & z_{bc} \\ z_{ca} & z_{cb} & z_{cc} \end{bmatrix} \Omega/\text{mile} \quad (4.54)$$

For a three-phase distribution line that is not transposed, the diagonal terms of Equation 4.54 will not be equal, and the off-diagonal terms will not be equal. However, the matrix will be symmetrical.

For two-phase (V-phase) and single-phase lines in grounded wye systems, the modified Carson's equations can be applied, which will lead to initial 3×3 and 2×2 primitive impedance matrices. Kron reduction will reduce the matrices to 2×2 and a single element. These matrices can be expanded to 3×3 "phase frame matrices" by the addition of rows and columns consisting of zero elements for the missing phases. For example, a V-phase line consisting of phases **a** and **c**, the phase impedance matrix would be

$$[z_{abc}] = \begin{bmatrix} z_{aa} & 0 & z_{ac} \\ 0 & 0 & 0 \\ z_{ca} & 0 & z_{cc} \end{bmatrix} \Omega/\text{mile} \quad (4.55)$$

The phase impedance matrix for a phase **b** single-phase line would be

$$[z_{abc}] = \begin{bmatrix} 0 & 0 & 0 \\ 0 & z_{bb} & 0 \\ 0 & 0 & 0 \end{bmatrix} \Omega/\text{mile} \quad (4.56)$$

The phase impedance matrix for a three-wire line is determined by the application of Carson's equations without the Kron reduction step.

The phase impedance matrix can be used to accurately determine the voltage drops on the feeder line segments once the currents have been determined. Since no approximations (transposition for example) have been made regarding the spacing between conductors, the effect of the mutual coupling between phases is accurately included. The application of the modified Carson's equations and the phase frame matrix leads to the most accurate model of a line segment. Figure 4.6 shows the

Series Impedance of Overhead and Underground lines

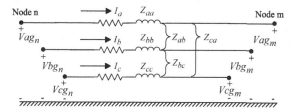

FIGURE 4.6 Three-phase line segment model.

general three-phase model of a line segment. Keep in mind that for V-phase and single-phase lines, some of the impedance values will be zero.

The voltage equation in matrix form for the line segment is

$$\begin{bmatrix} V_{ag} \\ V_{bg} \\ V_{cg} \end{bmatrix}_n = \begin{bmatrix} V_{ag} \\ V_{bg} \\ V_{cg} \end{bmatrix}_m + \begin{bmatrix} Z_{aa} & Z_{ab} & Z_{ac} \\ Z_{ba} & Z_{bb} & Z_{bc} \\ Z_{ca} & Z_{cb} & Z_{cc} \end{bmatrix} \cdot \begin{bmatrix} I_a \\ I_b \\ I_c \end{bmatrix}, \qquad (4.57)$$

where $Z_{ij} = z_{ij} \cdot length$.

Equation 4.57 can be written in "condensed" form as

$$[VLG_{abc}]_n = [VLG_{abc}]_m + [Z_{abc}] \cdot [I_{abc}] \qquad (4.58)$$

Example 4.1: An overhead three-phase distribution line is constructed, as shown in Figure 4.7. Determine the phase impedance matrix of the line. The phase conductors are 336,400 26/7 ACSR, and the neutral conductor is 4/0 6/1 ACSR.

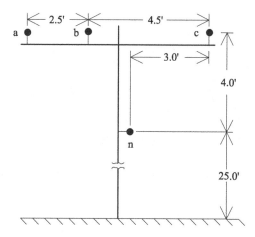

FIGURE 4.7 Three-phase distribution line spacings.

Solution:

From the table of standard conductor data (Appendix A), the following is found:

336,400 26/7 ACSR: GMR = 0.0244 ft
Resistance = 0.306 Ω/mile

4/0 6/1 ACSR: GMR = 0.00814 ft
Resistance = 0.5920 Ω/mile

An effective way of computing the distance between all conductors is to specify each position on the pole in Cartesian coordinates using complex number notation. The ordinate will be selected as a point on the ground directly below the left-most position. For the line in Figure 4.7, the positions are as follows:

$$d_1 = 0 + j29 \quad d_2 = 2.5 + j29 \quad d_3 = 7.0 + j29 \quad d_4 = 4.0 + j25$$

The distances between the positions can be computed as follows:

$$D_{12} = |d_1 - d_2| \quad D_{23} = |d_2 - d_3| \quad D_{31} = |d_3 - d_1|$$
$$D_{14} = |d_1 - d_4| \quad D_{24} = |d_2 - d_4| \quad D_{34} = |d_3 - d_4|$$

For this example, phase **a** is in position 1, phase **b** in position 2, phase **c** in position 3, and the neutral in position 4.

$$D_{ab} = 2.5' \quad D_{bc} = 4.5' \quad D_{ca} = 7.0'$$
$$D_{an} = 5.6569' \quad D_{bn} = 4.272' \quad D_{cn} = 5.0'$$

The diagonal terms of the distance matrix are the GMRs of the phase and neutral conductors.

$$D_{aa} = D_{bb} = D_{cc} = 0.0244 \quad D_{nn} = 0.00814$$

Applying the modified Carson's equation for self-impedance (Equation 4.41), the self-impedance for phase a is

$$\hat{z}_{aa} = 0.0953 + 0.306 + j0.12134 \cdot \left(\ln\frac{1}{0.0244} + 7.93402 \right) = 0.4013 + j1.4133 \, \Omega/\text{mile}$$

Applying Equation 4.42 for the mutual impedance between phases a and b,

$$\hat{z}_{ab} = 0.0953 + j0.12134 \cdot \left(\ln\frac{1}{2.5} + 7.93402 \right) = 0.0953 + j0.8515 \, \Omega/\text{mile}$$

Series Impedance of Overhead and Underground lines

Applying the equations for the other self- and mutual impedance terms results in the primitive impedance matrix.

$$[\hat{z}] = \begin{bmatrix} 0.4013 + j1.4133 & 0.0953 + j0.8515 & 0.0953 + j0.7266 & 0.0953 + j0.7524 \\ 0.0953 + j0.8515 & 0.4013 + j1.4133 & 0.0953 + j0.7802 & 0.0953 + j0.7865 \\ 0.0953 + j0.7266 & 0.0953 + j0.7802 & 0.4013 + j1.4133 & 0.0953 + j0.7674 \\ 0.0953 + j0.7524 & 0.0953 + j0.7865 & 0.0953 + j0.7674 & 0.6873 + j1.5465 \end{bmatrix}$$
$\times \Omega$/mile

The primitive impedance matrix in partitioned form is as follows:

$$[\hat{z}_{ij}] = \begin{bmatrix} 0.4013 + j1.4133 & 0.0953 + j0.8515 & 0.0953 + j0.7266 \\ 0.0953 + j0.8515 & 0.4013 + j1.4133 & j0.0943 + j0.7865 \\ 0.0953 + j0.7266 & 0.0953 + j0.7802 & 0.4013 + j1.4133 \end{bmatrix} \Omega/\text{mile}$$

$$[\hat{z}_{in}] = \begin{bmatrix} 0.0953 + j0.7524 \\ 0.0953 + j0.7865 \\ 0.0953 + j0.7674 \end{bmatrix} \Omega/\text{mile}$$

$$[\hat{z}_{nj}] = \begin{bmatrix} 0.0953 + j0.7524 & 0.0953 + j0.7865 & 0.0953 + j0.7674 \end{bmatrix} \Omega/\text{mile}$$

$$[\hat{z}_{nn}] = \begin{bmatrix} 0.6873 + j1.5465 \end{bmatrix} \Omega/\text{mile}$$

The Kron reduction of Equation 4.53 results in the "phase impedance matrix".

$$[z_{abc}] = [\hat{z}_{ij}] - [\hat{z}_{in}] \cdot [\hat{z}_{nn}]^{-1} \cdot [\hat{z}_{nj}]$$

$$[z_{abc}] = \begin{bmatrix} 0.4576 + j1.0780 & 0.1560 + j0.5017 & 0.1535 + j0.3849 \\ 0.1560 + j0.5017 & 0.4666 + j1.0482 & 0.1580 + j0.4236 \\ 0.1535 + j0.3849 & 0.1580 + j0.4236 & 0.4615 + j1.0651 \end{bmatrix} \Omega/\text{mile}$$

The neutral transformation matrix given by Equation 4.50 is as follows:

$$[t_n] = -\left([\hat{z}_{nn}]^{-1} \cdot [\hat{z}_{nj}]\right)$$

$$[t_n] = \begin{bmatrix} -0.4292 - j0.1291 & -0.4476 - j0.1373 & -0.4373 - j0.1327 \end{bmatrix}$$

Example 4.2: The length of the line of Example 4.1 is 5,000 feet and is connected to an ideal balanced 12.47 kV three-phase source. The line serves a balanced per-phase load of 5,000 kVA at a power factor of lagging 85%. Determine

- three-phase, line-to-neutral voltages;
- three-phase line currents; and
- three-phase voltage drops.

A Mathcad ladder technique computer program was written as shown in Figure 4.8.

$$\text{Tol} := .0001 \qquad \text{start} := \begin{pmatrix} 0 \\ 0 \\ 0 \end{pmatrix}$$

$$Y := \begin{vmatrix} I_{abc} \leftarrow \text{start} \\ V_{old} \leftarrow \text{start} \\ \text{for } n \in 1..30 \\ \quad \begin{pmatrix} VL_{abc} \leftarrow E_S - Z \cdot I_{abc} \\ \text{for } i \in 1..3 \\ \quad \text{Error}_i \leftarrow ||VL_{abc_i}| - |V_{old_i}|| \\ \text{Err} \leftarrow \max(\text{Error}) \\ \text{break if Err} < \text{Tol} \\ V_{old} \leftarrow VL_{abc} \\ \text{for } i \in 1..3 \\ \quad I_{abc_i} \leftarrow \dfrac{\overline{SL_i \cdot 1000}}{VL_{abc_i}} \\ \text{for } i \in 1..3 \\ \quad Vdrop_i \leftarrow \dfrac{||E_{S_i}| - |VL_{abc_i}||}{|E_{S_i}|} \cdot 100 \end{pmatrix} \\ \text{Out}_1 \leftarrow n \\ \text{Out}_2 \leftarrow VL_{abc} \\ \text{Out}_3 \leftarrow I_{abc} \\ \text{Out}_4 \leftarrow Vdrop \\ \text{Out} \end{vmatrix}$$

FIGURE 4.8 Mathcad Ladder technique for Example 4.2.

The results are

$$[VL_{abc}] = \begin{bmatrix} 6718.0\angle -1.9° \\ 6848.8\angle -122.0° \\ 6783.76\angle 117.7° \end{bmatrix} \quad [I_{abc}] = \begin{bmatrix} 744.3\angle -33.7° \\ 730.0\angle -153.8° \\ 737.1\angle 85.9° \end{bmatrix} \quad [V_{drop}] = \begin{bmatrix} 6.69 \\ 4.87 \\ 5.78 \end{bmatrix} \%.$$

4.1.7 PARALLEL OVERHEAD DISTRIBUTION LINES

It is common in a distribution system to find instances where two distribution lines are "physically" parallel. The parallel combination may have both distribution lines constructed on the same pole, or the two lines may run in parallel on separate poles but on the same right-of-way. For example, two different feeders leaving a substation may share a common pole or right-of-way before they branch out to their own service area. It is also possible that two feeders may converge and run in parallel until again they branch out into their own service areas. The lines could also be underground circuits sharing a common trench. In these cases, the question becomes, How should the parallel lines be modeled and analyzed?

Two parallel overhead lines on one pole are shown in Figure 4.9:
Note in Figure 4.9 the phasing of the two lines.

The phase impedance matrix for the parallel distribution lines is computed by the application of Carson's equations and the Kron reduction method. The first step is to number the phase positions as follows:

Position	1	2	3	4	5	6	7-
Line-Phase	1-a	1-b	1-c	2-a	2-b	2-c	neutral

With the phases numbered, the 7×7 primitive impedance matrix for one mile can be computed using the modified Carson's equations. If the two parallel lines are on different poles, most likely each pole will have a grounded neutral conductor. In this case, there will be eight positions and position 8 will correspond to the neutral on line 2. An 8×8 primitive impedance matrix will be developed for this case. The Kron reduction will reduce the matrix to a 6×6 phase impedance matrix. With reference to Figure 4.8, the voltage drops in the two lines are given by:

$$\begin{bmatrix} v1_a \\ v1_b \\ v1_c \\ v2_a \\ v2_b \\ v2_c \end{bmatrix} = \begin{bmatrix} z11_{aa} & z11_{ab} & z11_{ac} & z12_{aa} & z12_{ab} & z12_{ac} \\ z11_{ba} & z11_{bb} & z11_{bc} & z12_{ba} & z12_{bb} & z12_{bc} \\ z11_{ca} & z11_{cb} & z11_{cc} & z12_{ca} & z12_{cb} & z12_{cc} \\ z21_{aa} & z21_{ab} & z21_{ac} & z22_{aa} & z22_{ab} & z22_{ac} \\ z21_{ba} & z21_{bb} & z21_{bc} & z22_{ba} & z22_{bb} & z22_{bc} \\ z21_{ca} & z21_{cb} & z21_{cc} & z22_{ca} & z22_{cb} & z22_{cc} \end{bmatrix} \cdot \begin{bmatrix} I1_a \\ I1_b \\ I1_c \\ I2_a \\ I2_b \\ I2_c \end{bmatrix} \quad (4.59)$$

Partition Equation 4.59 between the third and fourth rows and columns so that series voltage drops for 1 mile of line are given by

$$[v] = [z] \cdot [I] = \begin{bmatrix} [v1] \\ [v2] \end{bmatrix} = \begin{bmatrix} [z11] & [z12] \\ [z21] & [z22] \end{bmatrix} \cdot \begin{bmatrix} [I1] \\ [I2] \end{bmatrix} V \quad (4.60)$$

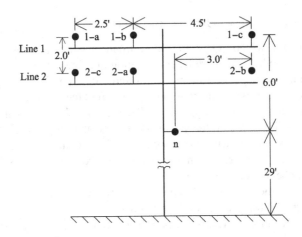

FIGURE 4.9 Parallel overhead lines.

Example 4.3: Two parallel distribution lines are on a single pole, as shown in Figure 4.10.

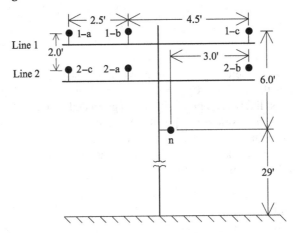

FIGURE 4.10 Example parallel OH lines.

The phase conductors are as follows:

Line 1: 336,400 26/7 ACSR: $GMR_1 = 0.0244'$ $r_1 = 0.306$ Ω/mile $d_1 = 0.721''$
Line 2: 250,000 AA: $GMR_2 = 0.0171'$ $r_2 = 0.41$ Ω/mile $d_2 = 0.567''$
Neutral: 4/0 6/1 ACSR: $GMR_n = 0.008814'$ $r_n = 0.592$ Ω/mile $d_n = 0.563''$

Determine the 6×6 phase impedance matrix.
Define the conductor positions according to the following phasing:

$d_1 = 0 + j35$ $d_2 = 2.5 + j35$ $d_3 = 7 + j35$
$d_4 = 2.5 + j33$ $d_5 = 7 + j33$ $d_6 = 0 + j33$
$d_7 = 4 + j29$

Series Impedance of Overhead and Underground lines

Using $D_{ij} = |d_i - d_j|$, the distances between all conductors can be computed. The diagonal terms of the resulting spacing matrix will be zero. It is convenient to define the diagonal terms of the spacing matrix as the GMR of the conductors occupying the position. The final spacing matrix is

$$[D] = \begin{bmatrix} 0.0244 & 2.5 & 7 & 3.2016 & 7.2801 & 2 & 7.2111 \\ 2.5 & 0.0244 & 4.5 & 2 & 4.9244 & 3.2016 & 6.1847 \\ 7 & 4.5 & 0.0244 & 4.9244 & 2 & 7.2801 & 6.7082 \\ 3.2016 & 2 & 4.9244 & 0.0171 & 4.5 & 2.5 & 4.2720 \\ 7.2801 & 4.9244 & 2 & 4.5 & 0.0171 & 7 & 5 \\ 2 & 3.2016 & 7.2801 & 2.5 & 7 & 0.0171 & 5.6869 \\ 7.2111 & 6.1847 & 6.7082 & 4.2720 & 5 & 5.6569 & 0.0081 \end{bmatrix} \quad (4.61)$$

The terms for the primitive impedance matrix can be computed using the modified Carson's equations. For this example, the subscripts *I* and *j* will run from 1 to 7. The 7×7 primitive impedance matrix is partitioned between rows and columns 6 and 7. The Kron reduction will now give the final phase impedance matrix. In partitioned form, the phase impedance matrices are as follows:

$$[z_{11}]_{abc} = \begin{bmatrix} 0.4502 + j1.1028 & 0.1464 + j0.5334 & 0.1452 + j0.4126 \\ 0.1464 + j0.5334 & 0.4548 + j1.0873 & 0.1475 + j0.4584 \\ 0.1452 + j0.4126 & 0.1475 + j0.4584 & 0.4523 + j1.0956 \end{bmatrix} \Omega/\text{mile}$$

$$[z_{12}]_{abc} = \begin{bmatrix} 0.1519 + j0.4848 & 0.1496 + j0.3931 & 0.1477 + j0.5560 \\ 0.1545 + j0.5336 & 0.1520 + j0.4323 & 0.1502 + j0.4909 \\ 0.1531 + j0.4287 & 0.1507 + j0.5460 & 0.1489 + j0.3955 \end{bmatrix} \Omega/\text{mile}$$

$$[z_{21}]_{abc} = \begin{bmatrix} 0.1519 + j0.4848 & 0.1545 + j0.5336 & 0.1531 + j0.4287 \\ 0.1496 + j0.3931 & 0.1520 + j0.4323 & 0.1507 + j0.5460 \\ 0.1477 + j0.5560 & 0.1502 + j0.4909 & 0.1489 + j0.3955 \end{bmatrix} \Omega/\text{mile}$$

$$[z_{22}]_{abc} = \begin{bmatrix} 0.5706 + j1.0913 & 0.1580 + j0.4236 & 0.1559 + j0.5017 \\ 0.1580 + j0.4236 & 0.5655 + j1.1082 & 0.1535 + j0.3849 \\ 0.1559 + j0.5017 & 0.1535 + j0.3849 & 0.5616 + j1.1212 \end{bmatrix} \Omega/\text{mile}$$

4.2 SERIES IMPEDANCE OF UNDERGROUND LINES

Figure 4.11 shows the general configuration of three underground cables (concentric neutral or tape shielded) with an additional neutral conductor.

The modified Carson's equations can be applied to underground cables in much the same manner as for overhead lines. The circuit of Figure 4.10 will result in a 7×7 primitive impedance matrix. For underground circuits that do not have the additional neutral conductor, the primitive impedance matrix will be 6×6.

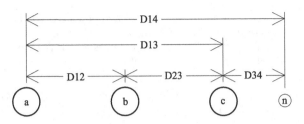

FIGURE 4.11 Three-phase underground with additional neutral.

Two popular types of underground cables are the "concentric neutral cable" and the "tape shield cable". To apply the modified Carson's equations the resistance and GMR of the phase conductor and the equivalent neutral must be known.

4.2.1 Concentric Neutral Cable

Figure 4.12 shows a simple detail of a concentric neutral cable. The cable consists of a central "phase conductor" covered by a thin layer of non-metallic semi-conducting screen to which is bonded to the insulating material. The insulation is then covered by a semi-conducting insulation screen. The solid strands of concentric neutral are spiraled around the semi-conducting screen with uniform spacing between strands. Some cables will also have an insulating "jacket" encircling the neutral strands.

In order to apply Carson's equations to this cable, the following data needs to be extracted from a table of underground cables (Appendices A and B).

d_c = phase conductor diameter (inches)
d_{od} = nominal diameter over the concentric neutrals of the cable (inches)
d_s = diameter of a concentric neutral strand (inches)
GMR_c = geometric mean radius of the phase conductor (feet)
GMR_s = geometric mean radius of a neutral strand (feet)
r_c = resistance of the phase conductor (Ω/mile)
r_s = resistance of a solid neutral strand (Ω/mile)
k = number of concentric neutral strands

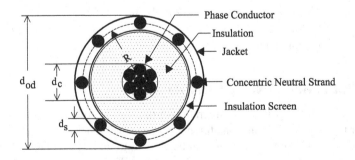

FIGURE 4.12 Concentric neutral cable.

Series Impedance of Overhead and Underground lines

The geometric mean radii of the phase conductor and a neutral strand are obtained from a standard table of conductor data (Appendix A). The equivalent geometric mean radius of the concentric neutral is computed using the equation for the geometric mean radius of bundled conductors used in high voltage transmission lines. [2]

$$GMR_{cn} = \sqrt[k]{GMR_s \cdot k \cdot R^{k-1}} \text{ ft,} \qquad (4.62)$$

where

R = (as shown in Figure 4.12) the radius of a circle passing through the center of the concentric neutral strands,

$$R = \frac{d_{od} - d_s}{24} \text{ ft} \qquad (4.63)$$

The equivalent resistance of the concentric neutral is

$$r_{cn} = \frac{r_s}{k} \text{ }\Omega/\text{mile} \qquad (4.64)$$

The various spacings between a concentric neutral and the phase conductors and other concentric neutrals are as follows:

Concentric neutral to its own phase conductor

$D_{ij} = R$ (Equation 4.63 above)

Concentric neutral to an adjacent concentric neutral

D_{ij} = center-to-center distance of the phase conductors

Concentric neutral to an adjacent phase conductor

Figure 4.13 shows the relationship between the distance between centers of concentric neutral cables and the radius of a circle passing through the centers of the neutral strands.

The geometric mean distance between a concentric neutral and an adjacent phase conductor is given by

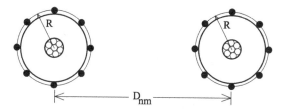

FIGURE 4.13 Distances between concentric neutral cables.

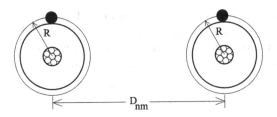

FIGURE 4.14 Equivalent neutral cables.

$$D_{ij} = \sqrt[k]{D_{nm}^k - R^k} \text{ ft.}, \tag{4.65}$$

where
D_{nm} = center-to-center distance between phase conductors.

The distance between cables will be much greater than the radius R, so a good approximation of modeling the concentric neutral cables is shown in Figure 4.14. In this figure, the concentric neutrals are modeled as one equivalent conductor (shown in black) directly above the phase conductor.

In applying the modified Carson's equations, the numbering of conductors and neutrals is important. For example, a three-phase underground circuit with an additional neutral conductor must be numbered as follows:

1 = phase **a** conductor #1
2 = phase **b** conductor #2
3 = phase **c** conductor #3
4 = neutral of conductor #1
5 = neutral of conductor #2
6 = neutral of conductor #3
7 = additional neutral conductor (if present)

Example 4.4: Three concentric neutral cables are buried in a trench with spacings, as shown in Figure 4.15.

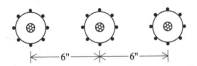

FIGURE 4.15 Three-phase concentric neutral cable spacing.

Series Impedance of Overhead and Underground lines

The concentric neutrals cables of Figure 4.15 can be modeled as shown in Figure 4.16. Notice the numbering of the phase conductors and the equivalent neutrals.

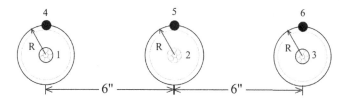

FIGURE 4.16 Three-phase equivalent concentric neutral cable spacing.

The cables are 15 kV, 250,000 CM stranded all aluminum with 13 strands of #14 annealed coated copper wires (one-third neutral). The outside diameter of the cable over the neutral strands is 1.29 inches (Appendix B). Determine the phase impedance matrix.

SOLUTION:

The data for the phase conductor and neutral strands from a conductor data table (Appendix A) are as follows:

250,000 AA phase conductor:

$GMR_c = 0.0171$ ft
Diameter = 0.567 in.
Resistance = 0.41 Ω/mile

14 copper neutral strands:

$GMR_s = 0.00208$ ft
Resistance = 14.8722 Ω/mile
Diameter $(d_s) = 0.0641$ in.

The radius of the circle passing through the center of the strands is

$$R = \frac{d_{od} - d_s}{24} = 0.0511 \text{ft}$$

The equivalent GMR of the concentric neutral is computed by

$$GMR_{cn} = \sqrt[k]{GMR_s \cdot k \cdot R^{k-1}} = \sqrt[13]{0.00208 \cdot 13 \cdot 0.0511^{13-1}} = 0.0486 \text{ ft}$$

The equivalent resistance of the concentric neutral is

$$r_{cn} = \frac{r_s}{k} = \frac{14.8722}{13} = 1.1440 \ \Omega/\text{mile}$$

The phase conductors are numbered 1, 2, and 3. The concentric neutrals are numbered 4, 5, and 6.

A convenient method of computing the various spacings is to define each conductor using Cartesian coordinates. Using this approach, the conductor coordinates are as follows:

$$d_1 = 0 + j0 \qquad d_2 = 0.5 + j0 \qquad d_3 = 1 + j0$$
$$d_4 = 0 + ji \qquad d_5 = 0.5 + jR \qquad d_6 = 1 + jR$$

The spacings of off-diagonal terms of the spacing matrix are computed by the following:

For $n = 1$ to 6 and $m = 1$ to 6

$$D_{n,m} = |d_n - d_m|$$

The diagonal terms of the spacing matrix are the GMRs of the phase conductors and the equivalent neutral conductors:

For $l = 1$ to 3 and $j = 4$ to 6

$$D_{i,i} = GMR_c$$

$$D_{j,j} = GMR_s$$

The resulting spacing matrix is

$$[D] = \begin{bmatrix} 0.0171 & 0.5 & 1 & 0.0511 & 0.5026 & 1.0013 \\ 0.5 & 0.0171 & 0.5 & 0.5026 & 0.0511 & 0.5026 \\ 1 & 0.5 & 0.0171 & 1.0013 & 0.5026 & 0.0511 \\ 0.0511 & 0.5026 & 1.0013 & 0.0486 & 0.5 & 1 \\ 0.5026 & 0.0511 & 0.5026 & 0.5 & 0.0486 & 0.5 \\ 1.0013 & 0.5026 & 0.0511 & 1 & 0.5 & 0.0486 \end{bmatrix} \text{ ft}$$

The self-impedance for the cable in position 1 is

$$\hat{z}_{11} = 0.0953 + 0.41 + j0.12134 \cdot \left(\ln \frac{1}{0.0171} + 7.93402 \right) = 0.5053 + j1.4564 \, \Omega/\text{mile}$$

The self-impedance for the concentric neutral for cable #1 is

$$\hat{z}_{44} = 0.0953 + 1.144 + j0.12134 \cdot \left(\ln \frac{1}{0.0486} + 7.93402 \right) = 1.2391 + j1.3296 \frac{\Omega}{\text{mile}}$$

The mutual impedance between cable #1 and cable #2 is

… Series Impedance of Overhead and Underground lines

$$\hat{z}_{12} = 0.0953 + j0.12134 \cdot \left(\ln \frac{1}{0.5} + 7.93402 \right) = 0.0953 + j1.0468 \, \Omega/\text{mile}$$

The mutual impedance between cable #1 and its concentric neutral is

$$\hat{z}_{14} = 0.0953 + j0.12134 \cdot \left(\ln \frac{1}{0.0511} + 7.93402 \right) = 0.0953 + j1.3236 \, \Omega/\text{mile}$$

The mutual impedance between the concentric neutral of cable #1 and the concentric neutral of cable #2 is

$$\hat{z}_{45} = 0.0953 + j0.12134 \cdot \left(\ln \frac{1}{0.5} + 7.93402 \right) = 0.0953 + j1.0468 \, \Omega/\text{mile}$$

Continuing the application of the modified Carson's equations results in a 6×6 primitive impedance matrix. This matrix in partitioned (Equation 4.33) form is as follows:

$$[\hat{z}_{ij}] = \begin{bmatrix} 0.5053 + j1.4564 & 0.0953 + j1.0468 & 0.0953 + j0.9627 \\ 0.0953 + j1.0468 & 0.5053 + j1.4564 & 0.0953 + j1.0468 \\ 0.0953 + j0.9627 & 0.0953 + j1.0468 & 0.5053 + j1.4564 \end{bmatrix} \Omega/\text{mile}$$

$$[\hat{z}_{in}] = \begin{bmatrix} 0.0953 + j1.3236 & 0.0953 + j1.0468 & 0.0953 + j0.9627 \\ 0.0953 + j1.0462 & 0.0953 + j1.3236 & 0.0953 + j1.0462 \\ 0.0953 + j.9626 & 0.0953 + j1.0462 & 0.0953 + j1.3236 \end{bmatrix} \Omega/\text{mile}$$

$$[\hat{z}_{nj}] = [\hat{z}_{in}]$$

$$[\hat{z}_{nn}] = \begin{bmatrix} 1.2393 + j1.3296 & 0.0953 + j1.0468 & 0.0953 + j.9627 \\ 0.0953 + j1.0468 & 1.2393 + j1.3296 & 0.0953 + j1.0468 \\ 0.0953 + j0.9627 & 0.0953 + j1.0468 & 1.2393 + j1.3296 \end{bmatrix} \Omega/\text{mile}$$

Using the Kron reduction results in the phase impedance matrix:

$$[z_{abc}] = [\hat{z}_{ij}] - [\hat{z}_{in}] \cdot [\hat{z}_{nn}]^{-1} \cdot [\hat{z}_{nj}]$$

$$[z_{abc}] = \begin{bmatrix} 0.7981 + j0.4467 & 0.3188 + j0.0334 & 0.2848 - j0.0138 \\ 0.3188 + j0.0334 & 0.7890 + j0.4048 & 0.3188 + j0.0334 \\ 0.2848 - j0.0138 & 0.3188 + j0.0334 & 0.7981 + j0.4467 \end{bmatrix} \Omega/\text{mile}$$

4.2.2 Tape Shielded Cables

Figure 4.17 shows a simple detail of a tape shielded cable. The cable consists

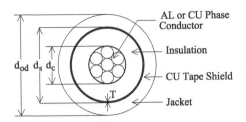

FIGURE 4.17 Tape shielded cable.

of a central "phase conductor" covered by a thin layer of non-metallic semi-conducting screen to which is bonded the insulating material. The insulation is covered by a semi-conducting insulation screen. The shield is bare copper tape helically applied around the insulation screen. An insulating "jacket" encircles the tape shield.

Parameters of the tape shielded cable are as follows:

d_c = diameter of phase conductor (inches): Appendix A
d_t = outside diameter of the tape shield (inches): Appendix B
d_{od} = outside diameter over jacket (inches): Appendix B
T = thickness of copper tape shield in mils: Appendix B

Once again, the modified Carson's equations will be applied to calculate the self-impedances of the phase conductor and the tape shield, as well as the mutual impedance between the phase conductor and the tape shield. The resistance and GMR of the phase conductor are found in a standard table of conductor data (Appendix A).

The resistance of the tape shield with d_s in inches and T in mils is given by

$$r_{tape} = \frac{2.49395 \cdot 10^{12} \cdot \rho m_{20}}{\pi \cdot T \cdot (d_t \cdot 1000 - T)} \; \Omega/\text{mile} \quad (4.66)$$

The resistance of the tape shield given in Equation 4.66 assumes a resistivity of $1.7721 \cdot 10^{-8}$ Ω-meter and a temperature of 20 degrees C. The outside diameter of the tape shield d_s is given in inches and the thickness of the tape shield T is in mils.

The GMR of the tape shield is given by

$$GMR_{shield} = \frac{Te^{-\frac{1}{4}}}{12000} \; \text{ft} \quad (4.67)$$

The geometric mean distance from the tape shield to its own phase conductor is the radius to the midpoint of the shield and is given by

$$D_{ij} = R = \frac{\dfrac{d_t}{2} - \dfrac{T}{2000}}{12} \tag{4.68}$$

The various spacings between a tape shield and the conductors and other tape shields are as follows:

Tape shield to its own phase conductor
  D_{ij} = radius to midpoint of the shield (Equation 4.68 above) feet.
Tape shield to an adjacent tape shield

$$D_{ij} = \text{center-to-center distance of the phase conductors' feet} \tag{4.69}$$

Tape shield to an adjacent phase or neutral conductor

$$D_{ij} = D_{nm} \text{ ft.,} \tag{4.70}$$

where D_{nm} = center-to-center distance between phase conductors.

Example 4.5: A single-phase circuit consists of a 1/0 AA, 220 mil insulation tape shielded cable, and a 1/0 CU neutral conductor, as shown in Figure 4.18. The single-phase line is connected to phase *b*. Determine the phase impedance matrix.

Cable Data: 1/0 AA
  Outside diameter of the tape shield = d_t = 0.88 in.
  Resistance = 0.97 Ω/mile
  GMR_c = 0.0111 ft
  Tape shield thickness = T = 5 mils
  Resistivity = ρm_{20} = 1.7721 · 10^{-8} Ω – meter
Neutral Data: 1/0 Copper, 7 strand

FIGURE 4.18 Single-phase tape shield with neutral.

Resistance = 0.607 Ω/mile
GMR_n = 0.01113 ft
Distance between cable and neutral = D_{nm} = 3 in.
The resistance of the tape shield is computed according to Equation 4.66:

$$r_{shield} = \frac{2.494 \cdot 10^{12} \, \rho m_{20}}{\pi T (d_t \cdot 1000 - T)} = \frac{2.494 \cdot 10^{12} \cdot 1.7721 \cdot 10^{-8}}{\pi \cdot 5 \cdot (1000 \cdot 0.88 - 5)} = 3.2156 \, \Omega/\text{mile}$$

The GMR of the tape shield is computed according to Equation 4.67:

$$GMR_{shield} = \frac{Te^{-\frac{1}{4}}}{12000} = \frac{5 \cdot e^{-\frac{1}{4}}}{12000} = 3.2450 \cdot 10^{-4} \text{ ft}$$

The conductors are numbered such that

#1 = 1/0 AA conductor,
#2 = tape shield,
#3 = 1/0 copper ground.

The spacings used in the modified Carson's equations are

$$D_{12} = \frac{\frac{d}{2} - \frac{T}{2000}}{12} = \frac{\frac{0.88}{2} - \frac{5}{2000}}{12} = 0.365,$$

$$D_{13} = \frac{3}{12} = 0.25.$$

The self-impedance of conductor #1 is

$$\hat{z}_{11} = 0.0953 + 0.97 + j0.12134 \cdot \left(\ln \frac{1}{0.0111} + 7.93402 \right)$$
$$= 1.0653 + j1.5088 \, \Omega/\text{mile}$$

The mutual impedance between conductor #1 and the tape shield (conductor #2) is

$$\hat{z}_{12} = 0.0953 + j0.12134 \cdot \left(\ln \frac{1}{0.0365} + 7.93402 \right)$$
$$= 0.0953 + j1.3645 \, \Omega/\text{mile}$$

The self-impedance of the tape shield (conductor #2) is

$$\hat{z}_{22} = 0.0953 + 3.2156 + j0.12134 \cdot \left(\ln \frac{1}{3.2450 \cdot 10^{-4}} + 7.93402 \right)$$
$$= 3.3109 + j1.9375 \, \Omega/\text{mile}$$

Series Impedance of Overhead and Underground lines

Continuing on the final primitive impedance matrix is

$$[\hat{z}] = \begin{bmatrix} 1.0653 + j1.5088 & 0.0953 + j1.3645 & 0.0953 + 1.1309 \\ 0.0953 + j1.3645 & 3.3109 + j1.9375 & 0.0953 + j1.1309 \\ 0.0953 + j1.1309 & 0.0953 + j1.1309 & 0.7023 + j1.5085 \end{bmatrix} \frac{\Omega}{mile}$$

In partitioned form, the primitive impedance matrix is as follows:

$$[\hat{z}_{ij}] = 1.0653 + j1.5088$$

$$[\hat{z}_{in}] = [0.0953 + j1.3645 \quad 0.0953 + j1.1309]$$

$$[\hat{z}_{nj}] = \begin{bmatrix} 0.0953 + j1.3645 \\ 0.0953 + j1.1309 \end{bmatrix}$$

$$[\hat{z}_{nn}] = \begin{bmatrix} 3.3109 + j1.9375 & 0.0953 + j1.1309 \\ 0.0953 + j1.1309 & 0.7023 + j1.5085 \end{bmatrix} \Omega/mile$$

Applying Kron's reduction method will result in a single impedance that represents the equivalent single-phase impedance of the tape shield cable and the neutral conductor.

$$z_{1p} = [\hat{z}_{ij}] - [\hat{z}_{in}] \cdot [\hat{z}_{nn}]^{-1} \cdot [\hat{z}_{nj}]$$

$$z_{1p} = 1.3127 + j0.6418 \, \Omega/mile$$

Since the single-phase line is on phase b, then the phase impedance matrix for the line is

$$[z_{abc}] = \begin{bmatrix} 0 & 0 & 0 \\ 0 & 1.3124 + j0.6425 & 0 \\ 0 & 0 & 0 \end{bmatrix} \Omega/mile$$

4.2.3 Parallel Underground Distribution Lines

The procedure for computing the phase impedance matrix for two overhead parallel lines has been presented in Section 4.1.7. Figure 4.19 shows two concentric neutral parallel lines, each with a separate grounded neutral conductor.

The process for computing the 6×6 phase impedance matrix follows exactly the same procedure as for the overhead lines. In this case, there are a total of 14 conductors (six phase conductors, six equivalent concentric neutral conductors, and two grounded neutral conductors). Applying Carson's equations will result in a 14×14 primitive impedance matrix. This matrix is partitioned between the sixth and seventh rows and columns. The Kron reduction is applied to form the final 6×6 phase impedance matrix.

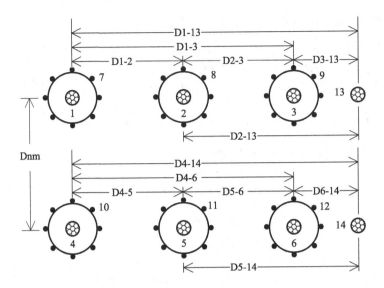

FIGURE 4.19 Parallel concentric neutral underground lines.

Example 4.6: Two concentric neutral three-phase underground parallel lines are shown in Figure 4.20.

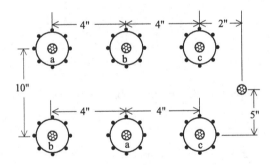

FIGURE 4.20 Parallel concentric neutral three-phase lines for Example 4.6.

Cables (both lines): 250 kcmil, 1/3 neutral
Extra neutral: 4/0, 7 strand copper
Determine the 6×6 phase impedance matrix.

Solution:

From Appendix B for the cables:

Outside diameter: $d_{od} = 1.29''$
Neutral strands: $k = 13$ #14 copper strands

Series Impedance of Overhead and Underground lines

From Appendix A for the conductors:

250 kcmil Al: $\quad GMR_c = 0.0171'$, $r_c = 0.41$ Ω/mile, $d_c = 0.567''$
#14 Copper: $\quad GMR_s = 0.00208'$, $r_s = 14.8722$ Ω/mile, $d_s = 0.0641''$
4/0 Copper: $\quad GMR_n = 0.01579''$, $r_n = 0.303$ Ω/mile, $d_n = 0.522''$

The radius of the circle to the center of the strands is

$$R = \frac{d_{od} - d_s}{24} = \frac{1.29 - 0.0641}{24} = 0.0511'$$

The equivalent geometric mean radius (GMR_{cn}) of the concentric neutral strands is computed as

$$GMR_{cn} = \sqrt[k]{GMR \cdot k \cdot R^{k-1}} = \sqrt[13]{0.00208 \cdot 13 \cdot 0.05111^{12}} = 0.0486'$$

The positions of the six cables and extra neutral using Cartesian coordinates with the phase *a* cable in line 1 (top line) as the ordinate. Note the phasing in both lines.

Phase a, line 1: $\quad d_1 = 0 + j0$
Phase b, line 1: $\quad d_2 = \frac{4}{12} + j0$
Phase c, line 1: $\quad d_3 = \frac{8}{12} + j0$

Phase a, line 2: $\quad d_4 = \frac{4}{12} - j\frac{10}{12}$
Phase b, line 1: $\quad d_5 = 0 - j\frac{10}{12}$
Phase c, line 1: $\quad d_6 = \frac{8}{12} - j\frac{10}{12}$

Equivalent neutrals:

Phase a, line 1: $\quad d_7 = d_1 + jR_b$
Phase b, line 1: $\quad d_8 = d_2 + jR_b$
Phase c, line 1: $\quad d_9 = d_3 + jR_b$

Phase a, line 2: $\quad d_{10} = d_4 + jR_b$
Phase b, line 2: $\quad d_{11} = d_5 + jR_b$
Phase c, line 2: $\quad d_{12} = d_6 + jR_b$

Extra neutral:

$$d_{13} = \frac{10}{12} - j\frac{5}{12}$$

The spacing matrix defining the distances between conductors can be computed by the following:

$$i = 1 \text{ to } 13 \; j = 1 \text{ to } 13$$

$$D_{i,j} = |d_i - d_j|$$

The diagonal terms of the spacing matrix are defined as the appropriate GMR:

$$D_{1,1} = D_{2,2} = D_{3,3} = D_{4,4} = D_{5,5} = D_{6,6} = GMR_c = 0.0171'$$

$$D_{7,7} = D_{8,8} = D_{9,9} = D_{10,10} = D_{11,11} = D_{12,12} = GMR_{cn} = 0.0486'$$

$$D_{13,13} = GMR_n = 0.01579'$$

The resistance matrix is defined as the following:

$$r_1 = r_2 = r_3 = r_4 = r_5 = r_6 = 0.41\,\Omega/\text{mile}$$

$$r_7 = r_8 = r_9 = r_{10} = r_{11} = r_{12} = \frac{r_s}{k} = \frac{14.8722}{13} = 1.144\,\Omega/\text{mile}$$

$$r_{13} = r_n = 0.303\,\Omega/\text{mile}$$

The primitive impedance matrix (13×13) is computed using Carson's equations:

$$i = 1 \text{ to } 13 \quad j = 1 \text{ to } 13$$

$$zp_{i,j} = 0.0953 + j0.12134 \cdot \left(\ln\left(\frac{1}{D_{i,j}}\right) + 7.93402 \right)$$

$$zp_{i,i} = r_i + 0.0953 + j0.12134 \cdot \left(\ln\left(\frac{1}{D_{i,i}}\right) + 7.93402 \right)$$

Once the primitive impedance matrix is developed, it is partitioned between the sixth and seventh rows and columns, and the Kron reduction method is applied to develop the 6×6 phase impedance matrix. The phase impedance matrix in partitioned form is as follows:

$$[z_{11}]_{abc} = \begin{bmatrix} 0.6423 + j0.4346 & 0.1174 + j0.0671 & 0.1352 + j0.0046 \\ 0.1174 + j0.0671 & 0.6239 + j0.3982 & 0.1601 + j0.0558 \\ 0.1352 + j0.0046 & 0.1601 + j0.0558 & 0.6094 + j0.4086 \end{bmatrix} \Omega/\text{mile}$$

$$[z_{12}]_{abc} = \begin{bmatrix} 0.1174 - j0.0155 & 0.1331 + j0.0058 & 0.1010 - j0.0254 \\ 0.1095 - j0.0239 & 0.1175 - j0.0164 & 0.0996 - j0.0268 \\ 0.0998 - j0.0273 & 0.1013 - j0.0268 & 0.0092 - j0.0200 \end{bmatrix} \Omega/\text{mile}$$

$$[z_{21}]_{abc} = \begin{bmatrix} 0.1174 - j0.0155 & 0.1095 - j0.0239 & 0.0998 - j0.0273 \\ 0.1331 + j0.0058 & 0.1175 - j0.0164 & 0.1023 - j0.0268 \\ 0.1010 - j0.0254 & 0.0996 - j0.0268 & 0.0992 - j0.0200 \end{bmatrix} \Omega/\text{mile}$$

$$[z_{22}]_{abc} = \begin{bmatrix} 0.6245 + j0.4057 & 0.1779 + j0.0770 & 0.1597 + j0.0661 \\ 0.1779 + j0.0770 & 0.6427 + j0.4440 & 0.1353 + j0.0142 \\ 0.1597 + j0.0661 & 0.1353 + j0.0142 & 0.6077 + j0.4185 \end{bmatrix} \Omega/\text{mile}$$

4.3 SUMMARY

This chapter has been devoted to presenting methods for computing the phase impedances of overhead lines and underground cables. Carson's equations have been modified to simplify the computation of the phase impedances. When using the modified Carson's equations, there is no need to make any assumptions, such as transposition of the lines. By assuming an untransposed line and including the actual phasing of the line, the most accurate values of the phase impedances, self and mutual, are determined. It is highly recommended that no assumptions be made in the computation of the impedances. Since voltage drop is a primary concern on a distribution line, the impedances used for the line must be as accurate as possible. This chapter also included the process of applying Carson's equations to two distribution lines that are physically parallel. This same approach would be taken when there are more than two lines physically parallel.

PROBLEMS

4.1 Modify the MATLAB scripts M0401.m and M0404.m to import the conductor data in the AppendexA.csv file. Your new script should import the data parameters for any conductor you wish to use.

4.2 The configuration of a three-phase overhead line is shown in Figure 4.21.

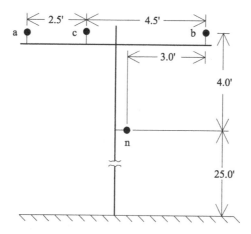

FIGURE 4.21 Three-phase configuration for Problem 4.2.

Using the MATLAB script from Problem 4.1, please import the following conductor data
Phase Conductors: 1/0 ACSR
Neutral Conductor: 1/0 ACSR
and use the script to
(a) Determine the phase impedance matrix $[z_{abc}]$ in Ohms/mile
(b) Determine the neutral transformation matrix $[t_n]$

4.3 The line of Problem 4.2 is connected to a 12.47 kV balanced three-phase source. The length of the line is 6,000 feet and serves a per-phase impedance load of 25 + j15 ohms. Determine
   (a) the three-phase line impedance matrix,
   (b) the three-phase load currents,
   (c) the three-phase line-to-neutral load voltages,
   (d) the per-phase load kW and kvar,
   (e) the percent voltage drop/phase.
4.4 Change Problem 4.3 for the line serving a balanced per-phase load of 5,000 kVA at 0.9 power factor. Modify the MATLAB script M0402.m to determine the required solutions listed in Problem 4.3.
4.5 Determine the phase impedance $[z_{abc}]$ matrix in Ω/mile for the two-phase configuration in Figure 4.22.

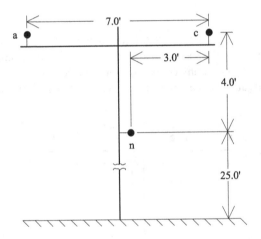

FIGURE 4.22 Two-phase configuration for Problem 4.5.

Phase Conductors: 336,400 26/7 ACSR
Neutral Conductor: 4/0 6/1 ACSR

4.6 Determine the phase impedance $[z_{abc}]$ matrix in Ω/mile for the single-phase configuration shown in Figure 4.23.

Phase and Neutral Conductors: 1/0   6/1 ACSR

4.7 Create the spacings of Problems 4.2, 4.5, and 4.6 in the distribution analysis program WindMil. Compare the phase impedance matrices to those computed in the previous problems.
4.8 Determine the phase impedance matrix $[z_{abc}]$ in Ω/mile for the three-phase pole configuration in Figure 4.24. The phase and neutral conductors are 250,000 all aluminum.

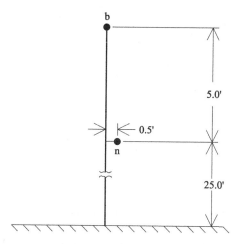

FIGURE 4.23 Single-phase pole configuration for Problem 4.3.

4.9 Determine the $[z_{abc}]$ matrix in Ω/mile for the three-phase configuration shown in Figure 4.25. The phase conductors are 350,000 all aluminum and the neutral conductor is 250,000 all aluminum.

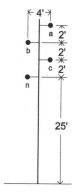

FIGURE 4.24 Three-phase pole configuration for Problem 4.8.

4.10 A 4/0 aluminum concentric neutral cable is to be used for a single-phase lateral. The cable has a full neutral (see Appendix B). Determine the impedance of the cable and the resulting phase impedance matrix in Ω/mile assuming the cable is connected to phase b.

4.11 Three 250,000 CM aluminum concentric cables with one-third neutral are buried in a trench in a horizontal configuration (see Figure 4.14). Determine the $[z_{abc}]$ the matrix in Ω/1,000 feet. assuming phasing of c-a-b.

4.12 Create the spacings and configurations of Problems 4.10 and 4.11 in WindMil. Compare the values of the phase impedance matrices to those

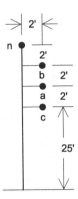

FIGURE 4.25 Three-phase pole configuration for Problem 4.9.

computed in the previous problems. In order to check the phase impedance matrix, it will be necessary for you to connect the line to a balanced three-phase source. A source of 12.47 kV works fine.

4.13 A single-phase underground line is composed of a 350,000 CM aluminum tape shielded cable. A 4/0 copper conductor is used as the neutral. The cable and neutral are separated by 4 inches. Determine the phase impedance matrix in Ω/mile for this single-phase cable line assuming phase c.

4.14 Three one-third neutral 2/0 aluminum jacketed concentric neutral cables are installed in a 6-inch conduit. Assume the cable jacket has a thickness of 0.2 inches and the cables lie in a triangular configuration inside the conduit. Compute the phase impedance matrix in Ω/mile for this cabled line.

4.15 Create the spacing and configuration of Problem 4.14 in WindMil. Connect a 12.47 kV source to the line and compare results to those of 4.14.

4.16 Two three-phase distribution lines are physically parallel as shown in Figure 4.26.

Use the code from Problem 4.1 to import the following conductor data.

Line # 1 (left side):	Phase conductors = 266,800 26/7 ACSR
	Neutral conductor = 3/0 6/1 ACSR
Line # 2 (right side):	Phase conductors = 300,000 CON LAY Aluminum
	Neutral conductor = 4/0 CLASS A Aluminum

Use this script to
(a) determine the 6×6 phase impedance matrix, and
(b) determine the neutral transform matrix.

Series Impedance of Overhead and Underground lines 91

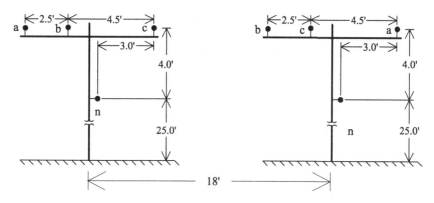

FIGURE 4.26 Parallel OH lines.

4.17 Two concentric neutral underground three-phase lines are physically parallel, as shown in Figure 4.27.

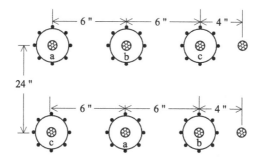

FIGURE 4.27 Parallel concentric neutral three-phase lines.

Line # 1 (top):	Cable = 250 kcmil, 1/3 neutral
	Additional neutral: 4/0 6/1 ACSR
Line #2 (bottom):	Cable = 2/0 kcmil, 1/3 neutral
	Additional neutral: 2/0 ACSR

(a) Determine the 6×6 phase impedance matrix
(b) Determine the neutral transform matrix

4.18 Many regions of the world use a frequency 50 Hz for their respective power systems. Modify the MATLAB script M0401.m to redo Example 4.1 for a 50 Hz system.

(a) Determine the phase impedance matrix
(b) Determine the neutral transformation matrix
(c) Comment on the effect that this change in frequency has on the transmission line's overall performance

WINDMIL ASSIGNMENT

Follow the method outlined in the user's manual to build a system called "System 1" in WindMil that will have the following components:

- 12.47 kV line-to-line source. The "Bus Voltage" should be set to 120 volts
- Connect to the node and call it Node 1
- A 10,000-foot-long overhead three distribution line as defined in Problem 4.1 Call this line OH-1
- Connect a node to the end of the line and call it Node 2
- A wye-connected, unbalanced, three-phase load is connected to Node 2 and is modeled as a constant PQ load with values of the following:
  - Phase a–g: 1,000 kVA, Power factor = 90% lagging
  - Phase b–g: 800 kVA, Power factor = 85% lagging
  - Phase c–g: 1,200 kVA, Power factor = 95% lagging

Determine the voltages on a 120-volt base at Node 2 and the current flowing on the OH-1 line.

REFERENCES

1. Glover, J. D., Sarma, M., *Power System Analysis and Design*, PWS-Kent Publishing, Boston, MA, 2nd Edition, 1994.
2. Carson, John R., "Wave propagation in overhead wires with ground return", *Bell System Technical Journal*, New York, Vol. 5, 539–554, 1926.
3. Kron, G., "Tensorial analysis of integrated transmission systems, part I, the six basic reference frames", *AIEE Transactions*, Vol. 71, 1239–1248, 1952.

5 Shunt Admittance of Overhead and Underground Lines

The shunt admittance of a line consists of the conductance and the capacitive susceptance. The conductance is usually ignored because it is very small compared to the capacitive susceptance. The capacitance of a line is the result of the potential difference between conductors. A charged conductor creates an electric field that emanates outward from the center of the conductor. Lines of equipotential are created that are concentric to the charged conductor. This is illustrated in Figure 5.1.

In Figure 5.1, a difference of potential between two points (P_1 and P_2) is a result of the electric field of the charged conductor. When the potential difference between the two points is known, then the capacitance between the two points can be computed. If there are other charged conductors nearby, the potential difference between the two points will be a function of the distance to the other conductors and the charge on each conductor. The principle of superposition is used to compute the total voltage drop between two points and then the resulting capacitance between the points. Understand that the points can be points in space or the surface of two conductors or the surface of a conductor and ground.

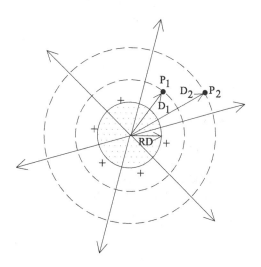

FIGURE 5.1 Electric field of a charged round conductor.

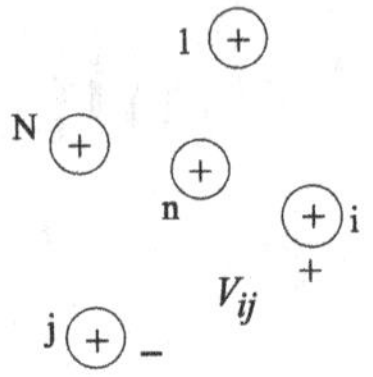

FIGURE 5.2 Array of round conductors.

5.1 GENERAL VOLTAGE DROP EQUATION

Figure 5.2 shows an array of N positively charged solid round conductors. Each conductor has a unique uniform charge density of q C/meter.

The voltage drop between conductor i and conductor j as a result of all of the charged conductors is given by

$$V_{ij} = \frac{1}{2\pi\epsilon}\left(q_1 \ln\frac{D_{1j}}{D_{1i}} + \ldots + q_i \ln\frac{D_{ij}}{RD_i} + \ldots + q_j \ln\frac{RD_j}{D_{ij}} + \ldots + q_N \ln\frac{D_{Nj}}{D_{Ni}}\right) \quad (5.1)$$

Equation 5.1 can be written in a general form as

$$V_{in} = \frac{1}{2\pi\epsilon}\sum_{n=1}^{N} q_n \ln\frac{D_{nj}}{D_{ni}}, \quad (5.2)$$

where
  $\epsilon = \epsilon_0 \epsilon_r$ = permittivity of the medium,
  ϵ_0 = permittivity of free space = 8.85×10^{-12} (F/meter),
  ϵ_r = relative permittivity of the medium,
  q_n = charge density on conductor n (C/m),
  D_{ni} = distance between conductor n and conductor i (feet),
  D_{nj} = distance between conductor n and conductor j (feet),
  D_{nn} = radius (RD_n) of conductor n (feet).

5.2 OVERHEAD LINES

The method of conductors and their images is employed in the calculation of the shunt capacitance of overhead lines. This is the same concept that was used in Chapter 4 in the general application of Carson's equations. Figure 5.3 illustrates the conductors and their images and will be used to develop a general voltage drop equation for overhead lines.

In Figure 5.3, it is assumed that

$$q_i' = -q_i$$

$$q_j' = -q_j \quad (5.3)$$

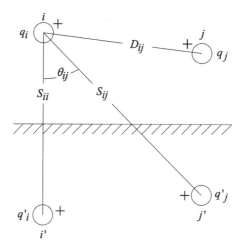

FIGURE 5.3 Conductors and images.

$$V_{ii} = \frac{1}{2\pi\epsilon}\left(q_i \ln\frac{S_{ii}}{RD_i} - q'_j \ln\frac{RD_i}{S_{ii}} + q_j \ln\frac{S_{ij}}{D_{ij}} - q'_j \ln\frac{D_{ij}}{S_{ij}}\right) \quad (5.4)$$

Because of this, the assumptions of Equation 5.3, Equation 5.4 can be simplified to:

$$V_{ii} = \frac{1}{2\pi\epsilon}\left(q_i \ln\frac{S_{ii}}{RD_i} - q_i \ln\frac{RD_i}{S_{ii}} + q_j \ln\frac{S_{ij}}{D_{ij}} - q_j \ln\frac{D_{ij}}{S_{ij}}\right)$$

$$V_{ii} = \frac{1}{2\pi\epsilon}\left(q_i \ln\frac{S_{ii}}{RD_i} + q_i \ln\frac{S_{ii}}{RD_i} + q_j \ln\frac{S_{ij}}{D_{ij}} + q_j \ln\frac{S_{ij}}{D_{ij}}\right)$$

$$V_{ii} = \frac{1}{2\pi\epsilon}\left(2q_i \ln\frac{S_{ii}}{RD_i} + 2q_j \ln\frac{S_{ij}}{D_{ij}}\right) \quad (5.5)$$

where
S_{ii} = distance from conductor i to its image i' (feet),
S_{ij} = distance from conductor i to the image of conductor j (feet),
D_{ij} = distance from conductor i to conductor j (feet),
RD_i = radius of conductor i in (feet).

Equation 5.5 gives the total voltage drop between conductor i and its image. The voltage drop between conductor i and ground will be one-half of that given in Equation 5.5.

$$V_{ig} = \frac{1}{2\pi\epsilon}\left(q_i \ln\frac{S_{ii}}{RD_i} + q_j \ln\frac{S_{ij}}{D_{ij}}\right) \quad (5.6)$$

Equation 5.6 can be written in general form as

$$V_{ig} = P_{ii} \cdot q_i + P_{ij} \cdot q_j, \tag{5.7}$$

where P_{ii} and P_{ij} are the self and mutual "potential coefficients".

For overhead lines, the relative permittivity of air is assumed to be 1.0 so that

$$\varepsilon_{air} = 1.0 \times 8.85 \times 10^{-12} \text{ F/meter},$$

$$\varepsilon_{air} = 1.4240 \times 10^{-2} \ \mu F/\text{mile} \tag{5.8}$$

Using the value of permittivity in μF/mile, the self and mutual potential coefficients are defined in Equations 5.9 and 5.10.

$$\hat{P}_{ii} = 11.17689 \cdot \ln \frac{S_{ii}}{RD_i} \tag{5.9}$$

$$\hat{P}_{ij} = 11.17689 \cdot \ln \frac{S_{ij}}{D_{ij}} \text{ mile}/\mu F \tag{5.10}$$

Note: The values of RD_i, S_{ii}, S_{ij}, and D_{ij} must all be in the same units. For overhead lines, the distances between conductors are typically specified in feet while the value of the conductor diameter from a table will typically be in inches. Care must be taken to assure that the radius in feet is used in applying the two equations.

For an overhead line of **ncond** conductors, the "primitive potential coefficient matrix" $\left[\hat{P}_{primitive}\right]$ can be constructed. The primitive potential coefficient matrix will be an **ncond × ncond** matrix. For a four-wire grounded wye line, the primitive coefficient matrix will be of the form

$$\left[\hat{P}_{primitive}\right] = \begin{bmatrix} \hat{P}_{aa} & \hat{P}_{ab} & \hat{P}_{ac} & \cdot & \hat{P}_{an} \\ \hat{P}_{ba} & \hat{P}_{bb} & \hat{P}_{bc} & \cdot & \hat{P}_{bn} \\ \hat{P}_{ca} & \hat{P}_{cb} & \hat{P}_{cc} & \cdot & \hat{P}_{cn} \\ \cdot & \cdot & \cdot & \cdot & \cdot \\ \hat{P}_{na} & \hat{P}_{nb} & \hat{P}_{nc} & \cdot & \hat{P}_{nn} \end{bmatrix} \tag{5.11}$$

Equation 5.11 shows the partition of the matrix between the third and fourth rows and columns. In partitioned from Equation 5.11 becomes

$$\left[\hat{P}_{primitve}\right] = \begin{bmatrix} \left[\hat{P}_{ij}\right] & \left[\hat{P}_{in}\right] \\ \left[\hat{P}_{nj}\right] & \left[\hat{P}_{nn}\right] \end{bmatrix} \tag{5.12}$$

Shunt Admittance of Overhead and Underground Lines

Because the neutral conductor is grounded, the matrix can be reduced using the "Kron reduction" method to an *n*-phase × *n*-phase phase potential coefficient matrix $[P_{abc}]$.

$$[P_{abc}] = [\hat{P}_{ij}] - [\hat{P}_{in}][\hat{P}_{nn}]^{-1}[\hat{P}_{jn}] \tag{5.13}$$

The inverse of the potential coefficient matrix will give the *n*-phase × *n*-phase capacitance matrix $[C_{abc}]$.

$$[C_{abc}] = [P_{abc}]^{-1} \; \mu F/\text{mile} \tag{5.14}$$

For a two-phase line, the capacitance matrix of Equation 5.14 will be 2×2. A row and column of zeros must be inserted for the missing phase. For a single-phase line, Equation 5.14 will result in a single element. Again, rows and columns of zero must be inserted for the missing phase. In the case of the single-phase line, the only non-zero term will be that of the phase in use.

Neglecting the shunt conductance, the phase shunt admittance matrix is given by

$$[y_{abc}] = 0 + j\omega[C_{abc}] \; \mu S/\text{mile}, \tag{5.15}$$

where

$$\omega = 2\pi f = 376.9911$$

Example 5.1: Determine the shunt admittance matrix for the overhead line of Example 4.1. Assume that the neutral conductor is 25 feet above ground.

The diameters of the phase and neutral conductors from the conductor table (Appendix A) are as follows:

Conductor:	336,400 26/7 ACSR	$d_c = 0.721$ in.,	$RD_c = 0.03004$ ft
	4/0 6/1 ACSR	$d_n = 0.563$ in.,	$RD_n = 0.02346$ ft

Using the Cartesian coordinated in Example 4.1, the image distance matrix is given by

$$S_{ij} = |d_i - d_j^*|$$

where

$$d_j^* = \text{the conjugate of } d_j$$

For the configuration, the distances between conductors and images in matrix form are

$$[S] = \begin{bmatrix} 58.0000 & 58.0539 & 58.4209 & 54.1479 \\ 58.0539 & 58.0000 & 58.1743 & 54.0208 \\ 58.4209 & 58.1743 & 58.0000 & 54.0833 \\ 54.1479 & 54.0208 & 54.0833 & 58.0000 \end{bmatrix} \text{ft}$$

The self-primitive potential coefficient for phase a and the mutual primitive potential coefficient between phases **a** and **b** are

$$\hat{P}_{aa} = 11.17689 \cdot \ln\frac{58}{0.03004} = 84.5600 \, \text{mile}/\mu F,$$

$$\hat{P}_{ab} = 11.17689 \cdot \ln\frac{58.0539}{2.5} = 35.1522 \, \text{mile}/\mu F$$

Using Equations 5.9 and 5.10, the total primitive potential coefficient matrix is computed to be

$$[\hat{P}_{primitive}] = \begin{bmatrix} 84.5600 & 35.1522 & 23.7174 & 25.2469 \\ 35.1522 & 84.5600 & 28.6058 & 28.3590 \\ 23.7174 & 28.6058 & 84.5600 & 26.6231 \\ 25.2469 & 28.3590 & 26.6231 & 84.5600 \end{bmatrix} \text{mile}/\mu F$$

Since the fourth conductor (neutral) is grounded, the Kron reduction method is used to compute the "phase potential coefficient matrix". Because only one row and column need to be eliminated, the $[\hat{P}_{nn}]$ term is a single element so that the Kron reduction equation for this case can be modified to

$$P_{ij} = \hat{P}_{ij} - \frac{\hat{P}_{ij} \cdot \hat{P}_{ij}}{\hat{P}_{ij}},$$

where $i = 1,2,3$ and $j = 1, 2, 3$.

For Example 5.1, the value of P_{cb} is computed to be

$$P_{cb} = \hat{P}_{32} - \frac{\hat{P}_{34} \cdot \hat{P}_{42}}{\hat{P}_{44}} = 19.7957$$

Following the Kron reduction, the phase potential coefficient matrix is

Shunt Admittance of Overhead and Underground Lines

$$[P_{abc}] = \begin{bmatrix} 77.1194 & 26.7944 & 15.8714 \\ 26.7944 & 75.1720 & 19.7957 \\ 15.8714 & 19.7957 & 76.2923 \end{bmatrix} \text{mile}/\mu F$$

Invert $[P_{abc}]$ to determine the shunt capacitance matrix:

$$[C_{abc}] = [P]^{-1} = \begin{bmatrix} 0.015 & -0.0049 & -0.0019 \\ -0.0049 & 0.0159 & -0.0031 \\ -0.0019 & -0.0031 & 0.0143 \end{bmatrix} \text{mile}/\mu F$$

Multiply $[C_{abc}]$ by the radian frequency to determine the final three-phase shunt admittance matrix.

$$[y_{abc}] = j376.9911 \cdot [C_{abc}] = \begin{bmatrix} j5.6711 & -j1.8362 & -j0.7033 \\ -j1.8362 & j5.9774 & -j1.1690 \\ -j0.7033 & -j1.1690 & j5.3911 \end{bmatrix} \mu S/\text{mile}$$

5.2.1 THE SHUNT ADMITTANCE OF OVERHEAD PARALLEL LINES

The development of the shunt admittance matrix for parallel overhead lines is similar to the steps taken to create the phase impedance matrix. The numbering of the conductors must be the same as was used in developing the phase impedance matrix. To develop the shunt admittance matrix for overhead lines, it is necessary to know the distance from each conductor to ground, and it will be necessary to know the radius in feet for each conductor.

The first step is to create the primitive potential coefficient matrix. This will be an *ncond* × *ncond* matrix where *ncond* is the total number of phase and neutral conductors. For the lines of Figure 4.9, *ncond* will be 7; for two lines each with its own grounded neutral, *ncond* will be 8.

The elements of the primitive potential coefficient matrix are given by

$$\hat{P}_{ii} = 11.17689 \cdot \ln \frac{S_{ii}}{RD_i}$$

$$\hat{P}_{ij} = 11.17689 \cdot \ln \frac{S_{ij}}{D_{ij}} \text{ mile}/\mu F \tag{5.16}$$

where
S_{ii} = distance in ft from a conductor to its image below ground,
S_{ij} = distance in feet from a conductor to the image of an adjacent conductor,
D_{ij} = distance in feet between two overhead conductors,
RD_i = radius in feet of conductor i.

The last one or two rows and columns of the primitive potential coefficient matrix are eliminated by using Kron reduction. The resulting voltage equation is

$$\begin{bmatrix} V1_{ag} \\ V1_{bg} \\ V1_{cg} \\ V2_{ag} \\ V2_{bg} \\ V2_{cg} \end{bmatrix} = \begin{bmatrix} P11_{aa} & P11_{ab} & P11_{ac} & P12_{aa} & P12_{ab} & P12_{ac} \\ P11_{ba} & P11_{bb} & P11_{bc} & P12_{ba} & P12_{bb} & P12_{bc} \\ P11_{ca} & P11_{cb} & P11\_cc & P12_{ca} & P12_{cb} & P12_{cc} \\ P21_{aa} & P21_{ab} & P21_{ac} & P22_{aa} & P22_{ab} & P22_{ac} \\ P21_{ba} & P21_{bb} & P21_{bc} & P22_{ba} & P22_{bb} & P22_{bc} \\ P21_{ca} & P21_{cb} & P21_{cc} & P22_{ca} & P22_{cb} & P2_{cc} \end{bmatrix} \cdot \begin{bmatrix} q1_a \\ q1_b \\ q1_c \\ q2_a \\ q2_b \\ q2_c \end{bmatrix} V \quad (5.17)$$

In shorthand form, Equation 5.17 is

$$[V_{LG}] = [P] \cdot [q] \quad (5.18)$$

The shunt capacitance matrix is determined by

$$[q] = [P]^{-1}[V_{LG}] = [C] \cdot [V_{LG}] \quad (5.19)$$

The resulting capacitance matrix is partitioned between the third and fourth rows and columns.

$$[C] = [P]^{-1} = \begin{bmatrix} [C11] & [C12] \\ [C21] & [C22] \end{bmatrix} \quad (5.20)$$

The shunt admittance matrix is given by

$$[y_{abc}] = j\omega \cdot [C] \cdot 10^{-6} = \begin{bmatrix} [y11] & [y12] \\ [y21] & [y22] \end{bmatrix} \mu S/\text{mile}, \quad (5.21)$$

where
$\omega = 2 \cdot \pi \cdot frequency$.

Shunt Admittance of Overhead and Underground Lines

Example 5.2: Determine the shunt admittance matrix for the parallel overhead lines of Example 4.2.

The position coordinates for the seven conductors and the distance matrix are defined in Example 4.4. The diagonal terms of the distance matrix (Example 4.4) matrix must be the radius in feet of the individual conductors. For this example,

$$D_{11} = D_{22} = D_{23} = \frac{d_1}{24} = \frac{0.721}{24} = 0.0300 \text{ ft.,}$$

$$D_{44} = D_{55} = D_{66} = \frac{d_2}{24} = \frac{0.567}{24} = 0.0236 \text{ ft.,}$$

$$D_{77} = \frac{d_n}{24} = \frac{0.721}{24} = 0.0235 \text{ ft}$$

The resulting distance matrix is

$$[D] = \begin{bmatrix} 0.0300 & 2.5000 & 7.0000 & 3.2016 & 7.2801 & 2.0000 & 7.2111 \\ 2.5000 & 0.0300 & 4.5000 & 2.0000 & 4.9244 & 3.2016 & 6.1847 \\ 7.0000 & 4.5000 & 0.0300 & 4.9244 & 2.0000 & 7.2801 & 6.7082 \\ 3.2016 & 2.0000 & 4.9244 & 0.0236 & 4.5000 & 2.5000 & 4.2720 \\ 7.2801 & 4.9244 & 2.0000 & 4.5000 & 0.0236 & 7.0000 & 5.0000 \\ 2.0000 & 3.2016 & 7.2801 & 2.5000 & 7.0000 & 0.0236 & 5.6569 \\ 7.2111 & 6.1847 & 6.7082 & 4.2720 & 5.0000 & 5.6569 & 0.0236 \end{bmatrix} \text{ft}$$

The distances between conductors and conductor images (image matrix) can be determined by

$$S_{ij} = |d_i - d_j^*|$$

For this example, the image matrix is

$$[S] = \begin{bmatrix} 70.000 & 70.045 & 70.349 & 68.046 & 68.359 & 68.000 & 64.125 \\ 70.045 & 70.000 & 70.145 & 68.000 & 68.149 & 68.046 & 64.018 \\ 70.349 & 70.145 & 70.000 & 68.149 & 68.000 & 68.359 & 64.070 \\ 68.046 & 68.000 & 68.149 & 66.000 & 66.153 & 66.047 & 62.018 \\ 68.359 & 68.149 & 68.000 & 66.153 & 66.000 & 66.370 & 66.370 \\ 68.000 & 68.046 & 68.359 & 66.047 & 66.370 & 66.000 & 62.129 \\ 64.125 & 64.018 & 64.070 & 62.018 & 62.073 & 66.370 & 60.000 \end{bmatrix} \text{ft}$$

The distance and image matrices are used to compute the 7 × 7 potential coefficient matrix by

$$Pp_{ij} = 11.17689 \cdot \ln \frac{S_{ij}}{D_{ij}}$$

The primitive potential coefficient matrix is partitioned between the sixth and seventh rows and columns and the Kron reduction method produces the 6 × 6 potential matrix. This matrix is then inverted and multiplied by $\omega = 376.9911$ to give the shunt admittance matrix. The final shunt admittance matrix in partitioned form is as follows:

$$[y11] = \begin{bmatrix} j6.2992 & -j1.3413 & -j0.4135 \\ -j1.3413 & j6.5009 & -j0.8038 \\ -j0.4135 & -j0.8038 & j6.0257 \end{bmatrix} \mu S/mile$$

$$[y12] = \begin{bmatrix} -j0.7889 & -j0.2992 & -j1.6438 \\ -j1.4440 & -j0.5698 & -j0.7988 \\ -j0.5553 & -j1.8629 & -j0.2985 \end{bmatrix} \mu S/mile$$

$$[y21] = \begin{bmatrix} -0.7889 & -1.4440 & -j0.5553 \\ -j0.2992 & -0.5698 & -j1.8629 \\ -j1.6438 & -j0.7988 & -j0.2985 \end{bmatrix} \mu S/mile$$

$$[y22] = \begin{bmatrix} j6.3278 & -j0.6197 & -j1.1276 \\ -0.2992 & -0.5698 & -j1.8629 \\ -j1.6438 & -j0.7988 & -j0.2985 \end{bmatrix} \mu S/mile$$

5.3 CONCENTRIC NEUTRAL CABLE UNDERGROUND LINES

Most underground distribution lines consist of one or more concentric neutral cables. Figure 5.4 illustrates a basic concentric neutral cable with the center conductor being the phase conductor and the concentric neutral strands displaced equally around a circle of radius R_b.

Referring to Figure 5.4, the following definitions apply:

R = radius of a circle passing through the centers of the neutral strands
d_c = diameter of the phase conductor

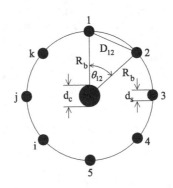

FIGURE 5.4 Basic concentric neutral cable.

Shunt Admittance of Overhead and Underground Lines

d_s = diameter of a neutral strand
k = total number of neutral strands

The concentric neutral strands are grounded so that they are all at the same potential. Because of the stranding, it is assumed that the electric field created by the charge on the phase conductor will be confined to the boundary of the concentric neutral strands. To compute the capacitance between the phase conductor and ground, the general voltage drop of Equation 5.2 will be applied. Since all the neutral strands are at the same potential, it is only necessary to determine the potential difference between the phase conductor p and strand **1**.

$$V_{p1} = \frac{1}{2\pi\varepsilon}\left(q_p \ln\frac{R_b}{RD_c} + q_1 \ln\frac{RD_s}{R_b} + q_2 \ln\frac{D_{12}}{R_b} + \ldots + q_i \ln\frac{D_{1i}}{R_b} + \ldots + q_k \ln\frac{D_{k1}}{R_b}\right), \quad (5.22)$$

where

$$RD_c = \frac{d_c}{2}$$

$$RD_s = \frac{d_s}{2}$$

It is assumed that each of the neutral strands carries the same charge such that

$$q_1 = q_2 = q_i = q_k = -\frac{q_p}{k} \quad (5.23)$$

Equation 5.22 can be simplified:

$$V_{p1} = \frac{1}{2\pi\varepsilon}\left[q_p \ln\frac{R_b}{RD_c} - \frac{q_p}{k}\left(\ln\frac{RD_s}{R_b} + \ln\frac{D_{12}}{R_b} + \ldots + \ln\frac{D_{1i}}{R_b} + \ldots + \ln\frac{D_{1k}}{R_b}\right)\right]$$

$$V_{p1} = \frac{q_p}{2\pi\varepsilon}\left[\ln\frac{R_b}{RD_c} - \frac{1}{k}\left(\ln\frac{RD_s \cdot D_{12} \cdot D_{1i} \cdot \ldots \cdot D_{1k}}{R_b^k}\right)\right] \quad (5.24)$$

The numerator of the second ln term in Equation 5.24 needs to be expanded. The numerator represents the product of the radius and the distances between strand i and all the other strands. Referring to Figure 5.4, the following relations apply:

$$\theta_{12} = \frac{2\pi}{k},$$

$$\theta_{13} = 2\cdot\theta_{12} = \frac{4\cdot\pi}{k}$$

In general, the angle between strand 1 and any other strand i is given by

$$\theta_{1i} = (i-1) \cdot \theta_{12} = \frac{(i-1) \cdot 2\pi}{k} \tag{5.25}$$

The distances between the various strands are given by the following:

$$D_{12} = 2 \cdot R_b \cdot \sin\left(\frac{\theta_{12}}{2}\right) = 2 \cdot R_b \cdot \sin\left(\frac{\pi}{k}\right)$$

$$D_{13} = 2 \cdot R_b \cdot \sin\left(\frac{\theta_{13}}{2}\right) = 2 \cdot R_b \cdot \sin\left(\frac{2\pi}{k}\right) \tag{5.26}$$

The distance between strand 1 and any other strand i is given by

$$D_{1i} = 2 \cdot R_b \cdot \sin\left(\frac{\theta_{1i}}{2}\right) = 2 \cdot R_b \cdot \sin\left[\frac{(i-1) \cdot \pi}{k}\right] \tag{5.27}$$

Equation 5.27 can be used to expand the numerator of the second log term of Equation 5.24.

$$RD_s \cdot D_{12} \cdot \ldots \cdot D_{1i} \cdot \ldots \cdot D_{1k}$$

$$RD_s \cdot R_b^{k-1} \cdot \left\{ 2\sin\left[\frac{\pi}{k}\right] \cdot 2 \right.$$

$$= \times \sin\left[\frac{2\pi}{k}\right] \cdot \ldots \cdot 2\sin\left[\frac{(i-1)\pi}{k}\right] \tag{5.28}$$

$$\left. \times \ldots \cdot 2\sin\left[\frac{(k-1)\pi}{k}\right] \right\}$$

The term inside the bracket in Equation 5.28 is a trigonometric identity that is merely equal to the number of strands k[1]. Using that identity, Equation 5.18 becomes

$$V_{p1} = \frac{q_p}{2\pi\varepsilon} \left[\ln\frac{R_b}{RD_c} - \frac{1}{k}\left(\ln\frac{k \cdot RD_s \cdot R_b^{k-1}}{R_b^k} \right) \right]$$

$$V_{p1} = \frac{q_p}{2\pi\varepsilon} \left[\ln\frac{R_b}{RD_c} - \frac{1}{k}\left(\ln\frac{k \cdot RD_s}{R_b} \right) \right] \tag{5.29}$$

Equation 5.29 gives the voltage drop from the phase conductor to neutral strand 1. Care must be taken that the units for the various radii are the same. Typically,

Shunt Admittance of Overhead and Underground Lines

underground spacings are given in inches, so the radii of the phase conductor (RD_c) and the strand conductor (RD_s) should be specified in inches.

Since the neutral strands are all grounded, Equation 5.29 gives the voltage drop between the phase conductor and ground. Therefore, the capacitance from phase to ground for a concentric neutral cable is given by

$$C_{pg} = \frac{q_p}{V_{p1}} = \frac{2\pi\varepsilon}{\ln\dfrac{R_b}{RD_c} - \dfrac{1}{k}\ln\dfrac{k \cdot RD_s}{R_b}} \;\mu F/\text{mile}, \tag{5.30}$$

where
$\varepsilon = \varepsilon_0 \varepsilon_r$ = permittivity of the medium,
ε_0 = permittivity of free space = 0.01420 μF/mile,
ε_r = relative permittivity of the medium.

The electric field of a cable is confined to the insulation material. Various types of insulation material are used, and each will have a range of values for the relative permittivity. Table 5.1 gives the range of values of relative permittivity for four common insulation materials [2].

Cross-linked polyethlyene is a very popular insulation material. If the minimum value of relative permittivity is assumed (2.3), the equation for the shunt admittance of the concentric neutral cable is given by

$$y_{ag} = 0 + j\frac{77.3619}{\ln\dfrac{R_b}{RD_c} - \dfrac{1}{k}\ln\dfrac{k \cdot RD_s}{R_b}} \;\mu S/\text{mile} \tag{5.31}$$

Example 5.3: Determine the three-phase shunt admittance matrix for the concentric neutral line of Example 4.3 in Chapter 4.

From Example 4.3,

$$R = 0.0511 \text{ft.} = 0.631 \text{in}$$

Diameter of the 250,000 AA phase conductor = 0.567 inches.

$$RD_c = \frac{0.567}{2} = 0.2835 \text{in}$$

Diameter of the #14 CU concentric neutral strand = 0.0641 inches.

$$RD_s = \frac{0.0641}{2} = 0.03205 \text{ in}$$

Substitute into Equation 5.31:

$$y_{ag} = j\frac{77.3619}{\ln\dfrac{R}{RD_c} - \dfrac{1}{k}\ln\dfrac{k\cdot RD_s}{R}}$$

$$y_{ab} = j\frac{77.3619}{\ln\left(\dfrac{0.6132}{0.2835}\right) - \dfrac{1}{13}\cdot\ln\left(\dfrac{13\cdot 0.03205}{0.6132}\right)} = j96.6098\,\mu S/\text{mile}$$

The phase admittance for this three-phase underground line is

$$[y_{abc}] = \begin{bmatrix} j96.6098 & 0 & 0 \\ 0 & j96.6098 & 0 \\ 0 & 0 & j96.6098 \end{bmatrix} \mu S/\text{mile}.$$

5.4 TAPE SHIELDED CABLE UNDERGROUND LINES

A tape shielded cable is shown in Figure 5.5.

Referring to Figure 5.5, R is the radius of a circle passing through the center of the tape shield. As with the concentric neutral cable, the electric field is confined to the insulation so that the relative permittivity of Table 5.1 will apply.

The tape shielded conductor can be visualized as a concentric neutral cable where the number of strands k has become infinite. When k in Equation 5.31 approaches infinity, the second term in the denominator approaches zero. Therefore, the equation for the shunt admittance of a tape shielded conductor becomes

$$y_{ag} = 0 + j\frac{77.3619}{\ln\dfrac{R}{RD_c}}\,\mu S/\text{mile}. \qquad (5.32)$$

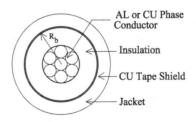

FIGURE 5.5 Tape shielded conductor.

TABLE 5.1
Typical Values of Relative Permittivity (ε_r)

Material	Range of Values of Relative Permittivity
Polyvinyl Chloride	3.4–8.0
Ethylene-Propylene Rubber	2.5–3.5
Polyethylene	2.5–2.6
Cross-Linked Polyethlyene	2.3–6.0

Example 5.4: Determine the shunt admittance of the single-phase tape shielded cable of Example 4.6 in Chapter 4. From Example 4.6, the outside diameter of the tape shield is 0.88 inches. The thickness of the tape shield (T) is 5 mils. Converting the radius of a circle passing through the center of the tape shield from Example 4.6 is

$$R_b = 0.4375 \text{ in.}$$

The diameter of the 1/0 AA phase conductor = 0.368 inches.

$$RD_c = \frac{d_c}{2} = \frac{0.368}{2} = 0.1840 \text{ in.}$$

Substitute into Equation 5.32:

$$y_{bg} = j\frac{77.3619}{\ln\left(\frac{R_b}{RD_c}\right)} = \frac{j77.3619}{\ln\left(\frac{0.4375}{0.184}\right)} = j89.3179 \frac{\mu S}{\text{mile}}$$

The line is on phase b so that the phase admittance matrix becomes

$$[y_{abc}] = \begin{bmatrix} 0 & 0 & 0 \\ 0 & j89.3179 & 0 \\ 0 & 0 & 0 \end{bmatrix} \mu S/\text{mile}$$

5.5 THE SHUNT ADMITTANCE OF PARALLEL UNDERGROUND LINES

For underground cable lines using either concentric neutral cables or tape shielded cables, the computation of the shunt admittance matrix is quite simple. The electric field created by the charged phase conductor does not link to adjacent conductors because of the presence of the concentric neutrals or the tape shield. As a result, the shunt admittance matrix for parallel underground lines will consist of diagonal terms only.

The diagonal terms for concentric neutral cables are given by

$$y_{ii} = 0 + j\frac{77.3619}{\ln\frac{R_b}{RD_i} - \frac{1}{k}\ln\frac{k \cdot RD_s}{R_b}} \mu S/\text{mile}, \quad (5.33)$$

where
 R_b = radius in feet of the circle going through the center of the neutral strands,
 RD_i = radius in feet of the center phase conductor,
 RD_s = radius in feet of the neutral strands,
 k = number of neutral strands.

The diagonal terms for tape shielded cables are given by

$$y_{ii} = 0 + j\frac{77.3619}{\ln\frac{R_b}{RD_i}} \mu S/mile, \qquad (5.34)$$

where
R_b = radius in feet of the circle passing through the center of the tape shield,
RD_i = radius in feet of the center phase conductor.

Example 5.5: Compute the shunt admittance matrix (6 × 6) for the concentric neutral underground configuration of Example 4.7.

From Example 4.7:

Diameter of the central conductor: $d_c = 0.567"$
Diameter of the strands: $d_s = 0.641"$
Outside diameters of concentric neutral strands: $d_{od} = 1.29"$
Radius of the circle passing through the strands: $R_b = \dfrac{d_{od} - d_s}{24} = 0.0511'$
Radius of the central conductor: $RD_c = \dfrac{d_c}{2} = \dfrac{0.567}{2} = 0.236'$
Radius of the strands: $RD_s = \dfrac{d_s}{24} = \dfrac{0.0641}{24} = 0.0027'$

Since all cables are identical, the shunt admittance of a cable is as follows:

$$y_c = 0 + j \cdot \frac{77.3619}{\ln\left(\dfrac{R_b}{RD_c}\right) - \dfrac{1}{k}\ln\left(\dfrac{k \cdot RD_s}{R_b}\right)} = 0 + j \cdot \frac{77.3619}{\ln\left(\dfrac{0.0511}{0.0236}\right) - \dfrac{1}{13}\ln\left(\dfrac{13 \cdot 0.0027}{0.0511}\right)}$$

$$y_c = 0 + j \cdot 96.6098 \, \mu S/mile$$

The phase admittance matrix is

$$[y_{abc}] = \begin{bmatrix} j96.6098 & 0 & 0 & 0 & 0 & 0 \\ 0 & j96.6098 & 0 & 0 & 0 & 0 \\ 0 & 0 & j96.6098 & 0 & 0 & 0 \\ 0 & 0 & 0 & j96.6098 & 0 & 0 \\ 0 & 0 & 0 & 0 & j96.6098 & 0 \\ 0 & 0 & 0 & 0 & 0 & j96.6098 \end{bmatrix}$$

× S/mile.

5.6 SUMMARY

Methods for computing the shunt capacitive admittance for overhead and underground lines have been presented in this chapter. Included is the development of computing the shunt admittance matrix for parallel overhead and underground lines.

Distribution lines are typically so short that the shunt admittance can be ignored. However, there are cases of long lightly loaded overhead lines where the shunt admittance should be included. Underground cables have a much higher shunt admittance per mile than overhead lines. Again, there will be cases where the shunt admittance of an underground cable should be included in the analysis process. When the analysis is being done using a computer, the approach to take is to model the shunt admittance for both overhead and underground lines. There is no need to make a simplifying assumption when it is not necessary.

PROBLEMS

5.1 Modify the MATLAB scripts M0501.m and M0502.m to import the conductor data in the AppendexA.csv file. Your new script should import the data parameters for any conductor you wish to use.

5.2 Modify the MATLAB M0502.m Determine the phase admittance matrix $[y_{abc}]$ in µS/mile for the three-phase overhead line of Problem 4.2.

5.3 Determine the phase admittance matrix in µS/mile for the two-phase line of Problem 4.5.

5.4 Determine the phase admittance matrix in µS/mile for the single-phase line of Problem 4.6.

5.5 Verify the results of Problems 5.2, 5.3, and 5.4 using WindMil.

5.6 Determine the phase admittance matrix in µS/mile for the three-phase line of Problem 4.8.

5.7 Modify the MATLAB script M0503.m to determine the phase admittance matrix in µS/mile for the single-phase concentric neutral cable of Problem 4.10.

5.8 Modify the MATLAB script M0503.m to determine the phase admittance matrix for the three-phase concentric neutral line of Problem 4.11.

5.9 Verify the results of Problems 5.7 and 5.8 using WindMil.

5.10 Determine the phase admittance matrix in µS/mile for the single-phase tape shielded cable line of Problem 4.13.

5.11 Determine the phase admittance for the three-phase tape shielded cable line of Problem 4.14.

5.12 Verify the results of Problems 5.10 and 5.11 using WindMil.

5.13 Modify the MATLAB script M0502.m to determine the shunt admittance matrix for the parallel overhead lines of Problem 4.16.

5.14 Modify the MATLAB script M0505.m to determine the shunt admittance matrix for the underground concentric neutral parallel lines of Problem 4.17.

WINDMIL ASSIGNMENT

Add to the WindMil System 1 a single-phase line connected to Node 2. Call this "System 2". The single-phase line is on phase b and is defined in Problem 4.5. Call this line OH-2. At the end of the line, connect a node and call it Node 3. The load at Node 3 is 200 kVA at a 90 % lagging power factor. The load is modeled as a constant impedance load.

Determine the voltages at the nodes on a 120-volt base and the currents flowing on the two lines.

REFERENCES

1. Glover, J. D., Sarma, M., *Power System Analysis and Design*, PWS-Kent Publishing, Boston, MA, 2nd Edition, 1995.
2. Arnold, T. P., Ed., *Power Cable Manual*, C.D. Mercier, Southwire Company, Carrollton, Georgia, 2nd Edition, 1997.

6 Distribution System Line Models

The modeling of distribution overhead and underground line segments is a critical step in the analysis of a distribution feeder. It is important in line modeling to include the actual phasing of the line and the correct spacing between conductors. Chapters 4 and 5 developed the method for the computation of the phase impedance and phase admittance matrices with no simplifying assumptions. Those matrices will be used in the models for overhead and underground line segments.

6.1 EXACT LINE SEGMENT MODEL

The model of a three-phase, two-phase, or single-phase overhead or underground line is shown in Figure 6.1.

When a line segment is two-phase (V-phase), or single-phase, some of the impedance and admittance values will be zero. Recall in Chapters 4 and 5 that in all cases, the phase impedance and phase admittance matrices were 3×3. Rows and columns of zeros for the missing phases represent two-phase and single-phase lines. Therefore, one set of equations can be developed to model all overhead and underground line segments. The values of the impedances and admittances in Figure 6.1 represent the total impedances and admittances for the line. That is, the phase impedance matrix, derived in Chapter 4, has been multiplied by the length of the line segment. The phase admittance matrix, derived in Chapter 5, has also been multiplied by the length of the line segment.

For the line segment of Figure 6.1, the equations relating the input (Node-n) voltages and currents to the output (Node-m) voltages and currents are developed as follows.

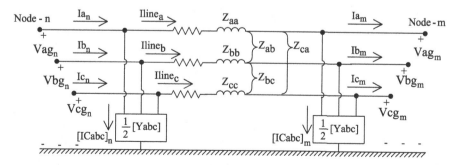

FIGURE 6.1 Three-phase line segment model.

KCL applied at Node-m:

$$\begin{bmatrix} Iline_a \\ Iline_b \\ Iline_c \end{bmatrix} = \begin{bmatrix} I_a \\ I_b \\ I_c \end{bmatrix}_m + \frac{1}{2} \cdot \begin{bmatrix} Y_{aa} & Y_{ab} & Y_{ac} \\ Y_{ba} & Y_{bb} & Y_{bc} \\ Y_{ca} & Y_{cb} & Y_{cc} \end{bmatrix} \cdot \begin{bmatrix} V_{ag} \\ V_{bg} \\ V_{cg} \end{bmatrix}_m \quad (6.1)$$

In condensed form Equation 6.1 becomes

$$[Iline_{abc}] = [I_{abc}]_m + \frac{1}{2} \cdot [Y_{abc}] \cdot [VLG_{abc}]_m \quad (6.2)$$

KVL applied to the model gives

$$\begin{bmatrix} V_{ag} \\ V_{bg} \\ V_{cg} \end{bmatrix}_n = \begin{bmatrix} V_{ag} \\ V_{bg} \\ V_{cg} \end{bmatrix}_m + \frac{1}{2} \cdot \begin{bmatrix} Z_{aa} & Z_{ab} & Z_{ac} \\ Z_{ba} & Z_{bb} & Z_{bc} \\ Z_{ca} & Z_{cb} & Z_{cc} \end{bmatrix} \cdot \begin{bmatrix} Iline_a \\ Iline_b \\ Iline_c \end{bmatrix} \quad (6.3)$$

In condensed form, Equation 6.3 becomes

$$[VLG_{abc}]_n = [VLG_{abc}]_m + \frac{1}{2} \cdot [Z_{abc}] \cdot [Iline_{abc}] \quad (6.4)$$

Substituting Equation 6.2 into Equation 6.4,

$$[VLG_{abc}]_n = [VLG_{abc}]_m + \frac{1}{2} \cdot [Z_{abc}] \cdot \left([I_{abc}]_m + \frac{1}{2} \cdot [Y_{abc}] \cdot [VLG_{abc}]_m \right) \quad (6.5)$$

Collecting terms,

$$[VLG_{abc}]_n = \left([U] + \frac{1}{2} \cdot [Z_{abc}] \cdot [Y_{abc}]_m \right) \cdot [VLG_{abc}]_m + [Z_{abc}] \cdot [I_{abc}]_m, \quad (6.6)$$

where

$$[U] = \begin{bmatrix} 1 & 0 & 0 \\ 0 & 1 & 0 \\ 0 & 0 & 1 \end{bmatrix} \quad (6.7)$$

Distribution System Line Models

Equation 6.6 is of the general form

$$[VLG_{abc}]_n = [a] \cdot [VLG_{abc}]_m + [b] \cdot [I_{abc}]_m, \qquad (6.8)$$

where

$$[a] = [U] + \frac{1}{2} \cdot [Z_{abc}] \cdot [Y_{abc}]_m, \qquad (6.9)$$

$$[b] = [Z_{abc}] \qquad (6.10)$$

The input current to the line segment at Node-n is

$$\begin{bmatrix} I_a \\ I_b \\ I_c \end{bmatrix}_n = \begin{bmatrix} Iline_a \\ Iline_b \\ Iline_c \end{bmatrix} + \frac{1}{2} \cdot \begin{bmatrix} Y_{aa} & Y_{ab} & Y_{ac} \\ Y_{ba} & Y_{bb} & Y_{bc} \\ Y_{ca} & Y_{cb} & Y_{cc} \end{bmatrix} \cdot \begin{bmatrix} Vag \\ Vbg \\ Vcg \end{bmatrix}_n \qquad (6.11)$$

In condensed form, Equation 3.11 becomes

$$[I_{abc}]_n = [Iline_{abc}] + \frac{1}{2} \cdot [Y_{abc}] \cdot [VLG_{abc}]_n \qquad (6.12)$$

Substitute Equation 6.2 into Equation 6.12:

$$[I_{abc}]_n = [I_{abc}]_m + \frac{1}{2}[Y_{abc}] \cdot [VLG_{abc}]_m + \frac{1}{2} \cdot [Y_{abc}] \cdot [VLG_{abc}]_n \qquad (6.13)$$

Substitute Equation 6.6 into Equation 6.13:

$$\begin{aligned}[I_{abc}]_n = &[I_{abc}]_m + \frac{1}{2}[Y_{abc}] \cdot [VLG_{abc}]_m \\ &+ \frac{1}{2} \cdot [Y_{abc}] \cdot \left(\left\{ [U] + \frac{1}{2} \cdot [Z_{abc}] \cdot [Y_{abc}] \right\} \cdot [VLG_{abc}]_m \right. \\ &\left. + [Z_{abc}] \cdot [I_{abc}]_m \right) \end{aligned} \qquad (6.14)$$

Collecting terms in Equation 6.14,

$$[I_{abc}]_n = \left\{[Y_{abc}] + \frac{1}{4}\cdot[Y_{abc}]\cdot[Z_{abc}]\cdot[Y_{abc}]\right\}\cdot[VLG_{abc}]_m$$
$$+ \left\{[U] + \frac{1}{2}\cdot[Y_{abc}]\cdot[Z_{abc}]\right\}[I_{abc}]_m \quad (6.15)$$

Equation 6.15 is of the form

$$[I_{abc}]_n = [c]\cdot[VLG_{abc}]_m + [d]\cdot[I_{abc}]_m, \quad (6.16)$$

where

$$[c] = [Y_{abc}] + \frac{1}{4}\cdot[Y_{abc}]\cdot[Z_{abc}]\cdot[Y_{abc}], \quad (6.17)$$

$$[d] = [U] + \frac{1}{2}\cdot[Y_{abc}]\cdot[Z_{abc}] \quad (6.18)$$

Equations 6.8 and 6.16 can be put into partitioned matrix form:

$$\begin{bmatrix} [VLG_{abc}]_n \\ [I_{abc}]_n \end{bmatrix} = \begin{bmatrix} [a] & [b] \\ [c] & [d] \end{bmatrix} \cdot \begin{bmatrix} [VLG_{abc}]_m \\ [I_{abc}]_m \end{bmatrix} \quad (6.19)$$

Equation 6.19 is very similar to the equation used in transmission line analysis when the ABCD parameters have been defined [1]. In this case, the ***abcd*** parameters are 3×3 matrices rather than single variables and will be referred to as the "generalized line matrices."

Equation 6.19 can be turned around to solve for the voltages and currents at Node-*m* in terms of the voltages and currents at Node-*n*.

$$\begin{bmatrix} [VLG_{abc}]_m \\ [I_{abc}]_m \end{bmatrix} = \begin{bmatrix} [a] & [b] \\ [c] & [d] \end{bmatrix}^{-1} \cdot \begin{bmatrix} [VLG_{abc}]_n \\ [I_{abc}]_n \end{bmatrix} \quad (6.20)$$

The inverse of the ***abcd*** matrix is simple because the determinant is

$$[a]\cdot[d] - [b]\cdot[c] = [U] \quad (6.21)$$

Using the relationship of Equations 6.21 and 6.20 becomes

Distribution System Line Models

$$\begin{bmatrix} [VLG_{abc}]_m \\ [I_{abc}]_m \end{bmatrix} = \begin{bmatrix} [d] & -[b] \\ -[c] & [a] \end{bmatrix} \cdot \begin{bmatrix} [VLG_{abc}]_n \\ [I_{abc}]_n \end{bmatrix} \qquad (6.22)$$

Since the matrix $[a]$ is equal to the matrix $[d]$, Equation 6.22 in expanded form becomes

$$[VLG_{abc}]_m = [a] \cdot [VLG_{abc}]_n - [b] \cdot [I_{abc}]_n, \qquad (6.23)$$

$$[I_{abc}]_m = -[c] \cdot [VLG_{abc}]_n + [d] \cdot [I_{abc}]_n \qquad (6.24)$$

Sometimes it is necessary to compute the voltages at Node-m as a function of the voltages at Node-n and the currents entering Node-m. This is true in the iterative technique that is developed in Chapter 10.

Solving Equation 6.8 for the Node-m voltages gives the following:

$$[VLG_{abc}]_m = [a]^{-1} \cdot \{[VLG_{abc}]_n - [b] \cdot [I_{abc}]_m\}$$

$$[VLG_{abc}]_m = [a]^{-1} \cdot [VLG_{abc}]_n - [a]^{-1} \cdot [b] \cdot [I_{abc}]_m \qquad (6.25)$$

Equation 6.25 is of the form

$$[VLG_{abc}]_m = [A] \cdot [VLG_{abc}]_n - [B] \cdot [I_{abc}], \qquad (6.26)$$

where

$$[A] = [a]^{-1}, \qquad (6.27)$$

$$[B] = [a]^{-1} \cdot [b] \qquad (6.28)$$

The line-to-line voltages are computed by

$$\begin{bmatrix} V_{ab} \\ V_{bc} \\ V_{ca} \end{bmatrix}_m = \begin{bmatrix} 1 & -1 & 0 \\ 0 & 1 & -1 \\ -1 & 0 & 1 \end{bmatrix} \cdot \begin{bmatrix} V_{ag} \\ V_{bg} \\ V_{cg} \end{bmatrix}_m = [Dv] \cdot [VLG_{abc}]_m, \qquad (6.29)$$

where

$$[Dv] = \begin{bmatrix} 1 & -1 & 0 \\ 0 & 1 & -1 \\ -1 & 0 & 1 \end{bmatrix} \quad (6.30)$$

Because the mutual couplings between phases for the line segments are not equal, there will be different values of voltage drop on each of the three phases. As a result, the voltages on a distribution feeder become unbalanced even when the loads are balanced. A common method of describing the degree of unbalance is to use the National Electrical Manufacturers Association (NEMA) definition of voltage unbalance, as given in Equation 3.31 [2].

$$V_{unbalance} = \frac{|Maximum\ Deviation\ of\ Voltages\ from\ Average|}{V_{average}} \cdot 100\%$$

$$V_{unbalance} = \frac{|dV|}{V_{average}} \cdot 100\% \quad (3.31)$$

Example 6.1: A balanced three-phase load of 6,000 kVA, 12.47 kV, 0.9 lagging power factor is being served at Node-*m* of a 10,000-foot, three-phase line segment. The load voltages are rated and balanced 12.47 kV. The configuration and conductors of the line segment are those of Example 4.1.

Determine:

- generalized line matrices: [a], [b], [c], [d], [A], and [B];
- line-to-ground load voltages and currents;
- line currents at the source end;
- voltage unbalance in percent;
- complex power at the source and the load; and
- total three-phase power loss.

Solution:

The phase impedance matrix and the shunt admittance matrix for the line segment as computed in Examples 4.1 and 5.1 are

$$[z_{abc}] = \begin{bmatrix} 0.4576 + j1.0780 & 0.1560 + j.5017 & 0.1535 + j0.3849 \\ 0.1560 + j0.5017 & 0.4666 + j1.0482 & 0.1580 + j0.4236 \\ 0.1535 + j0.3849 & 0.1580 + j0.4236 & 0.4615 + j1.0651 \end{bmatrix} \Omega/\text{mile},$$

Distribution System Line Models

$$[Y_{abc}] = j \cdot 376.9911 \cdot [C_{abc}] = \begin{bmatrix} j5.6711 & -j1.8362 & -j0.7033 \\ -j1.8362 & j5.9774 & -j1.169 \\ -j0.7033 & -j1.169 & j5.3911 \end{bmatrix} \mu S/\text{mile}$$

For the 10,000-foot line segment, the total phase impedance matrix and shunt admittance matrix are

$$[Z_{abc}] = \begin{bmatrix} 0.8667 + j2.0417 & 0.2955 + j0.9502 & 0.2907 + j0.7290 \\ 0.2955 + j0.9502 & 0.8837 + j1.9852 & 0.2992 + j0.8023 \\ 0.2907 + j0.7290 & 0.2992 + j0.8023 & 0.8741 + j2.0172 \end{bmatrix} \Omega,$$

$$[Y_{abc}] = \begin{bmatrix} j10.7409 & -j3.4777 & -j1.3322 \\ -j3.4777 & j11.3208 & -j2.2140 \\ -j1.3322 & -j2.2140 & j10.2104 \end{bmatrix} \mu S$$

It should be noted that the elements of the phase admittance matrix are very small.

The generalized matrices computed according to Equations 6.9, 6.10, 6.17, and 6.18 are as follows:

$$[a] = [U] + \frac{1}{2} \cdot [Z_{abc}] \cdot [Y_{abc}] = \begin{bmatrix} 1.0 & 0 & 0 \\ 0 & 1.0 & 0 \\ 0 & 0 & 1.0 \end{bmatrix}$$

Rounded to four significant figures

$$[b] = [Z_{abc}] = \begin{bmatrix} 0.8667 + j2.0417 & 0.2955 + j0.9502 & 0.2907 + j0.7290 \\ 0.2955 + j0.9502 & 0.8837 + j1.9852 & 0.2992 + j0.8023 \\ 0.2907 + j0.7290 & 0.2992 + j0.8023 & 0.8741 + j2.0172 \end{bmatrix} \Omega$$

$$[c] = \begin{bmatrix} 0 & 0 & 0 \\ 0 & 0 & 0 \\ 0 & 0 & 0 \end{bmatrix} S$$

Rounded to four significant figures,

$$[d] = [U] + \frac{1}{2} \cdot [Y_{abc}] \cdot [Z_{abc}] = \begin{bmatrix} 1.0 & 0 & 0 \\ 0 & 1.0 & 0 \\ 0 & 0 & 1.0 \end{bmatrix},$$

$$[A] = [a]^{-1} = \begin{bmatrix} 1.0 & 0 & 0 \\ 0 & 1.0 & 0 \\ 0 & 0 & 1.0 \end{bmatrix},$$

$$[B] = [a]^{-1} \cdot [b] = \begin{bmatrix} 0.8667 + j2.0417 & 0.2955 + j0.9502 & 0.2907 + j0.7290 \\ 0.2955 + j0.9502 & 0.8837 + j1.9852 & 0.2992 + j0.8023 \\ 0.2907 + j0.7290 & 0.2992 + j0.8023 & 0.8741 + j2.0172 \end{bmatrix} \Omega$$

Because the elements of the phase admittance matrix are so small, the [a], [A], and [d] matrices appear to be the unity matrix. If more significant figures are displayed, the 1,1 element of these matrices is

$$a_{1,1} = A_{1,1} = 0.99999117 + j0.00000395$$

Also, the elements of the [c] matrix appear to be zero. Again, if more significant figures are displayed the 1,1 term is

$$c_{1,1} = -.0000044134 + j.0000127144$$

The point here is that for all practical purposes, the phase admittance matrix can be neglected.

The magnitude of the line-to-ground voltages at the load is

$$[VLG_{abc}]_m = \begin{bmatrix} 7200\underline{/0} \\ 7200\underline{/-120} \\ 7200\underline{/120} \end{bmatrix} V$$

For a 0.9 lagging power factor, the load current matrix is

$$[I_{abc}]_m = \begin{bmatrix} 277.79\underline{/-25.84} \\ 277.79\underline{/-145.84} \\ 277.79\underline{/94.16} \end{bmatrix} A$$

The line-to-ground voltages at Node-n are computed to be

$$[VLG_{abc}]_n = [a] \cdot [VLG_{abc}]_m + [b] \cdot [I_{abc}]_m = \begin{bmatrix} 7538.70\underline{/1.57} \\ 7451.25\underline{/-118.30} \\ 7485.11\underline{/121.93} \end{bmatrix} V$$

It is important to note that the voltages at Node-n are unbalanced, even though the voltages and currents at the load (Node-m) are perfectly balanced. This is a result of the unequal mutual coupling between phases. The degree of voltage unbalance is of

Distribution System Line Models

concern since, for example, the operating characteristics of a three-phase induction motor are very sensitive to voltage unbalance. Using the NEMA definition for voltage unbalance (Equation 6.29), the voltage unbalance is as follows:

$$V_{average} = \frac{1}{3} \cdot \sum_{k=1}^{3} |VLG_n|_k = 7491.69$$

For : $i = 1, 2, 3$

$$dV_i = |V_{average} - |VLG_n|_i| = \begin{bmatrix} 47.01 \\ 40.44 \\ 6.57 \end{bmatrix}$$

$$V_{unbalance} = \frac{dV_{max}}{V_{average}} = \frac{47.01}{7491.70} \cdot 100\% = 0.6275\%$$

Although this may not seem like a large unbalance, it does give an indication of how the unequal mutual coupling can generate an unbalance. It is important to know that NEMA standards require that induction motors be derated when the voltage unbalance exceeds 1.0%.

Selecting rated line-to-ground voltage as the base (7199.56), the per-unit voltages at Node-n are

$$\begin{bmatrix} V_{ag} \\ V_{bg} \\ V_{cg} \end{bmatrix}_n = \frac{1}{7199.56} \begin{bmatrix} 7538.70 \underline{/1.57} \\ 7451.25 \underline{/-118.30} \\ 7485.11 \underline{/121.93} \end{bmatrix} = \begin{bmatrix} 1.0471 \underline{/1.57} \\ 1.0350 \underline{/-118.30} \\ 1.0397 \underline{/121.93} \end{bmatrix} \text{per unit}$$

By converting the voltages to per unit, it is easy to see that the voltage drop by phase is 4.71% for phase a, 3.50% for phase b, and 3.97% for phase c.

The line currents at Node-n are computed to be

$$[I_{abc}]_n = [c] \cdot [VLG_{abc}]_m + [d] \cdot [I_{abc}]_m = \begin{bmatrix} 277.71 \underline{/-25.83} \\ 277.73 \underline{/-148.82} \\ 277.73 \underline{/94.17} \end{bmatrix} A$$

Comparing the computed line currents at Node-n to the balanced load currents at Node-m, a very slight difference is noted that is another result of the unbalanced voltages at Node-n and the shunt admittance of the line segment.

The complex power at the Node-n and node are

$$[S_n] = \begin{bmatrix} 1858.71 + j963.40 \\ 1835.23 + j956.22 \\ 1839.61 + j968.17 \end{bmatrix} \quad [S_m] = \begin{bmatrix} 1800 + j871.78 \\ 1800 + j871.78 \\ 1800 + j871.78 \end{bmatrix} \text{kVA}$$

The complex power loss is

$$[S_{loss}] = [S_n] - [S_m] = \begin{bmatrix} 58.71 + j91.62 \\ 35.23 + j84.44 \\ 39.61 + j96.39 \end{bmatrix} \text{kVA}$$

The three-phase power loss is

$$Ploss_{total} = \text{Real}\left(\sum_{k=1}^{k=3}[S_{loss}]_k\right) = 133.57 \text{ kW}$$

6.2 THE MODIFIED LINE MODEL

It was demonstrated in Example 6.1 that the shunt admittance of an overhead line is so small that it can be neglected. Figure 6.2 shows the modified line segment model with the shunt admittance neglected.

When the shunt admittance is neglected, the generalized matrices become the following:

$$[a] = [U] \tag{6.32}$$

$$[b] = [Z_{abc}] \tag{6.33}$$

$$[c] = [0] \tag{6.34}$$

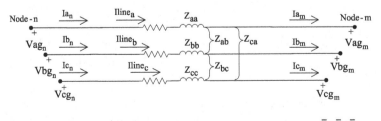

FIGURE 6.2 Modified line segment model.

Distribution System Line Models

$$[d] = [U] \tag{6.35}$$

$$[A] = [U] \tag{6.36}$$

$$[B] = [Z_{abc}] \tag{6.37}$$

6.2.1 THE THREE-WIRE LINE

If the line is three wire, then the voltage drops down the line must be in terms of the line-to-line voltages and line currents. However, it is possible to use "equivalent" line-to-neutral voltages so that the equations derived to this point will still apply. Writing the voltage drops in terms of line-to-line voltages for the line in Figure 6.2 results in

$$\begin{bmatrix} V_{ab} \\ V_{bc} \\ V_{ca} \end{bmatrix}_n = \begin{bmatrix} V_{ab} \\ V_{bc} \\ V_{ca} \end{bmatrix}_m + \begin{bmatrix} vdrop_a \\ vdrop_b \\ vdrop_c \end{bmatrix} - \begin{bmatrix} vdrop_b \\ vdrop_c \\ vdrop_a \end{bmatrix}, \tag{6.38}$$

where

$$\begin{bmatrix} vdrop_a \\ vdrop_b \\ vdrop_c \end{bmatrix} = \begin{bmatrix} Z_{aa} & Z_{ab} & Z_{ac} \\ Z_{ba} & Z_{bb} & Z_{bc} \\ Z_{ca} & Z_{cb} & Z_{cc} \end{bmatrix} \cdot \begin{bmatrix} Iline_a \\ Iline_b \\ Iline_c \end{bmatrix}$$

$$[vdrop_{abc}] = [Z_{abc}] \cdot [Iline_{abc}] \tag{6.39}$$

Expanding Equation 6.38 for the Phase a–b:

$$Vab_n = Vab_m + vdrop_a - vdrop_b, \tag{6.40}$$

but

$$Vab_n = Van_n - Vbn_n$$

$$Vab_m = Van_m - Vbn_m \tag{6.41}$$

Substitute Equation 6.41 into Equation 6.40:

$$Van_n - Vbn_n = Van_m - Vbn_m + vdrop_a - vdrop_b$$

or

$$Van_n = Van_m + vdrop_a$$

$$Vbn_n = Vbn_m + vdrop_b \tag{6.42}$$

In general, Equation 6.42

$$[A] = [a]^{-1}$$

$[B] = [a]^{-1} \cdot [b]''$ line-to-neutral voltages.

$$[VLN]_n = [VLN]_m + [vdrop_{abc}]$$

$$[VLN]_n = [VLN]_m + [Z_{abc}] \cdot [Iline_{abc}] \tag{6.43}$$

The conclusion is that it is possible to work with "equivalent" line-to-neutral voltages in a three-wire line. This is very important since it makes the development of general analyses techniques the same for four-wire and three-wire systems.

6.2.2 THE COMPUTATION OF NEUTRAL AND GROUND CURRENTS

In Chapter 4, the Kron reduction method was used to reduce the primitive impedance matrix to the 3×3 phase impedance matrix. Figure 6.3 shows a three-phase line with grounded neutral that is used in the Kron reduction. Note in Figure 6.3 that the direction of the current flowing in the ground is shown.

In the development of the Kron reduction method, Equation 4.50 defined the "neutral transform matrix" $[t_n]$. That equation is shown as Equation 6.44.

$$t_n = -[\hat{z}_{nn}]^{-1} \cdot [\hat{z}_{nj}] \tag{6.44}$$

The matrices $[\hat{z}_{nn}]$ and $[\hat{z}_{nj}]$ are the partitioned matrices in the primitive impedance matrix.

When the currents flowing in the lines have been determined, Equation 6.45 is used to compute the current flowing in the grounded neutral wire(s).

Distribution System Line Models

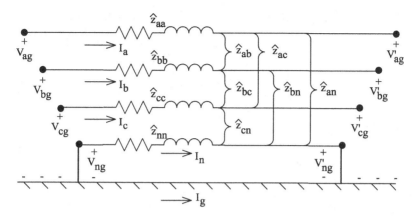

FIGURE 6.3 Three-phase line with neutral and ground currents.

$$[I_n] = [t_n] \cdot [I_{abc}] \quad (6.45)$$

In Equation 6.45, the matrix $[I_n]$ for an overhead line with one neutral wire will be a single element. However, in the case of an underground line consisting of concentric neutral cables or taped shielded cables with or without a separate neutral wire, $[I_n]$ will be the currents flowing in each of the cable neutrals and the separate neutral wire if present. Once the neutral current(s) have been determined, KCL is used to compute the current flowing in ground.

$$I_g = -(I_a + I_b + I_c + In_1 + In_2 + \ldots\ldots + In_k) \quad (6.46)$$

Example 6.2: The line of Example 6.1 will be used to supply an unbalanced load at Node-*m*. Assume that the voltages at the source end (Node-*n*) are balanced three phase at 12.47 kV line-to-line. The balanced line-to-ground voltages are

$$[VLG]_n = \begin{bmatrix} 7199.56 \underline{/0} \\ 7199.56 \underline{/-120} \\ 7199.56 \underline{/120} \end{bmatrix} V$$

The unbalanced currents measured at the source end are given by

$$\begin{bmatrix} I_a \\ I_b \\ I_c \end{bmatrix}_n = \begin{bmatrix} 249.97/-24.5 \\ 277.56/-145.8 \\ 305.54/95.2 \end{bmatrix} A$$

Determine:

- the line-to-ground and line-to-line voltages at the load end (Node-*m*) using the modified line model,
- the voltage unbalance,
- the complex powers of the load, and
- the currents flowing in the neutral wire and ground.

Solution:

The [A] and [B] matrices for the modified line model are as follows:

$$[A] = [U] = \begin{bmatrix} 1 & 0 & 0 \\ 0 & 1 & 0 \\ 0 & 0 & 1 \end{bmatrix}$$

$$[B] = [Z_{abc}] = \begin{bmatrix} 0.8666 + j2.0417 & 0.2955 + j0.9502 & 0.2907 + j0.7290 \\ 0.2955 + j0.9502 & 0.8837 + j1.9852 & 0.2992 + j0.8023 \\ 0.2907 + j0.7290 & 0.2992 + j0.8023 & 0.8741 + j2.0172 \end{bmatrix} \Omega$$

Since this is the approximate model, $[I_{abc}]_m$ is equal to $[I_{abc}]_n$. Therefore,

$$\begin{bmatrix} I_a \\ I_b \\ I_c \end{bmatrix}_m = \begin{bmatrix} 249.97/-24.5 \\ 277.56/-145.8 \\ 305.54/95.2 \end{bmatrix} A$$

The line-to-ground voltages at the load end are

$$[VLG]_m = [A] \cdot [VLG]_n - [B] \cdot [I_{abc}]_m = \begin{bmatrix} 6942.53/-1.47 \\ 6918.35/-121.55 \\ 6887.71/117.31 \end{bmatrix} V$$

The line-to-line voltages at the load end are

$$[Dv] = \begin{bmatrix} 1 & -1 & 0 \\ 0 & 1 & -1 \\ -1 & 0 & 1 \end{bmatrix},$$

Distribution System Line Models

$$[VLL]_m = [Dv] \cdot [VLG]_m = \begin{bmatrix} 12{,}008\underline{/28.4} \\ 12{,}025\underline{/-92.2} \\ 11{,}903\underline{/148.1} \end{bmatrix}$$

For this condition, the average load voltage is

$$V_{average} = \frac{1}{3} \cdot \sum_{k=1}^{3} |VLG_m|_k = 6916.20$$

The maximum deviation from the average is on phase c so that

for : $i = 1, 2, 3$;

$$dV_i = \left| V_{average} - |VLG_m|_i \right| = \begin{bmatrix} 26.33 \\ 2.15 \\ 28.49 \end{bmatrix};$$

$$V_{unbalance} = \frac{dV_{max}}{V_{average}} \cdot 100 = \frac{28.49}{6916.20} \cdot 100 = 0.4119\%$$

The complex powers of the load are

$$\begin{bmatrix} S_a \\ S_b \\ S_c \end{bmatrix}_m = \frac{1}{1000} \cdot \begin{bmatrix} V_{ag} \cdot I_a^* \\ V_{bg} \cdot I_b^* \\ V_{cg} \cdot I_c^* \end{bmatrix}_m = \begin{bmatrix} 1597.2 + j678.8 \\ 1750.8 + j788.7 \\ 1949.7 + j792.0 \end{bmatrix} \text{ kVA.}$$

The "neutral transformation matrix" from Example 4.1 is

$$[t_n] = [-0.4291 - j0.1291 \quad -0.4476 - j0.1273 \quad -0.4373 - j0.1327]$$

The neutral current is

$$[I_n] = [t_n] \cdot [I_{abc}]_m = 26.2\underline{/-29.5} \text{ A}$$

The ground current is

$$I_g = -(I_a + I_b + I_c + I_n) = 32.5\underline{/-77.6} \text{ A}$$

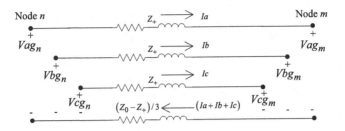

FIGURE 6.4 Equivalent source impedance.

6.3 SOURCE IMPEDANCES

When the analysis of a distribution system starts, it is important to include (and required) to input the 3×3 phase impedance matrix of the equivalent system serving the distribution substation. A typical system is shown in Figure 6.4

The source impedance can be called the Thevenin impedance between Nodes n and m in Figure 6.4. The distribution engineer usually does not have the necessary information to determine the impedance matrix but the transmission engineer will. For a transmission system, it is assumed that all of the transmission lines are transposed and any transformers have equal phases. With those assumptions, the short-circuit analysis of a transmission system will use the method of symmetrical components to analyze the system for all types of faults. In Figure 6.4, the three-phase and line-to-ground, short-circuit currents will be computed for faults at Node-m. Given the three-phase, short-circuit current, the "positive sequence" impedance (Z_1) computed up to Node-m is

$$Z_1 = \frac{VLN_{source}}{I_{three-phase}} \; \Omega \tag{6.47}$$

The usual assumption is that the "negative sequence" impedance (Z_2) is equal to the positive sequence impedance; that is, $Z_2 = Z_1$. The line-to-ground short current is computed as

$$I_{1p} = \frac{3 \cdot VLN_{source}}{Z_0 + 2 \cdot Z_1} = \frac{3 \cdot VLN_{source}}{Z_{eq}} \; \text{amps,}$$

where

$$Z_{eq} = Z_0 + 2 \cdot Z_1. \tag{6.48}$$

Equation 6.49 is used to compute the zero-sequence impedance (Z_0).

$$Z_0 = Z_{eq} - 2 \cdot Z_1 \tag{6.49}$$

Distribution System Line Models

The sequence impedance matrix between nodes n and m is

$$[Z_{012}] = \begin{bmatrix} Z_0 & 0 & 0 \\ 0 & Z_1 & 0 \\ 0 & 0 & Z_2 \end{bmatrix} \Omega \qquad (6.50)$$

The transformation matrix is defined as

$$[A_s] = \begin{bmatrix} 1 & 1 & 1 \\ 1 & a^2 & a \\ 1 & a & a^2 \end{bmatrix}, \qquad (6.51)$$

where

$$a = 1/\underline{120}$$

The sequence impedance matrix is converted to the actual a-b-c impedance matrix is

$$[Z_{system}] = [A_s] \cdot [Z_{012}] \cdot [A_s]^{-1} \qquad (6.52)$$

With the system impedance matrix computed, the new figure for the system is:
It must be pointed out that this is the **only** time that symmetrical components are used in distribution system analysis.

Example 6.3: The short-circuit study on a 115 kV system hass been completed. At the node feeding a distribution substation the short-circuit currents are as follows:

$$If_{3-phase} = 5677.7/\underline{-82.7} \text{ A},$$

$$If_{1-phase} = 4430.0/\underline{-80.2} \text{ A}.$$

The line-to-ground nominal voltage is $VLG = 66{,}395.3$ volts.
The positive sequence impedance is

$$Z_1 = \frac{66{,}395.3/\underline{0}}{5677.7/\underline{-82.7}} = 1.4859 + j11.5993 \,\Omega$$

The zero-sequence impedance is

$$Z_{eq} = \frac{3 \cdot 66395.281 \underline{/0}}{4430 \cdot \underline{/-80.2}} = 7.6531 + j44.3069$$

$$Z_0 = Z_{eq} - 2 \cdot Z_1 = 4.6931 + j21.1069$$

The sequence impedance matrix is

$$[Z_{012}] = \begin{bmatrix} Z_0 & 0 & 0 \\ 0 & Z_1 & 0 \\ 0 & 0 & Z_1 \end{bmatrix}$$

$$[Z_{012}] = \begin{bmatrix} 4.6931 + j21.1069 & 0 & 0 \\ 0 & 1.4859 + j11.5993 & 0 \\ 0 & 0 & 1.4859 + j11.5993 \end{bmatrix} \text{Ohms}$$

The phase impedance matrix is computed by

$$[Z_{abc}] = [A_s] \cdot [Z_{012}] \cdot [A_s]^{-1},$$

$$[Z_{abc}] = \begin{bmatrix} 2.5633 + j14.7667 & 1.0833 + j3.1667 & 1.0833 + j3.1667 \\ 1.0833 + j3.1667 & 2.5633 + j14.7667 & 1.0833 + j3.1667 \\ 1.0833 + j3.1667 & 1.0833 + j3.1667 & 2.5633 + j14.7667 \end{bmatrix} \text{ohm}$$

In a distribution system analysis program (like Milsoft's WindMil), the positive and zero-sequence impedances are input for the distribution system source.

6.4 THE LIT

The previous example problems have assumed a linear system. Unfortunately, that will not be the usual case for distribution feeders. When the source voltages are specified and the loads are specified as constant kW and kvar (constant PQ), the system becomes non-linear and the LIT **must** be used as developed in Chapter 3. Everything in the ladder technique must use phase components. Symmetrical components and the sequence impedances for the lines are never used in the ladder technique.

The ladder technique is composed of two parts:

1. Forward sweep
2. Backward sweep

The forward sweep computes the downstream voltages from the source by applying Equation 6.53.

$$[VLG_{abc}]_m = [A] \cdot [VLG_{abc}]_n - [B] \cdot [I_{abc}]_m \tag{6.53}$$

Distribution System Line Models

$$Y := \begin{vmatrix} \text{Start} := \begin{pmatrix} 0 \\ 0 \\ 0 \end{pmatrix} \quad \text{Tol} := .00001 \quad kV_{LN} := 7.2 \\ I_{abc} \leftarrow \text{Start} \\ V_{old} \leftarrow \text{Start} \\ \text{for } n \in 1..200 \\ \quad \begin{vmatrix} VLG_{abc} \leftarrow A \cdot E_{abc} - B \cdot I_{abc} \\ \text{for } i \in 1..3 \\ \quad I_{abc_i} \leftarrow \dfrac{SL_i \cdot 1000}{VLG_{abc_i}} \\ \text{for } j \in 1..3 \\ \quad \text{Error}_j \leftarrow \dfrac{|VLG_{abc_i} - V_{old_j}|}{kV_{LN} \cdot 1000} \\ \text{Err}_{max} \leftarrow \max(\text{Error}) \\ \text{break if } \text{Err}_{max} < \text{Tol} \\ V_{old} \leftarrow VLG_{abc} \end{vmatrix} \\ \text{Out}_1 \leftarrow VLG_{abc} \\ \text{Out}_2 \leftarrow I_{abc} \\ \text{Out}_3 \leftarrow n \\ \text{Out} \end{vmatrix}$$

FIGURE 6.5 Iterative power flow for equivalent system.

To start the process, the load currents $[I_{abc}]_m$ are assumed to be equal to zero, and the load voltages are computed. In the first iteration, the load voltages will be the same as the source voltages, and the load currents are computed.

The backward sweep computes the currents from the load back to the source using the most recently computed voltages from the forward sweep. Equation 6.54 is applied for this sweep.

$$[I_{abc}]_n = [c] \cdot [VLG_{abc}]_m + [d] \cdot [I_{abc}]_m$$

since

$$[c] = [0]$$

$$[I_{abc}]_n = [d] \cdot [I_{abc}]_m \quad (6.54)$$

After the first forward and backward sweeps, the new load voltages are computed using the most recent currents. The forward and backward sweeps continue until the error between the new and previous load voltages are within a specified tolerance. Using the matrices computed in Example 6.1, a very simple computer program that applies the ladder iterative technique is demonstrated in Example 6.5.

Example 6.4: The line of Example 6.1 serves an unbalanced three-phase load of the following:

Phase a: 2,500 kVA and PF = 0.9 lagging
Phase b: 2,000 kVA and PF = 0.85 lagging
Phase c: 1,500 kVA and PF = 0.95 lagging

The source voltages are balanced 12.47 kV.

$$ELN = \frac{12,470}{\sqrt{3}} = 7199.6 \text{ V}$$

$$[VLG_{abc}] = \begin{bmatrix} 7199.6 \underline{/0} \\ 7199.6 \underline{/-120} \\ 7199.6 \underline{/120} \end{bmatrix} \text{V}$$

The ladder technique is used to compute the load voltages and currents. The matrices [A] and [B] from Example 6.1 are used. The program is shown in Figure 6.6.

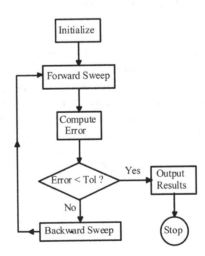

FIGURE 6.6 Ladder technique for Example 6.4.

Distribution System Line Models

After seven iterations, the load voltages and currents are computed to be

$$[VLG_{abc}] = \begin{bmatrix} 6678.2/-2.3 \\ 6972.8/-122 \\ 7055.5/118.7 \end{bmatrix} V [I_{abc}] = \begin{bmatrix} 374.4/-28.17 \\ 286.8/-153.9 \\ 212.6/100.5 \end{bmatrix} A$$

Example 6.4 demonstrates the application of the ladder iterative technique, which is used as models of other distribution feeder elements are developed. A simple flow chart of the program and one that will be used in other chapters is shown in Figure 6.7.

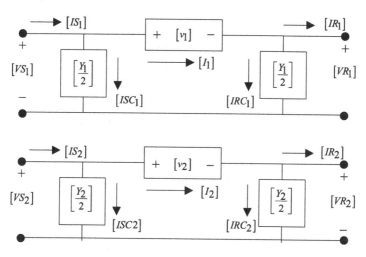

FIGURE 6.7 Simple ladder technique flow chart.

6.5 THE GENERAL MATRICES FOR PARALLEL LINES

The equivalent pi circuits for two parallel three-phase lines are shown in Figure 6.8.

The 6×6 phase impedance and shunt admittance matrices for parallel three-phase lines were developed in Chapters 4 and 5. These matrices are used in the development of the general matrices used in modeling parallel three-phase lines.

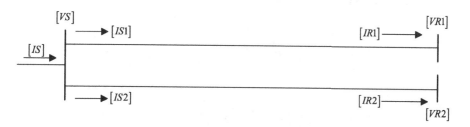

FIGURE 6.8 Equivalent pi parallel lines.

The first step in computing the ***abcd*** matrices is to multiply the 6×6 phase impedance matrix from Chapter 4 and the 6×6 shunt admittance matrix from Chapter 5 by the distance that the lines are parallel.

$$\begin{bmatrix} [v1] \\ [v2] \end{bmatrix} = \begin{bmatrix} [z11] & [z12] \\ [z21] & [z22] \end{bmatrix} \cdot length \cdot \begin{bmatrix} [I1] \\ [I2] \end{bmatrix} = \begin{bmatrix} [Z11] & [Z12] \\ [Z21] & [Z22] \end{bmatrix} \begin{bmatrix} [I1] \\ [I2] \end{bmatrix}$$

$$[v] = [Z] \cdot [I] \text{ V} \tag{6.55}$$

$$\begin{bmatrix} [y11] & [y12] \\ [y21] & [y22] \end{bmatrix} \cdot length = \begin{bmatrix} [Y11] & [Y12] \\ [Y21] & [Y22] \end{bmatrix} \text{ S} \tag{6.56}$$

Referring to Figure 6.7, the line currents in the two circuits are given by the following:

$$\begin{bmatrix} [I_1] \\ [I_2] \end{bmatrix} = \begin{bmatrix} [IR_1] \\ [IR_2] \end{bmatrix} + \frac{1}{2} \cdot \begin{bmatrix} [Y_{11}] & [Y_{12}] \\ [Y_{21}] & [Y_{22}] \end{bmatrix} \begin{bmatrix} [VR_1] \\ [VR_2] \end{bmatrix}$$

$$[I] = [IR] + \frac{1}{2} \cdot [Y] \cdot [VR] \text{ A} \tag{6.57}$$

The sending end voltages are given by $\begin{bmatrix} [VS_1] \\ [VS_2] \end{bmatrix} = \begin{bmatrix} [VR_1] \\ [VR_2] \end{bmatrix} + \begin{bmatrix} [Z_{11}] & [Z_{12}] \\ [Z_{21}] & [Z_{22}] \end{bmatrix} \begin{bmatrix} [I_1] \\ [I_2] \end{bmatrix}$,

$$[VS] = [VR] + [Z] \cdot [I] \text{ A} \tag{6.58}$$

Substitute Equation 6.57 into Equation 6.58.

$$\begin{bmatrix} [VS_1] \\ [VS_2] \end{bmatrix} = \begin{bmatrix} [VR_1] \\ [VR_2] \end{bmatrix} + \begin{bmatrix} [Z_{11}] & [Z_{12}] \\ [Z_{21}] & [Z_{22}] \end{bmatrix} \cdot \left(\begin{bmatrix} [IR_1] \\ [IR_2] \end{bmatrix} + \frac{1}{2} \cdot \begin{bmatrix} [Y_{11}] & [Y_{12}] \\ [Y_{21}] & [Y_{22}] \end{bmatrix} \begin{bmatrix} [VR_1] \\ [VR_2] \end{bmatrix} \right)$$

$$[VS] = [VR] + [Z] \cdot \left([IR] + \frac{1}{2} \cdot [Y] \cdot [VR] \right) \tag{6.59}$$

Distribution System Line Models

Combine terms in Equation 6.59.

$$\begin{bmatrix} [VS_1] \\ [VS_2] \end{bmatrix} = \left(\begin{bmatrix} [U] \\ [U] \end{bmatrix} + \frac{1}{2} \cdot \begin{bmatrix} [Z_{11}] & [Z_{12}] \\ [Z_{21}] & [Z_{22}] \end{bmatrix} \begin{bmatrix} [Y_{11}] & [Y_{12}] \\ [Y_{21}] & [Y_{22}] \end{bmatrix} \right) \begin{bmatrix} [VR_1] \\ [VR_2] \end{bmatrix}$$
$$+ \begin{bmatrix} [Z_{11}] & [Z_{12}] \\ [Z_{21}] & [Z_{22}] \end{bmatrix} \cdot \begin{bmatrix} [IR_1] \\ [IR_2] \end{bmatrix}$$

$$[VS] = \left([U] + \frac{1}{2} \cdot [Z] \cdot [Y] \right) \cdot [VR] + [Z] \cdot [IR] \qquad (6.60)$$

Equation 6.60 is of the form

$$[VS] = [a] \cdot [VR] + [b] \cdot [IR], \qquad (6.61)$$

where

$$[a] = [U] + \frac{1}{2} \cdot [Z] \cdot [Y],$$

$$[b] = [Z] \qquad (6.62)$$

The sending end currents are given by

$$\begin{bmatrix} [IS_1] \\ [IS_2] \end{bmatrix} = \begin{bmatrix} [I_1] \\ [I_2] \end{bmatrix} + \frac{1}{2} \cdot \begin{bmatrix} [Y_{11}] & [Y_{12}] \\ [Y_{21}] & [Y_{22}] \end{bmatrix} \cdot \begin{bmatrix} [VS_1] \\ [VS_2] \end{bmatrix},$$

$$[IS] = [I] + \frac{1}{2} \cdot [Y] \cdot [VS] \qquad (6.63)$$

Substitute Equations 6.57 and 6.61 into Equation 6.63 using the shorthand form.

$$[IS] = [IR] + \frac{1}{2} \cdot [Y] \cdot [VR] + \frac{1}{2} \cdot [Y] \cdot ([a] \cdot [VR] + [b] \cdot [IR]) \text{A} \qquad (6.64)$$

Combine terms in Equation 6.64.

$$[IS] = \frac{1}{2} \cdot ([Y] + [Y] \cdot [a]) \cdot [VR] + \left([U] + \frac{1}{2} \cdot [Y] \cdot [b] \right) \cdot [IR] \qquad (6.65)$$

Equation 6.65 is of the form

$$[IS] = [c] \cdot [VR] + [d] \cdot [IR], \qquad (6.66)$$

where

$$[c] = \frac{1}{2} \cdot ([Y] + [Y] \cdot [a]) = \left(\frac{1}{2} \cdot \left([Y] + [Y] \cdot \left([U] + \frac{1}{2} \cdot [Z] \cdot [Y]\right)\right)\right),$$

$$[c] = [Y] + \frac{1}{4} \cdot [Y] \cdot [Z] \cdot [Y],$$

$$[d] = [U] + \frac{1}{2} \cdot [Y] \cdot [b] = [U] + \frac{1}{2} \cdot [Y] \cdot [Z] \qquad (6.67)$$

The derived matrices [a], [b], [c], [d] will be 6×6 matrices. These four matrices can all be partitioned between the third and fourth rows and columns. The final voltage equation in partitioned form is given by

$$\begin{bmatrix}[VS_1]\\ [VS_2]\end{bmatrix} = \begin{bmatrix}[a_{11}] & [a_{12}]\\ [a_{21}] & [a_{22}]\end{bmatrix} \cdot \begin{bmatrix}[VR_1]\\ [VR_2]\end{bmatrix} + \begin{bmatrix}[b_{11}] & [b_{12}]\\ [b_{21}] & [b_{22}]\end{bmatrix} \cdot \begin{bmatrix}[IR_1]\\ [IR_2]\end{bmatrix} \qquad (6.68)$$

The final current equation in partitioned form is given by

$$\begin{bmatrix}[IS_1]\\ [IS_2]\end{bmatrix} = \begin{bmatrix}[c_{11}] & [c_{12}]\\ [c_{21}] & [c_{22}]\end{bmatrix} \cdot \begin{bmatrix}[VR_1]\\ [VR_2]\end{bmatrix} + \begin{bmatrix}[d_{11}] & [d_{12}]\\ [d_{21}] & [d_{22}]\end{bmatrix} \cdot \begin{bmatrix}[IR_1]\\ [IR_2]\end{bmatrix} \qquad (6.69)$$

Equations 6.68 and 6.69 are used to compute the sending end voltages and currents of two parallel lines. The matrices [A] and [B] are used to compute the receiving end voltages when the sending end voltages and receiving end currents are known. Solving Equation 6.61 for [VR],

$$[VR] = [a]^{-1} \cdot ([VS] - [b] \cdot [IR]),$$

$$[VR] = [a]^{-1} \cdot [VS] - [a]^{-1} \cdot [b] \cdot [IR],$$

$$[VR] = [A] \cdot [VS] - [B] \cdot [IR], \qquad (6.70)$$

where
[A] = [a]⁻¹
[B] = [a]⁻¹ · [b]

In expanded form, Equation 6.70 becomes

$$\begin{bmatrix} VR_1 \\ VR_2 \end{bmatrix} = \begin{bmatrix} [A_{11}] & [A_{12}] \\ [A_{21}] & [A_{22}] \end{bmatrix} \cdot \begin{bmatrix} VS_1 \\ VS_2 \end{bmatrix} - \begin{bmatrix} [B_{11}] & [B_{12}] \\ [B_{21}] & [B_{22}] \end{bmatrix} \cdot \begin{bmatrix} IR_1 \\ IR_2 \end{bmatrix} \quad (6.71)$$

6.5.1 Physically Parallel Lines

Two distribution lines can be physically parallel in two different ways in a radial system. Figure 6.9 illustrates two lines connected to the same sending end node, but the receiving ends of the lines do not share a common node.

The physically parallel lines of Figure 6.8 represent the common practice of two feeders leaving a substation on the same poles or right of ways and then branching in different directions downstream. Equations 6.68 and 6.69 are used to compute the sending end node voltages and currents using the known line current flows and node voltages at the receiving end. For this special case, the sending end node voltages must be the same at the end of the two lines so that Equation 6.68 is modified to reflect that [VS₁] = [VS₂]. A modified ladder iterative technique is used to force the two sending end voltages to be equal. In Chapter 10, the "ladder" iterative technique will be used to adjust the receiving end voltages in such a manner that the sending end voltages will be the same for both lines.

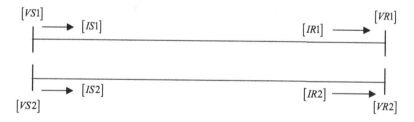

FIGURE 6.9 Physically parallel lines with a common sending end node.

Example 6.5: The parallel lines of Examples 4.4 and 5.2 are connected as shown in Figure 6.8 and are parallel to each other for 10 miles.

(1) Determine the *abcd* and *AB* matrices for the parallel lines.
From Examples 4.4 and 5.2, the per-mile values of the phase impedance and shunt admittance matrices in partitioned form are shown. The first step is to multiply these matrices by the length of the line. Note that the units for the shunt admittance matrix in Example 5.2 are in µS/mile.

$$dist = 10$$

$$Z = z \cdot dist$$

$$Y = y \cdot 10^{-6} \cdot dist$$

The unit matrix [U] must be defined as 6×6, and then the abcd matrices are computed using the equations developed in this chapter. The final results in partitioned form are as follows:

$$[a_{11}] = [a_{22}] = \begin{bmatrix} 0.9998 + j0.0001 & 0 & 0 \\ 0 & 0.9998 + j0.0001 & 0 \\ 0 & 0 & 0.9998 + j0.0001 \end{bmatrix}$$

$$[a_{12}] = [a_{21}] = \begin{bmatrix} 0 & 0 & 0 \\ 0 & 0 & 0 \\ 0 & 0 & 0 \end{bmatrix}$$

$$[b_{11}] = \begin{bmatrix} 4.5015 + j11.0285 & 1.4643 + j5.3341 & 1.4522 + j4.1255 \\ 1.4643 + j5.3341 & 4.5478 + j10.8726 & 1.4754 + j4.5837 \\ 1.4522 + j4.1255 & 1.4754 + j4.5837 & 4.5231 + j10.9556 \end{bmatrix}$$

$$[b_{12}] = \begin{bmatrix} 1.5191 + j4.8484 & 1.4958 + j3.9305 & 1.4775 + j5.5601 \\ 1.5446 + j5.3359 & 1.5205 + j4.3234 & 1.5015 + j4.9093 \\ 1.5311 + j4.2867 & 1.5074 + j5.4599 & 1.4888 + j3.9548 \end{bmatrix}$$

$$[b_{21}] = \begin{bmatrix} 1.5191 + j4.8484 & 1.5446 + j5.3359 & 1.5311 + j4.2867 \\ 1.4958 + j3.9305 & 1.5205 + j4.3234 & 1.5074 + j5.4599 \\ 1.4775 + j5.5601 & 1.5015 + j4.9093 & 1.4888 + j3.9548 \end{bmatrix}$$

$$[b_{22}] = \begin{bmatrix} 5.7063 + j10.9130 & 1.5801 + j4.2365 & 1.5595 + j5.0167 \\ 1.5801 + j4.2365 & 5.6547 + j11.0819 & 1.5348 + j3.8493 \\ 1.5595 + j5.0167 & 1.5348 + j3.8493 & 5.6155 + j11.2117 \end{bmatrix}$$

$$[c_{11}] = [c_{12}] = \begin{bmatrix} j.0001 & 0 & 0 \\ 0 & j.0001 & 0 \\ 0 & 0 & j.0001 \end{bmatrix}$$

$$[c_{21}] = [c_{22}] = \begin{bmatrix} 0 & 0 & 0 \\ 0 & 0 & 0 \\ 0 & 0 & 0 \end{bmatrix}$$

$$[d_{11}] = [a_{11}] \quad [d_{12}] = [a_{12}] \quad [d_{21}] = [a_{21}] \quad [d_{22}] = [a_{22}]$$

Distribution System Line Models

$$[A_{11}] = [A_{22}] = [a_{11}]^{-1} = \begin{bmatrix} 1.0002 - j0.0001 & 0 & 0 \\ 0 & 1.0002 - j0.0001 & 0 \\ 0 & 0 & 1.0002 - j0.0001 \end{bmatrix}$$

$$[A_{12}] = [A_{21}] = \begin{bmatrix} 0 & 0 & 0 \\ 0 & 0 & 0 \\ 0 & 0 & 0 \end{bmatrix}$$

$$[B_{11}] = [a_{11}]^{-1} \cdot [b_{11}] = \begin{bmatrix} 4.5039 + j11.031 & 1.4653 + j5.3357 & 1.4533 + j4.1268 \\ 1.4653 + j5.3357 & 4.5502 + j10.8751 & 1.4764 + j4.5852 \\ 1.4533 + j4.1268 & 1.4764 + j4.5852 & 4.5255 + j10.9580 \end{bmatrix}$$

$$[B_{12}] = [a_{12}]^{-1} \cdot [b_{12}] = \begin{bmatrix} 1.5202 + j4.8499 & 1.4969 + j3.9318 & 1.4786 + j5.5618 \\ 1.4969 + j3.9318 & 1.5216 + j4.3248 & 1.5026 + j4.9108 \\ 1.4786 + j5.5618 & 1.5026 + j4.9108 & 1.4899 + j3.9560 \end{bmatrix}$$

$$[B_{21}] = [a_{21}]^{-1} \cdot [b_{21}] = \begin{bmatrix} 1.5202 + j4.8499 & 1.5457 + j5.3375 & 1.5322 + j4.2881 \\ 1.5457 + j5.3375 & 1.5216 + j4.3248 & 1.5058 + j5.4615 \\ 1.5322 + j4.2881 & 1.5058 + j5.4615 & 1.4899 + j3.9560 \end{bmatrix}$$

$$[B_{22}] = [a_{22}]^{-1} \cdot [b_{22}] = \begin{bmatrix} 5.7092 + j10.9152 & 1.5812 + j4.2378 & 1.5606 + j5.0183 \\ 1.5812 + j4.2378 & 5.6577 + j11.0842 & 1.5360 + j3.8506 \\ 1.5606 + j5.0183 & 1.5360 + j3.8506 & 5.6184 + j11.2140 \end{bmatrix}$$

The loads at the ends of the two lines are treated as constant current loads with values of the following:

$$\text{Line 1:} \quad [IR1] = \begin{bmatrix} 102.6 /\underline{-20.4} \\ 82.1 /\underline{-145.2} \\ 127.8 /\underline{85.2} \end{bmatrix}$$

$$\text{Line 2:} \quad [IR2] = \begin{bmatrix} 94.4 /\underline{-27.4} \\ 127.4 /\underline{-152.5} \\ 100.2 /\underline{99.8} \end{bmatrix}$$

The voltages at the sending end of the lines are

$$[VS] = \begin{bmatrix} 14,400 /\underline{0} \\ 14,400 /\underline{-120} \\ 14,400 /\underline{120} \end{bmatrix} V$$

(2) Determine the receiving end voltages for the two lines.

Since the common sending end voltages are known and the receiving end line currents are known, Equation 6.73 is used to compute the receiving end voltages:

Line 1: $[VR1] = ([A_{11}] + [A_{12}]) \cdot [VS] - [B_{11}] \cdot [IR1] - [B_{12}] \cdot [IR2]$

$$[VR1] = \begin{bmatrix} 14{,}119/\underline{-2.3} \\ 14{,}022/\underline{-120.4} \\ 13{,}686/\underline{117.4} \end{bmatrix}$$

Line 2: $[VR2] = ([A_{21}] + [A_{22}]) \cdot [VS] - [B_{21}] \cdot [IR1] - [B_{22}] \cdot [IR2]$

$$[VR2] = \begin{bmatrix} 13{,}971/\underline{-1.6} \\ 13{,}352/\underline{-120.8} \\ 13{,}566/\underline{118.1} \end{bmatrix}$$

The second way in which two lines can be physically parallel in a radial feeder is to have neither the sending nor receiving ends common to both lines. This is shown in Figure 6.10.

FIGURE 6.10 Physically parallel lines without common nodes.

Equations 6.68 and 6.69 are again used for the analysis of this special case. Since neither the sending end nor receiving end nodes are common, no adjustments need to be made to Equation 6.68. Typically, these lines will be part of a large distribution feeder in which case an iterative process will be used to arrive at the final values of the sending and receiving end voltages and currents.

Example 6.6: The parallel lines of Examples 4.4 and 5.2 are connected as shown in Figure 6.9. The lines are parallel to each other for 10 miles.

The complex power flowing out of each line is as follows:

Line 1: $S1_a = 1450$ kVA, $PF_a = 0.95$
$S1_b = 1150$ kVA, $PF_b = 0.90$
$S1_c = 1750$ kVA, $PF_c = 0.85$

Distribution System Line Models

Line 2:
$$S2_a = 1320 \text{ kVA}, PF_a = 0.90$$
$$S2_b = 1700 \text{ kVA}, PF_b = 0.85$$
$$S2_c = 1360 \text{ kVA}, PF_c = 0.95$$

The line-to-neutral voltages at the receiving end nodes are as follows:

Line 1:
$$VR1_{an} = 13,430\underline{/-33.1}$$
$$VR1_{bn} = 13,956\underline{/-151.3}$$
$$VR1_{cn} = 14,071\underline{/86.0}$$

Line 2:
$$VR2_{an} = 14,501\underline{/-29.1}$$
$$VR2_{bn} = 13,932\underline{/-154.8}$$
$$VR2_{cn} = 12,988\underline{/90.3}$$

Determine the sending end voltages of the two lines.
The currents leaving the two lines are as follows:

Line 1: For $i = a, b, c$
$$IR1_i = \left(\frac{S1_i \cdot 1000}{V1_i}\right)^* = \begin{bmatrix} 108.0\underline{/-51.3} \\ 82.4\underline{/-177.1} \\ 124.4\underline{/54.2} \end{bmatrix}$$

Line 2: For $i = a, b, c$
$$IR2_i = \left(\frac{S2_i \cdot 1000}{V2_i}\right)^* = \begin{bmatrix} 91.0\underline{/-54.9} \\ 122.0\underline{/173.5} \\ 104.7\underline{/72.1} \end{bmatrix}$$

The sending end voltages of the two lines are computed using Equation 6.68.

Line 1:
$$[VS1] = \begin{bmatrix} 13,673\underline{/-30.5} \\ 14,361\underline{/-151.0} \\ 14,809\underline{/88.7} \end{bmatrix}$$

Line 2:
$$[VS2] = \begin{bmatrix} 14,845\underline{/-27.5} \\ 14,973\underline{/-154.3} \\ 13,898\underline{/92.5} \end{bmatrix}$$

The sending end currents are

$$[IS1] = \begin{bmatrix} 107.7\underline{/-50.8} \\ 82.0\underline{/-176.2} \\ 124.0\underline{/54.7} \end{bmatrix},$$

$$[IS2] = \begin{bmatrix} 90.5/\underline{-54.3} \\ 121.4/\underline{173.8} \\ 104.2/\underline{72.6} \end{bmatrix}$$

Note in this example the very slight difference between the sending and receiving end currents. The very small difference is due to the shunt admittance. It is seen that very little error will be made if the shunt admittance of the two lines is ignored. This will be the usual case. Exceptions will be for very long distribution lines (50 miles or more) and for underground concentric neutral lines that are in parallel for 10 miles or more.

A third option for physically parallel lines in a radial feeder might be considered with the receiving end nodes common to both lines and the sending end nodes not common. However, this would violate the "radial" nature of the feeder since the common receiving end nodes would constitute the creation of a loop.

6.5.2 Electrically Parallel Lines

Figure 6.11 shows two distribution lines that are electrically parallel.

The analysis of the electrically parallel lines requires an extra step from that of the physically parallel lines since the individual line currents are not known. In this case, only the total current leaving the parallel lines is known.

In the typical analysis, the receiving end voltages will either have been assumed or computed, and the total phase currents $[I_R]$ will be known. With $[V_S]$ and $[V_R]$ common to both lines, the first step must be to determine how much of the total current $[I_R]$ flows on each line. Since the lines are electrically parallel, Equation 6.68 can be modified to reflect this condition.

$$\begin{bmatrix} [VS] \\ [VS] \end{bmatrix} = \begin{bmatrix} [a_{11}] & [a_{12}] \\ [a_{21}] & [a_{22}] \end{bmatrix} \cdot \begin{bmatrix} [VR] \\ [VR] \end{bmatrix} + \begin{bmatrix} [b_{11}] & [b_{12}] \\ [b_{21}] & [b_{22}] \end{bmatrix} \cdot \begin{bmatrix} [IR1] \\ [IR2] \end{bmatrix} \quad (6.72)$$

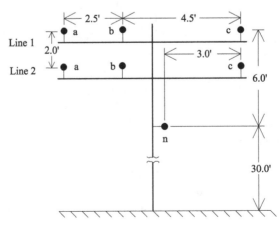

FIGURE 6.11 Electrically parallel lines.

Distribution System Line Models

The current in line 2 as a function of the total current and the current in line 1 is given by

$$[IR2] = [IR] - [IR1] \quad (6.73)$$

Substitute Equation 6.73 into Equation 6.72:

$$\begin{bmatrix}[VS]\\[VS]\end{bmatrix} = \begin{bmatrix}[a_{11}] & [a_{12}]\\ [a_{21}] & [a_{22}]\end{bmatrix} \cdot \begin{bmatrix}[VR]\\[VR]\end{bmatrix} + \begin{bmatrix}[b_{11}] & [b_{12}]\\ [b_{21}] & [b_{22}]\end{bmatrix} \cdot \begin{bmatrix}[IR1]\\ [IR]-[IR1]\end{bmatrix} \quad (6.74)$$

Since the sending end voltages are equal, Equation 6.74 is modified to reflect this.

$$\left([a_{11}]+[a_{12}]\right)\cdot[VR] + \left([b_{11}]-[b_{12}]\right)\cdot[IR1] + [b_{12}]\cdot[IR]$$

$$= \left([a_{21}]+[a_{22}]\right)\cdot[VR] + \left([b_{21}]-[b_{22}]\right)\cdot[IR1] + [b_{22}]\cdot[IR] \quad (6.75)$$

Collect terms in Equation 6.75:

$$\left([a_{11}]+[a_{12}]-[a_{21}]-[a_{22}]\right)\cdot[VR] + \left([b_{12}]-[b_{22}]\right)\cdot[IR]$$

$$= \left([b_{21}]-[b_{22}]-[b_{11}]+[b_{12}]\right)\cdot[IR1] \quad (6.76)$$

Equation 6.76 is in the form of

$$[Aa]\cdot[VR] + [Bb]\cdot[IR] = [Cc]\cdot[IR1], \quad (6.77)$$

where

$$[Aa] = [a_{11}]+[a_{12}]-[a_{21}]-[a_{22}]$$

$$[Bb] = [b_{12}]-[b_{22}]$$

$$[Cc] = [b_{21}]-[b_{22}]-[b_{11}]+[b_{12}] \quad (6.78)$$

Equation 6.77 can be solved for the receiving end current in line 1.

$$[IR1] = [Cc]^{-1}\cdot\left([Aa]\cdot[VR] + [Bb]\cdot[IR]\right) \quad (6.79)$$

Equation 6.73 can be used to compute the receiving end current in line 2.

With the two receiving end line currents known, Equations 6.68 and 6.69 are used to compute the sending end voltages. As with the physically parallel lines, an iterative process (Chapter 10) will have to be used to assure the sending end voltages for each line are equal.

Example 6.7: The two lines of Example 6.5 are electrically parallel as shown in Figure 6.10. The receiving end voltages are given by

$$[VR] = \begin{bmatrix} 13,280/-33.1 \\ 14,040/-151.7 \\ 14,147/86.5 \end{bmatrix} V$$

The complex power out of the parallel lines is the sum of the complex power of the two lines in Example 6.6.

$$S_a = 2,763.8 \, kVA \, at \, 0.928 \, PF$$

$$S_b = 2,846.3 \, kVA \, at \, 0.872 \, PF$$

$$S_c = 3,088.5 \, kVA \, at \, 0.90 \, PF$$

The first step in the solution is to determine the total current leaving the two lines.

$$IR_i = \left(\frac{S_i \cdot 1000}{VR_i}\right)^* = \begin{bmatrix} 208.1/-54.9 \\ 202.7/179.0 \\ 218.8/60.7 \end{bmatrix} A$$

Equation 6.83 is used to compute the current in line 1. Before that can be done, the matrices of Equation 6.82 must be computed.

$$[Aa] = [a_{11}] + [a_{12}] - [a_{21}] - [a_{22}] = \begin{bmatrix} 0 & 0 & 0 \\ 0 & 0 & 0 \\ 0 & 0 & 0 \end{bmatrix}$$

$$[Bb] = [b_{12}] - [b_{22}] = \begin{bmatrix} -4.1872 - j6.0646 & -0.0843 - j0.3060 & -0.0820 + j0.5434 \\ -0.0354 + j1.0995 & -4.1342 - j6.7585 & -0.0333 + j1.0599 \\ -0.0284 - j0.7300 & -0.0274 + j1.6105 & -4.1267 - j7.2569 \end{bmatrix}$$

$$[Cc] = [b_{21}] - [b_{22}] - [b_{11}] + [b_{12}]$$
$$= \begin{bmatrix} -7.1697 - j12.2446 & -0.0039 - j0.3041 & -0.0032 + j0.7046 \\ -0.0039 - j0.3041 & -7.1616 - j13.3077 & -0.0013 + j1.9361 \\ -0.0032 + j0.7046 & -0.0013 + j1.9361 & -7.1610 - j14.2577 \end{bmatrix}$$

Distribution System Line Models

The current in line 1 is now computed by

$$[IR1] = [Cc]^{-1} \cdot ([Aa] \cdot [VR] + [Bb] \cdot [IR]) = \begin{bmatrix} 110.4/-59.7 \\ 119.3/172.6 \\ 121.7/50.2 \end{bmatrix}$$

The current in line 2 is

$$[IR2] = [IR] - [IR1] = \begin{bmatrix} 98.5/-49.6 \\ 85.2/-172.2 \\ 101.1/73.3 \end{bmatrix}$$

The sending end voltages are

$$[VS1] = ([a_{11}] + [a_{12}]) \cdot [VR] + [b_{11}] \cdot [IR1] + [b_{12}] \cdot [IR2] = \begin{bmatrix} 13,738/-30.9 \\ 14,630/-151.0 \\ 14,912/88.3 \end{bmatrix},$$

$$[VS2] = ([a_{21}] + [a_{22}]) \cdot [VR] + [b_{21}] \cdot [IR1] + [b_{22}] \cdot [IR2] = \begin{bmatrix} 13,738/-30.9 \\ 14,630/-151.0 \\ 14,912/88.3 \end{bmatrix}$$

It is satisfying that the two equations give us the same results for the sending end voltages.

The sending end currents are

$$[IS1] = ([c_{11}] + [c_{12}]) \cdot [VR] + [d_{11}] \cdot 1[IR1] + [d_{12}] \cdot [IR2] = \begin{bmatrix} 110.0/-59.2 \\ 118.7/173.1 \\ 121.2/50.6 \end{bmatrix},$$

$$[IS2] = ([c_{21}] + [c_{22}]) \cdot [VR] + [d_{21}] \cdot [IR1] + [d_{22}] \cdot [IR2] = \begin{bmatrix} 98.2/-49.0 \\ 84.7/-171.7 \\ 100.7/73.9 \end{bmatrix}$$

When the shunt admittance of the parallel lines is ignored, a parallel equivalent 3×3 phase impedance matrix can be determined. Since very little error is made ignoring the shunt admittance on most distribution lines, the equivalent parallel phase impedance matrix can be very useful in distribution power flow programs that are not designed to model electrically parallel lines.

Since the lines are electrically parallel, the voltage drops in the two lines must be equal. The voltage drop in the two parallel lines is given by

$$\begin{bmatrix} [v_{abc}] \\ [v_{abc}] \end{bmatrix} = \begin{bmatrix} [Z_{11}] & [Z_{12}] \\ [Z_{21}] & [Z_{22}] \end{bmatrix} \cdot \begin{bmatrix} IR1 \\ IR2 \end{bmatrix} \quad (6.80)$$

Substitute Equation 6.73 into Equation 6.80.

$$\begin{bmatrix}[v_{abc}]\\[v_{abc}]\end{bmatrix} = \begin{bmatrix}[Z_{11}] & [Z_{12}]\\[Z_{21}] & [Z_{22}]\end{bmatrix} \cdot \begin{bmatrix}[IR1]\\[IR]-[IR1]\end{bmatrix} \quad (6.81)$$

Expand Equation 6.85 to solve for the voltage drops.

$$[v_{abc}] = [Z_{11}]\cdot[IR1]+[Z_{12}]\cdot([IR]-[IR1])[v_{abc}] = [Z_{21}]\cdot[IR1]+[Z_{22}]\cdot([IR]-[IR1])$$

$$[v_{abc}] = ([Z_{11}]-[Z_{12}])\cdot[IR1]+[Z_{12}]\cdot[IR]$$

$$[v_{abc}] = ([Z_{21}]-[Z_{22}])\cdot[IR1]+[Z_{22}]\cdot[IR] \quad (6.82)$$

Collect terms in Equation 6.82.

$$([Z_{11}]-[Z_{12}]-[Z_{21}]+[Z_{22}])\cdot[IR1] = ([Z_{22}]-[Z_{12}])\cdot[IR] \quad (6.83)$$

Let

$$[ZX] = ([Z_{11}]-[Z_{12}]-[Z_{21}]+[Z_{22}]) \quad (6.84)$$

Substitute Equation 6.84 into Equation 6.83 and solve for the current in line 1.

$$[IR1] = [ZX]^{-1}\cdot([Z_{22}]-[Z_{12}])\cdot[IR] \quad (6.85)$$

Substitute Equation 6.85 into the top line of Equation 6.81.

$$[v_{abc}] = (([Z_{11}-Z_{12}])\cdot[ZX]^{-1}\cdot([Z_{22}]-[Z_{12}])+[Z_{12}])\cdot[IR] \quad (6.86)$$

$$[v_{abc}] = [Z_{eq}]\cdot[IR]$$

$$[Z_{eq}] = (([Z_{11}-Z_{12}])\cdot[ZX]^{-1}\cdot([Z_{22}]-[Z_{12}])+[Z_{12}]) \quad (6.87)$$

The equivalent impedance of Equation 6.87 is the 3×3 equivalent for the two lines that are electrically parallel. This is the phase impedance matrix that can be used in conventional distribution power flow programs that cannot model electrically parallel lines.

Example 6.8: The same two lines are electrically parallel, but the shunt admittance is neglected. Compute the equivalent 3×3 impedance matrix using the impedance partitioned matrices of Example 6.6.

$$[ZX] = [Z_{11}] - [Z_{12}] - [Z_{21}] + [Z_{22}]$$

$$= \begin{bmatrix} 7.1697 + j12.2446 & 0.0039 + j0.3041 & 0.0032 - j0.7046 \\ 0.0039 + j0.3041 & 7.1616 + j13.3077 & 0.0013 - j1.9361 \\ 0.0032 - j0.7046 & 0.0013 - j1.9361 & 7.1610 + j14.2577 \end{bmatrix}$$

$$[Z_{eq}] = ([Z_{11}] - [Z_{12}]) \cdot [ZX]^{-1}$$
$$\cdot ([Z_{22}] - [Z_{12}]) + [Z_{12}]$$

$$= \begin{bmatrix} 3.3677 + j7.796 & 1.5330 + j4.7717 & 1.4867 + j4.7304 \\ 1.5330 + j4.7717 & 3.3095 + j7.6459 & 1.5204 + j4.7216 \\ 1.4867 + j4.7304 & 1.5204 - j4.7216 & 3.2662 + j7.5316 \end{bmatrix}$$

The sending end voltages are

$$[VS] = [VR] + [Z_{eq}] \cdot [IR] = \begin{bmatrix} 13,740/\underline{-31.0} \\ 14,634/\underline{-151.0} \\ 14,916/\underline{88.3} \end{bmatrix} V$$

6.6 SUMMARY

This chapter has developed the "exact" and "modified" line segment models. The exact model uses no approximations. That is, the phase impedance and shunt admittance matrices are developed assuming no transposition is used. The modified model ignores the shunt admittance. For the three-line models, generalized matrix equations have been developed. The equations utilize the generalized matrices [a], [b], [c], [d], [A], and [B]. The example problems demonstrate that because the shunt admittance is very small, the generalized matrices can be computed neglecting the shunt admittance with very little error. In most cases, the shunt admittance can be neglected; however, there are situations where the shunt admittances should not be neglected. This is particularly true for long rural lightly loaded lines and for many underground lines.

A method for computing the current flowing in the neutral and ground was developed. The only assumption used that can make a difference on the computing currents is that the resistivity of the earth was assumed to be 100 meters.

A simple version of the LIT was introduced and applied in Example 6.5. The ladder method will be used in future chapters and is fully developed in Chapter 10.

It must be pointed out that the method symmetrical components are never used in the analysis of a distribution system. The only time sequence impedances are used is in the development of the source phase impedance matrix of the high voltage system serving the distribution substation.

The generalized matrices for two lines in parallel have been derived. The analysis of physically parallel and electrically parallel lines were developed with examples to demonstrate the analysis process.

PROBLEMS

6.1 A two-mile-long, three-phase line uses the configuration of Problem 4.1. The phase impedance matrix and shunt admittance matrix for the configuration are as follows:

$$[z_{abc}] = \begin{bmatrix} 0.3375 + j1.0478 & 0.1535 + j0.3849 & 0.1559 + j0.5017 \\ 0.1535 + j0.3849 & 0.3414 + j1.0348 & 0.1580 + j0.4236 \\ 0.1559 + j0.5017 & 0.1580 + j0.4236 & 0.3465 + j1.0179 \end{bmatrix} \Omega/\text{mile}$$

$$[y_{abc}] = \begin{bmatrix} j5.9540 & -j0.7471 & -j2.0030 \\ -j0.7471 & j5.6322 & -j1.2641 \\ -j2.0030 & -j1.2641 & j6.3962 \end{bmatrix} \mu S/\text{mile}$$

The line is serving a balanced three-phase load of 10,000 kVA, with balanced voltages of 13.2 kV line to line and a power factor of 0.85 lagging. Alter the MATLAB script M0601.m to perform the following.
(a) Determine the generalized matrices.
(b) For the given load, compute the line-to-line and line-to-neutral voltages at the source end of the line.
(c) Compute the voltage unbalance at the source end.
(d) Compute the source end complex power per phase.
(e) Compute the power loss by phase over the line. (Hint: Power loss is defined as power in minus power out).

6.2 Use the line of Problem 6.1. For this problem, the source voltages are specified as

$$[VS_{LN}] = \begin{bmatrix} 7620/\underline{0} \\ 7620/\underline{-120} \\ 7620/\underline{120} \end{bmatrix}$$

The three-phase load is unbalanced, connected in wye, and given by

$$[kVA] = \begin{bmatrix} 2500 \\ 3500 \\ 1500 \end{bmatrix} \quad [PF] = \begin{bmatrix} 0.90 \\ 0.85 \\ 0.95 \end{bmatrix}$$

Write a MATLAB script that employs ladder iterative technique and determine
(a) the load line-to-neutral voltages,
(b) power at the source, and
(c) the voltage unbalance at the load.

6.3 Use WindMil for Problem 6.2.

6.4 The line of Problem 6.1 serves an unbalanced, grounded, wye-connected constant impedance load of

$$Z_{ag} = 15/\underline{30}\ \Omega, \quad Z_{bg} = 17/\underline{36.87}\ \Omega, \quad Z_{cg} = 20/\underline{25.84}\ \Omega.$$

The line is connected to a balanced three-phase 13.2 kV source. Alter the MATLAB script M061.m so that the load served is the constant impedance load shown earlier and use to perform the following.
(a) Determine the load currents.
(b) Determine the load line-to-ground voltages.
(c) Determine the complex power of the load by phase.
(d) Determine the source complex power by phase.
(e) Determine the power loss by phase and the total three-phase power loss.
(f) Determine the current flowing in the neutral and ground.

6.5 Repeat Problem 6.3, only the load on phase b is changed to $50/\underline{36.87}\ \Omega$.

6.6 The two-phase line of Problem 4.2 has the following phase impedance matrix:

$$[z_{abc}] = \begin{bmatrix} 0.4576 + j1.0780 & 0 & 0.1535 + j0.3849 \\ 0 & 0 & 0 \\ 0.1535 + j0.3849 & 0 & 0.4615 + j1.0651 \end{bmatrix} \Omega/\text{mile}$$

The line is 2 miles long and serves a two-phase load such that

S_{ag} = 2000 kVA at 0.9 lagging power factor and voltage of 7,620/$\underline{0}$ V,
S_{cg} = 1500 kVA at 0.95 lagging power factor and voltage of 7,620/$\underline{120}$ V.

Neglect the shunt admittance and determine the following:
(a) The source line-to-ground voltages using the generalized matrices. (Hint: Even though phase b is physically not present, assume that it is with a value of 7,620/$\underline{-120}$ V and is serving a 0 kVA load).
(b) The complex power by phase at the source.
(c) The power loss by phase on the line.
(d) The current flowing in the neutral and ground.

6.7 The single-phase line of Problem 4.3 has the following phase impedance matrix.

$$[z_{abc}] = \begin{bmatrix} 0 & 0 & 0 \\ 0 & 1.3292 + j1.3475 & 0 \\ 0 & 0 & 0 \end{bmatrix} \Omega/\text{mile}$$

The line is one mile long and is serving a single-phase load of 2,000 kVA, 0.95 lagging power factor at a voltage of 7,500/−120 V. Determine the source voltage and power loss on the line. (Hint: As in the previous problem, even though phases a and c are not physically present, assume they are, and, along with phase b, make up a balanced three-phase set of voltages).

6.8 The three-phase concentric neutral cable configuration of Problem 4.10 is 2 miles long and serves a balanced three-phase load of 10,000 kVA, 13.2 kV, 0.85 lagging power factor. The phase impedance and shunt admittance matrices for the cable line are

$$[z_{abc}] = \begin{bmatrix} 0.7891 + j0.4041 & 0.3192 + j0.0328 & 0.3192 + j0.0328 \\ 0.3192 + j0.0328 & 0.7982 + j0.4463 & 0.2849 - j0.0143 \\ 0.3192 + j0.0328 & 0.2849 - j0.0143 & 0.8040 + j0.4381 \end{bmatrix}, \Omega/\text{mile}$$

$$[y_{abc}] = \begin{bmatrix} j96.61 & 0 & 0 \\ 0 & j96.61 & 0 \\ 0 & 0 & j96.61 \end{bmatrix} \mu S/\text{mile}.$$

(a) Determine the generalized matrices.
(b) For the given load, compute the line-to-line and line-to-neutral voltages at the source end of the line.
(c) Compute the voltage unbalance at the source end.
(d) Compute the source end complex power per phase.
(e) Compute the power loss by **phase** over the line. (Hint: Power loss is defined as power in minus power out).

6.9 The line of Problem 6.9 serves an unbalanced, grounded, wye-connected constant impedance load of

$$Z_{ag} = 15/30 \,\Omega, \quad Z_{bg} = 50/36.87 \,\Omega, \quad Z_{cg} = 20/25.84 \,\Omega.$$

The line is connected to a balanced three-phase 13.2 kV source.
(a) Determine the load currents.
(b) Determine the load line-to-ground voltages.
(c) Determine the complex power of the load by phase.
(d) Determine the source complex power by phase.
(e) Determine the power loss by phase and the total three-phase power loss.
(f) Determine the current flowing in each neutral and ground.

6.10 The tape shielded cable single-phase line of Problem 4.12 is 2 miles long and serves a single-phase load of 3,000 kVA at 8.0 kV and 0.9 lagging power factor. The phase impedance and shunt admittances for the line are

$$[z_{abc}] = \begin{bmatrix} 0 & 0 & 0 \\ 0 & 0 & 0 \\ 0 & 0 & 0.5291 + j0.5685 \end{bmatrix} \Omega/\text{mile},$$

$$[y_{abc}] = \begin{bmatrix} 0 & 0 & 0 \\ 0 & 0 & 0 \\ 0 & 0 & j140.39 \end{bmatrix} \mu S/\text{mile}.$$

Determine the source voltage and the power loss for the loading condition.

6.11 Two distribution lines constructed on one pole are shown in Figure 6.12.

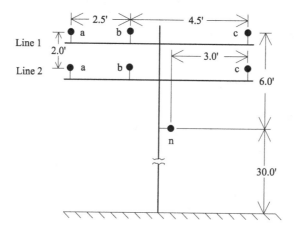

FIGURE 6.12 Two parallel lines on one pole.

Line #1 Data:
Conductors: 336,400 26/7 ACSR
GMR = 0.0244 ft., Resistance = 0.306 Ω/mile, Diameter = 0.721 in.
Line # 2 Data:
Conductors: 250,000 AA
GMR = 0.0171 ft., Resistance = 0.41 Ω/mile, Diameter = 0.574 in.
Neutral Conductor Data:
Conductor: 4/0 6/1 ACSR
GMR = 0.00814 ft., Resistance = 0.592 Ω/mile, Diameter = 0.563 in.
Length of lines is 10 miles
Balanced load voltages of 24.9 kV line to line
Unbalanced Loading:

Load #1:   Phase a: 1,440 kVA at 0.95 lagging power factor
Phase b: 1,150 kVA at 0.9 lagging power factor
Phase c: 1,720 kVA at 0.85 lagging power factor

Load #2: Phase a: 1,300 kVA at 0.9 lagging power factor
Phase b: 1,720 kVA at 0.85 lagging power factor
Phase c: 1,370 kVA at 0.95 lagging power factor

The two lines have a common sending end node (Figure 6.6).
Modify the MATLAB script M0606 to determine
(a) the total phase impedance matrix (6×6) and total phase admittance matrix (6×6),
(b) the abcd and AB matrices,
(c) the sending end node voltages and currents for each line for the specified loads,
(d) the sending end complex power for each line,
(e) the real power loss of each line, and
(f) determine the current flowing in the neutral conductor and ground.

6.12 The lines of Problem 6.11 do not share a common sending or receiving end node (Figure 6.7). Determine
(a) the sending end node voltages and currents for each line for the specified loads,
(b) the sending end complex power for each line, and
(c) the real power loss of each line.

6.13 The lines of Problem 6.11 are electrically parallel (Figure 6.8). Modify the MATLAB script M0608.m to compute the equivalent 3×3 impedance matrix and determine
(a) the sending end node voltages and currents for each line for the specified loads,
(b) the sending end complex power for each line, and
(c) the real power loss of each line.

WINDMIL ASSIGNMENT

Use System 2 and add a two-phase concentric neutral cable line connected to Node 2. Call this "System 3". The line uses phases *a* and *c* and is 300 feet long and consists of two 1/0 AA one-third neutral concentric neutral cables. The cables are 40 inches below ground and 6 inches apart. There is no additional neutral conductor. Call this line UG-1. At the end of UG-1, connect a node and call it Node 4. The load at Node 4 is a delta-connected load modeled as constant current. The load is 250 kVA at a 95% lagging power factor.

Determine the voltages at all nodes on a 120-volt base and all line currents.

REFERENCES

1. Glover, J.D., Sarma, M., *Power System Analysis and Design*, PWS-Kent Publishing, Boston, MA, 2nd Edition, 1995.
2. "ANSI/NEMA Standard Publication No. MG1-1978", National Electrical Manufactures Association, Washington, DC.

7 Voltage Regulation

The regulation of voltages is an important function on a distribution feeder. As the loads on the feeders vary, there must be some means of regulating the voltage so that every customer's voltage remains within an acceptable level. Common methods of regulating the voltage are the application of step-type voltage regulators, load tap changing transformers (LTC), and shunt capacitors.

7.1 STANDARD VOLTAGE RATINGS

The American National Standards Institute' (ANSI) standard ANSI C84.1-1995 for "Electric Power Systems and Equipment Voltage Ratings (60 Hertz)" provides the following definitions for system voltage terms [1]:

- System Voltage: The root mean square phasor voltage of a portion of an alternating-current electric system. Each system voltage pertains to a portion of the system that is bounded by transformers or utilization equipment.
- Nominal System Voltage: The voltage by which a portion of the system is designated, and to which certain operating characteristics of the system are related. Each nominal system voltage pertains to a portion of the system bounded by transformers or utilization equipment.
- Maximum System Voltage: The highest system voltage that occurs under normal operating conditions and the highest system voltage for which equipment and other components are designed for satisfactory continuous operation without a derating of any kind.
- Service Voltage: The voltage at the point where the electrical system of the supplier and the electrical system of the user are connected.
- Utilization Voltage: The voltage at the line terminals of utilization equipment.
- Nominal Utilization Voltage: The voltage rating of certain utilization equipment used on the system.

The ANSI standard specifies two voltage ranges. An oversimplification of the voltage ranges is as follows:

- Range A: Electric supply systems shall be so designated and operated that most service voltages will be within the limits specified for Range A. The occurrence of voltages outside of these limits should be infrequent.
- Range B: Voltages above and below Range A. When these voltages occur, corrective measures shall be undertaken within a reasonable time to improve voltages to meet Range A.

For a normal three-wire, 120-/240-volt service to a user, the Range A and Range B voltages are as follows:

- Range A:
  - Nominal Utilization Voltage = 115 volts
  - Maximum Utilization and Service Voltage = 126 volts
  - Minimum Service Voltage = 114 volts
  - Minimum Utilization Voltage = 110 volts
- Range B:
  - Nominal Utilization Voltage = 115 volts
  - Maximum Utilization and Service Voltage = 127 volts
  - Minimum Service Voltage = 110 volts
  - Minimum Utilization Voltage = 107 volts

These ANSI standards give the distribution engineer a range of "normal steady-state" voltages (Range A) and a range of "emergency steady-state" voltages (Range B) that must be supplied to all users.

In addition to the acceptable voltage magnitude ranges, the ANSI standard recommends that the "electric supply systems should be designed and operated to limit the maximum voltage unbalance to 3 percent when measured at the electric-utility revenue meter under a no-load condition". Voltage unbalance is defined as

$$Voltage_{unbalance} = \frac{max.deviation\ from\ average\ voltage}{Average\ voltage} \cdot 100\% \qquad (7.1)$$

The task for the distribution engineer is to design and operate the distribution system so that under normal steady-state conditions the voltages at the meters of all users will lie within Range A and the voltage unbalance will not exceed 3%.

A common device used to maintain system voltages is the step-voltage regulator. Step-voltage regulators can be single phase or three phase. Single-phase regulators can be connected in wye, delta, or open delta in addition to operating as a single-phase device. The regulators and their controls allow the voltage output to vary as the load varies.

A step-voltage regulator is basically an autotransformer with a load tap changing mechanism on the "series" winding. The voltage change is obtained by changing the number of turns (tap changes) of the series winding of the autotransformer.

An autotransformer can be visualized as a two-winding transformer with a solid connection between a terminal on the primary side of the transformer to a terminal on the secondary. Before proceeding to the autotransformer, a review of two-winding transformer theory and the development of generalized constants will be presented.

Voltage Regulation

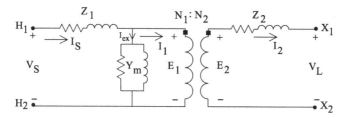

FIGURE 7.1 Two-winding transformer exact equivalent circuit.

7.2 TWO-WINDING TRANSFORMER THEORY

The exact equivalent circuit for a two-winding transformer is shown in Figure 7.1.

In Figure 7.1, the high-voltage transformer terminals are denoted by H_1 and H_2, and the low-voltage terminals are denoted by X_1 and X_2. The standards for these markings are such that at no load the voltage between H_1 and H_2 will be in phase with the voltage between X_1 and X_2. The currents I_1 and I_2 will also be in phase.

Without introducing a significant error, the exact equivalent circuit of Figure 7.1 is modified by referring the primary impedance (Z_1) to the secondary side, as shown in Figure 7.2.

To better understand the model for the step-regulator, a model for the two-winding transformer will first be developed. With reference to Figure 7.2, the turns ratio is defined as

$$n_t = \frac{N_1}{N_2} \tag{7.2}$$

The basic equations for the ideal voltages and currents are

$$E_1 = n_t \cdot E_2 \text{ and } I_1 = \frac{1}{n_t} \cdot I_2,$$

$$E_2 = \frac{1}{n_t} \cdot E_1 \text{ and } I_2 = n_t \cdot I_1 \tag{7.3}$$

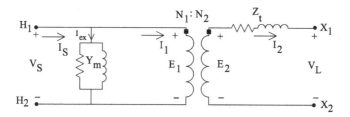

FIGURE 7.2 Two-winding transformer approximate equivalent circuit.

The ideal primary and secondary voltages are

$$E_1 = n_t \cdot E_2,$$

$$\text{but } E_2 = V_L + Z_2 \cdot I_2;$$

$$\text{therefore, } E_1 = n_t \cdot (V_L + Z_2 \cdot I_2) \qquad (7.4)$$

The source voltage is

$$V_S = E_1 + Z_1 \cdot I_1,$$

$$\text{but } I_1 = \frac{1}{n_t} \cdot I_2;$$

$$\text{therefore, } V_S = n_t \cdot (V_L + Z_2 \cdot I_2) + Z_1 \cdot I_1 \qquad (7.5)$$

Substitute into Equation 7.5 the expression for I_1 and collect terms.

$$V_S = n_t \cdot (V_L + Z_2 \cdot I_2) + Z_1 \cdot \frac{1}{n_t} \cdot I_2$$

$$V_S = n_t \cdot (V_L + Z_2 \cdot I_2) + Z_1 \cdot \frac{n_t}{n_t^2} \cdot I_2$$

$$V_S = n_t \cdot V_L + n_t \cdot \left(Z_2 + \frac{Z_1}{n_t^2}\right) \cdot I_2 \qquad (7.6)$$

The equivalent secondary impedance is defined as

$$Z_t = Z_2 + \frac{Z_1}{n_t^2}. \qquad (7.7)$$

Using the equivalent impedance of Equation 7.7, the source voltage is given as

$$V_S = n_t \cdot V_L + n_t \cdot Z_t \cdot I_2. \qquad (7.8)$$

The general terms for the source voltage are

$$V_S = a \cdot V_L + b \cdot I_2, \qquad (7.9)$$

Voltage Regulation

where

$a = n_t$,
$b = n_t \cdot Z_t$.

The input current to the two-winding transformer is given by

$$I_S = Y_m \cdot V_S + I_1. \tag{7.10}$$

Substitute Equations 7.3 and 7.8 into Equation 7.10.

$$I_S = Y_m \cdot (n_t \cdot V_L + n_t \cdot Z_t \cdot I_2) + \frac{1}{n_t} \cdot I_2,$$

$$I_S = n_t \cdot Y_m \cdot V_L + n_t \cdot \left(\frac{1}{n_t^2} + Y_m \cdot Z_t\right) \cdot I_2,$$

$$I_S = c \cdot V_L + d \cdot I_2, \tag{7.11}$$

where

$$c = n_t \cdot Y_m,$$

$$d = n_t \cdot \left(\frac{1}{n_t^2} + Y_m \cdot Z_t\right)$$

Equations 7.9 and 7.11 are used to compute the input voltage and current for a two-winding transformer when the load voltage and current are known. These two equations are of the same form as Equations 6.8 and 6.16 that were derived in Chapter 6 for the three-phase line models. The only difference at this point is that only a single-phase, two-winding transformer is being modeled. Later, in this chapter, the terms *a*, *b*, *c*, and *d* will be expanded to 3 × 3 matrices for all possible three-phase regulator connections.

Sometimes, particularly in the ladder iterative process, the output voltage needs to be computed knowing the input voltage and the load current. Solving Equation 7.9 for the load voltage yields

$$V_L = \frac{1}{a} \cdot V_S - \frac{b}{a} \cdot Z_t. \tag{7.12}$$

Substituting Equations 7.8 and 7.9 into Equation 7.15 results in

$$V_L = A \cdot V_S - B \cdot I_2, \tag{7.13}$$

where

$$A = \frac{1}{a} = \frac{1}{n_t},$$

$$B = \frac{b}{a} = \frac{n_t \cdot Z_t}{n_t} = Z_t$$

Again, Equation 7.13 is of the same form as Equation 6.26. Later, in this chapter, the expressions for A and B will be expanded to 3 × 3 matrices for all possible three-phase transformer connections.

Example 7.1: A single-phase transformer is rated 75 kVA, 2,400–240 volts. The transformer has the following impedances and shunt admittance:

$Z_1 = 0.612 + j1.2\ \Omega$ (high-voltage winding impedance)
$Z_2 = 0.0061 + j0.0115\ \Omega$ (low-voltage winding impedance)
$Y_m = 1.92 \times 10^{-4} - j8.52 \times 10^{-4}$ S (referred to the high-voltage winding)

Determine the generalized a, b, c, d constants and the A and B constants. The transformer "turns ratio" is

$$n_t = \frac{N_1}{N_2} = \frac{V_{1-rated}}{V_{2-rated}} = \frac{2400}{240} = 10$$

The equivalent transformer impedance referred to the low-voltage side:

$$Z_t = Z_2 + \frac{Z_1}{n_t^2} = 0.0122 + j0.0235$$

The generalized constants are as follows:

$$a = n_t = 10$$

$$b = n_t \cdot Z_t = 0.1222 + j0.2350$$

$$c = n_t \cdot Y_m = 0.0019 - j0.0085$$

$$d = n_t \cdot \left(Y_m \cdot Z_t + \frac{1}{n_t^2}\right) = 0.1002 - j0.0001$$

$$A = \frac{1}{n_t} = 0.1$$

$$B = Z_t = 0.0122 + j0.0235$$

Voltage Regulation

Assume that the transformer is operated at rated load (75 kVA) and rated voltage (240 volts) with a power factor of 0.9 lagging. Determine the source voltage and current using the generalized constants.

$$V_L = 240 \underline{/0}$$

$$I_2 = \frac{75 \cdot 1000}{240} \underline{/-\cos^{-1}(0.9)} = 312.5 \underline{/-25.84}$$

Applying the values of the previously computed a, b, c, and d parameters,

$$V_S = a \cdot V_L + b \cdot I_2 = 2466.9 \underline{/1.15} \, \text{V},$$

$$I_S = c \cdot V_L + d \cdot I_2 = 32.67 \underline{/-28.75} \, \text{A}$$

Using the computed source voltage and the load current, determine the load voltage.

$$V_L = A \cdot V_S - B \cdot I_2 = (0.1) \cdot (2466.9 \underline{/1.15}) - (0.0122 + j0.0235) \cdot (312.5 \underline{/-25.84})$$

$$V_L = 240.0 \underline{/0} \, \text{V}$$

For future reference, the per-unit impedance of the transformer is computed by

$$Z_{base} = \frac{kV_2^2 \cdot 1000}{kVA} = \frac{.240^2 \cdot 1000}{75} = 0.768 \, \Omega,$$

$$Z_{pu} = \frac{Z_t}{Z_{base}} = \frac{0.0122 + j0.0115}{0.768} = 0.0345 \underline{/62.5} \, \text{per-unit}$$

The per-unit shunt admittance is computed by

$$Y_{base} = \frac{kVA}{kV_1^2 \cdot 1000} = \frac{75}{2.4^2 \cdot 1000} = 0.013 \, \text{S},$$

$$Y_{pu} = \frac{Y_m}{Y_{base}} = \frac{1.92 \cdot 10^{-4} - j8.52 \cdot ^{-4}}{0.013} = 0.0147 - j0.0654 \, \text{per-unit}$$

Example 7.1 demonstrates that the generalized constants provide a quick method for analyzing the operating characteristics of a two-winding transformer.

7.3 TWO-WINDING AUTOTRANSFORMER

A two-winding transformer can be connected as an autotransformer. Connecting the high-voltage terminal H1 to the low-voltage terminal X2, as shown in Figure 7.3, can create a "step-up" autotransformer. The source is connected to terminals H1 and H2, while the load is connected between the X1 terminal and the extension of H2.

In Figure 7.3, V_S is the "source" voltage, and V_L is the "load" voltage. The low-voltage winding of the two-winding transformer (E_2) will be referred to as the "series" winding of the autotransformer and the high-voltage winding (E_1) of the two-winding transformer will be referred to as the "shunt" winding of the autotransformer.

Generalized constants like those of the two-winding transformer can be developed for the autotransformer. The total equivalent transformer impedance is referred to as the "series" winding. The "ideal" transformer Equations of 7.2 and 7.3 still apply.

Apply KVL in the secondary circuit, recognizing that the secondary current $I_2 = I_L$ and the input voltage $V_S = E_1$.

$$E_1 + E_2 = V_L + Z_t \cdot I_2$$

$$E_1 + \frac{E_1}{n_t} = \left(\frac{n_t + 1}{n_t}\right) \cdot E_1 = V_L + Z_t \cdot I_2$$

But $I_2 = I_L$ and $E_1 = V_s$

$$V_S + \frac{V_S}{n_t} = \left(\frac{n_t + 1}{n_t}\right) \cdot V_S = V_L + Z_t \cdot I_L \qquad (7.14)$$

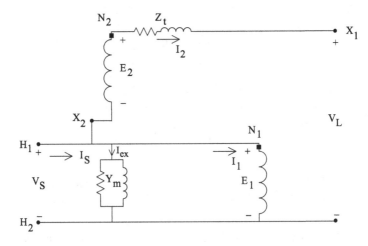

FIGURE 7.3 Step-up autotransformer.

Voltage Regulation

Solve Equation 7.14 for the source voltage.

$$V_S = \left(\frac{n_t}{n_t+1}\right) \cdot V_L + \left(\frac{n_t \cdot Z_t}{n_t+1}\right) \cdot I_L \tag{7.15}$$

Equation 7.15 is of the form

$$V_S = a \cdot V_L + b \cdot I_2, \tag{7.16}$$

where

$$a = \left(\frac{n_t}{n_t+1}\right), \tag{7.17}$$

$$b = \left(\frac{n_t \cdot Z_t}{n_t+1}\right) \tag{7.18}$$

Applying KCL at input Node H_1,

$$I_S = I_1 + I_2 + I_{ex},$$

$$I_S = Y_m \cdot V_S + \left(\frac{1}{n_t}+1\right) \cdot I_2 = Y_m \cdot V_S + \left(\frac{n_t+1}{n_t}\right) \cdot I_2 \tag{7.19}$$

In Equation 7.19, the secondary current is recognized as the load current.

$$I_2 = I_L$$

$$I_S = Y_m \cdot V_S + \left(\frac{n_t+1}{n_t}\right) \cdot I_L \tag{7.20}$$

Substitute Equation 7.15 into Equation 7.20 and simplify.

$$I_S = Y_m \cdot \left(\left(\frac{n_t}{n_t+1}\right) \cdot V_L + \left(\frac{n_t}{n_t+1}\right) \cdot Z_t \cdot I_L\right) + \left(\frac{n_t+1}{n_t}\right) \cdot I_L$$

$$I_S = \frac{n_t \cdot Y_m}{n_t+1} \cdot V_L + \left(\frac{n_t \cdot Y_m \cdot Z_t}{n_t} + \frac{n_t+1}{n_t}\right) \cdot I_L \tag{7.21}$$

Equation 7.21 can be written as

$$I_S = c \cdot V_L + d \cdot I_L, \quad (7.22)$$

where

$$c = \frac{n_t \cdot Y_m}{n_t + 1}, \quad (7.23)$$

$$d = \frac{n_t \cdot Y_m \cdot Z_t}{n_t + 1} + \frac{n_t + 1}{n_t} \quad (7.24)$$

Equations 7.17, 7.18, 7.22, and 7.23 define the generalized constants for the "step-up" autotransformer.

The two-winding transformer can also be connected in the "step-down" connection by reversing the connection between the shunt and series winding, as shown in Figure 7.4.

Generalized constants can be developed for the "step-down" connection following the same procedure as that for the step-up connection.

Apply KVL in the secondary circuit.

$$E_1 - E_2 = V_L + Z_t \cdot I_2 \quad (7.25)$$

Using the "ideal" transformer relationship of Equation 7.2,

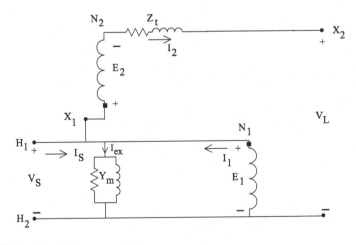

FIGURE 7.4 Step-down autotransformer.

Voltage Regulation

$$V_S - \frac{1}{n_t} \cdot V_S = \left(\frac{n_t - 1}{n_t}\right) \cdot V_S = V_L + Z_t \cdot I_L \quad (7.26)$$

Solve Equation 7.26 for the source voltage.

$$V_S = \left(\frac{n_t}{n_t - 1}\right) \cdot V_L + \left(\frac{n_t}{n_t - 1}\right) \cdot Z_t \cdot I_L, \quad (7.27)$$

$$V_S = a \cdot V_L + b \cdot I_L, \quad (7.28)$$

where

$$a = \left(\frac{n_t}{n_t - 1}\right), \quad (7.29)$$

$$b = \left(\frac{n_t}{n_t - 1}\right) \cdot Z_t \quad (7.30)$$

It is observed at this point that the only difference between the a and b constants of Equations 7.17 and 7.18 for the step-up connection and Equations 7.29 and 7.30 for the step-down connection is the sign in front of the number 1. This will also be the case for the c and d constants. Therefore, for the step-down connection, the c and d constants are defined by

$$c = \frac{n_t \cdot Y_m}{n_t - 1}, \quad (7.31)$$

$$d = \frac{n_t \cdot Y_m \cdot Z_t}{n_t - 1} + \frac{n_t - 1}{n_t} \quad (7.32)$$

The only difference between the definitions of the generalized constants is the sign of the number 1. In general, then, the generalized constants can be defined by the following:

$$a = \frac{n_t}{n_t \pm 1} \quad (7.33)$$

$$b = \frac{n_t \cdot Z_t}{n_t \pm 1} \quad (7.34)$$

$$c = \frac{n_t \cdot Y_m}{n_t \pm 1} \qquad (7.35)$$

$$d = \frac{n_t \cdot Y_m \cdot Z_t}{n_t \pm 1} + \frac{n_t \pm 1}{n_t} \qquad (7.36)$$

In Equations 7.33 through 7.36, the sign of the number 1 will be positive for the step-up connection and negative for the step-down connection.

As with the two-winding transformer, it is sometimes necessary to relate the output voltage as a function of the source voltage and the output current. Solving Equation 7.28 for the output voltage,

$$V_L = \frac{1}{a} \cdot V_S - \frac{b}{a} \cdot Z_t \cdot I_L, \qquad (7.37)$$

$$V_L = A \cdot V_S - B \cdot I_L, \qquad (7.38)$$

where

$$A = \frac{1}{a} = \frac{n_t \pm 1}{n_t}, \qquad (7.39)$$

$$B = \frac{b}{a} = Z_t \qquad (7.40)$$

The generalized equations for the step-up and step-down autotransformers have been developed. They are of the form that was derived for the two-winding transformer and for the line segment in Chapter 6. For the single-phase autotransformer, the generalized constants are single values but will be expanded later to 3 × 3 matrices for three-phase autotransformers.

7.3.1 Autotransformer Ratings

The kVA rating of the autotransformer is the product of the rated input voltage V_S times the rated input current I_S or the rated load voltage V_L times the rated load current I_L. Define the rated kVA and rated voltages of the two-winding transformer and autotransformer as follows:

kVA_{xfm} = kVA rating of the two-winding transformer
kVA_{auto} = kVA rating of the autotransformer
$V_{1-rated} = E_1$ = rated source voltage of the two-winding transformer
$V_{2-rated} = E_2$ = rated load voltage of the two-winding transformer
V_{S-auto} = rated source voltage of the autotransformer

Voltage Regulation

V_{L-auto} = rated load voltage of the autotransformer
I_1 = rated source current
I_2 = rated secondary current

For the following derivation neglect, the voltage drops through the series winding impedance.

$$V_{L-auto} = E_1 \pm E_2 = \left(1 \pm \frac{1}{n_t}\right) \cdot E_1 = \left(\frac{1 \pm n_t}{n_t}\right) \cdot E_1 \quad (7.41)$$

The rated output kVA is then

$$kVA_{auto} = \left(\frac{1 \pm n_t}{n_t}\right) \cdot \frac{E_1 \cdot I_2}{1000}, \quad (7.42)$$

$$\text{but } I_2 = n_t \cdot I_1 \quad (7.43)$$

Therefore,

$$kVA_{auto} = \left(\frac{1 \pm n_t}{n_t}\right) \cdot \frac{E_1}{1000} \cdot n_t \cdot I_1 = \left(1 \pm n_t\right) \cdot \frac{E_1 \cdot I_1}{1000} = \left(1 \pm n_t\right) \cdot kVA_{xfm} \quad (7.44)$$

Equation 7.44 gives the kVA_{auto} as a function of the kVA_{xfm} of the two-winding transformer when connected as an autotransformer. For the step-up connection, the sign of n_t will be positive, while the step-down will use the negative sign. The kVA_{auto} rating of the autotransformer will be considerably greater than the kVA_{xfm} rating of the two-winding transformer.

Example 7.2: The two-winding transformer of Example 7.1 is connected as a "step-up" autotransformer. Determine the kVA and voltage ratings of the autotransformer.

From Example 7.1, the turns ratio was determined to be $n_t = 10$.
The rated kVA of the step-up autotransformer using Equation 7.44 is:

$$kVA_{auto} = \left(1 + n_t\right) \cdot kVA_{xfm} = \left(1 + 10\right) \cdot 75 = 825 \, kVA$$

The voltage ratings are as follows:

$$V_{S-auto} = V_{1-rated} = 2400 \, V$$

$$V_{L-auto} = V_{1-rated} + V_{2-rated} = 2400 + 240 = 2640 \, V$$

The autotransformer is rated at 825 kVA, 2,400–2,640 volts

Suppose now that the autotransformer is supplying rated kVA at rated voltage with a power factor of 0.9 lagging; determine the source voltage and current.

$$V_L = V_{L-auto} = 2640\underline{/0}\,\text{V}$$

$$I_2 = \frac{kVA_{auto} \cdot 1000}{V_{L-auto}} = \frac{825{,}000}{2640}\underline{/-\cos^{-1}(0.9)} = 312.5\underline{/-25.84}\,\text{A}$$

Determine the generalized constants.

$$a = \frac{n_t}{n_t + 1} = 0.9091$$

$$b = \frac{n_t \cdot Z_t}{n_t + 1} = 0.0111 + j0.0214$$

$$c = \frac{n_t \cdot Y_m}{n_t + 1} = 0.00017455 - j0.0007455$$

$$d = \frac{n_t \cdot Y_m \cdot Z_t}{n_t + 1} + \frac{n_t + 1}{n_t} = 1.1 + j0$$

Applying the generalized constants.

$$V_S = a \cdot 2640\underline{/0} + b \cdot 312.5\underline{/-25.84} = 2406.0\underline{/0.1}\,\text{V}$$

$$I_S = c \cdot 2640\underline{/0} + d \cdot 312.5\underline{/-25.84} = 345.06\underline{/-26.11}\,\text{A}$$

When the load-side voltage is determined knowing the source voltage and load current, the A and B parameters are needed.

$$A = \frac{1}{a} = 1.1$$

$$B = \frac{b}{a} = 0.0122 + j0.0235$$

The load voltage is then

$$V_L = A \cdot 2406.04\underline{/0.107} - B \cdot 312.5\underline{/-25.84} = 2640.00\underline{/0}\,\text{V}$$

Voltage Regulation

Rework this example by setting the transformer impedances and shunt admittance to zero.

When this is done, the generalized matrices are as follows:

$$a = \frac{1}{1+n_t} = 0.9091$$

$$b = \frac{1}{1+n_t} \cdot Z_t = 0$$

$$c = \frac{Y_m}{1+n_t} = 0$$

$$d = \frac{Y_m \cdot Z_t}{1+n_t} + n_t + 1 = 1.1$$

Using these matrices, the source voltages and currents are

$$V_S = a \cdot V_L + b \cdot I_L = 2400\underline{/0},$$

$$I_S = c \cdot V_L + d \cdot I_L = 343.75\underline{/-25.8}$$

The "errors" for the source voltages and currents by ignoring the impedances and shunt admittance are

$$Error_V = \left(\frac{2406.0 - 2400}{2406}\right) \cdot 100 = 0.25\%,$$

$$Error_I = \left(\frac{345.07 - 343.75}{345.07}\right) \cdot 100 = 0.38\%$$

By ignoring the transformer impedances and shunt admittance, very little error has been made. This example demonstrates why, for all practical purposes, the impedance and shunt admittance of an autotransformer can be ignored. This idea will be carried forward for the modeling of voltage regulators.

7.3.2 PER-UNIT IMPEDANCE

The per-unit impedance of the autotransformer based upon the autotransformer kVA and kV ratings can be developed as a function of the per-unit impedance of the two-winding transformer based upon the two-winding transformer ratings.

Let:

Zpu_{xfm} = the per-unit impedance of the two-winding transformer based upon the two-winding kVA and kV ratings.

$V_{2-rated}$ = rated secondary voltage of the two-winding transformer

The base impedance of the two-winding transformer that referred to the low-voltage winding (series winding of the autotransformer) is:

$$Zbase_{xfm} = \frac{V_{rated\_2}^2}{kVA_{xfm} \cdot 1000} \quad (7.45)$$

The actual impedance of the transformer that referred to the low-voltage (series) winding is

$$Zt_{actual} = Zpu_{xfm} \cdot Zbase_{xfm} = Zpu_{xfm} \cdot \frac{V_{rated\_2}^2}{kVA_{xfm} \cdot 1000} \quad (7.46)$$

The rated load voltage of the autotransformer as a function of the rated low-side voltage of the transformer is

$$V_{2-auto} = \left(\frac{1 \pm n_t}{n_t}\right) \cdot V_{2-rated} \quad (7.47)$$

The base impedance for the autotransformer referenced to the load side is

$$Zbase_{auto} = \frac{V_{auto\_2}^2}{kVA_{auto} \cdot 1000} \quad (7.48)$$

Substitute Equation 7.44 and Equation 7.47 into Equation 7.48.

$$Zbase_{auto} = \frac{\left[(1 \pm n_t) \cdot V_{2-rated}\right]^2}{(1 \pm n_t) \cdot kVA_{xfm} \cdot 1000}$$

$$Zbase_{auto} = (1 \pm n_t) \cdot Zbase_{xfm} \quad (7.49)$$

The per-unit impedance of the autotransformer based upon the rating of the autotransformer is

$$Zauto_{pu} = \frac{Zt_{actual}}{Zbase_{auto}} \quad (7.50)$$

Voltage Regulation

Substitute Equations 7.51 and 7.54 into Equation 7.50.

$$Zauto_{pu} = Zpu_{xfm} \cdot \frac{Zbase_{xfm}}{\left(\frac{1 \pm n_t}{n_t}\right) \cdot Zbase_{xfm}} = \left(\frac{n_t}{1 \pm n_t}\right) \cdot Zpu_{xfm} \qquad (7.51)$$

Equation 7.56 gives the relationship between the per-unit impedance of the autotransformer and the per-unit impedance of the two-winding transformer with the point being that the per-unit impedance of the autotransformer is very small compared to that of the two-winding transformer. When the autotransformer is connected to boost the voltage 10%, the value of n_t is 0.1, and Equation 7.58 becomes

$$Zpu_{auto} = \frac{0.1}{1+0.1} \cdot Zpu_{xfm} = 0.0909 \cdot Zpu_{xfm} \qquad (7.52)$$

The per-unit shunt admittance of the autotransformer can be developed as a function of the per-unit shunt admittance of the two-winding transformer. Recall that the shunt admittance is represented on the source side of the two-winding transformer.

Example 7.3: The shunt admittance referred to the source side of the two-winding transformer of Example 7.2 is

$$Yt_{actual} = Y_m = 1.92 \cdot 10^{-4} - j8.52 \cdot 10^{-4} \text{ S}$$

a. Determine the per-unit shunt admittance based upon the two-winding transformer ratings.

$$Ybase_{xfm} = \frac{75}{2.4^2 \cdot 1000} = 0.013$$

$$Ypu_{xfm} = \frac{1.92 \cdot 10^{-4} - j8.52 \cdot 10^{-4}}{0.013} = 0.014746 - j0.065434$$

b. In Example 7.2, the kVA rating of the two-winding transformer connected as an autotransformer was computed to be 825 kVA, and the voltage ratings 2,640–2,400 volts. Determine the per-unit admittance based upon the autotransformer kVA rating and a nominal voltage of 2,400 volts and the ratio of the per-unit admittance of the autotransformer to the per-unit admittance of the two-winding transformer.

$$Ybase_{auto} = \frac{825 \cdot 1000}{2400^2} = 0.1432$$

$$Ypu_{auto} = \frac{1.92 \cdot 10^{-4} - j8.52 \cdot 10^{-4}}{0.1432} = 0.001341 - j0.005949$$

$$\text{Ratio} = \left|\frac{0.001341 - j0.005949}{0.014746 - j0.065434}\right| = 0.909$$

In this section, the equivalent circuit of an autotransformer has been developed for the "raise" and "lower" connections. These equivalent circuits included the series impedance and shunt admittance. If a detailed analysis of the autotransformer is desired, the series impedance and shunt admittance should be included. However, it has been shown in Example 7.2 that these values are very small, and when the autotransformer is to be a component of a system, very little error will be made by neglecting both the series impedance and shunt admittance of the equivalent circuit.

Let

Ypu_{xfm} = per-unit admittance of the two-winding transformer based upon the transformer ratings,

Ypu_{auto} = per-unit admittance of the autotransformer based upon the autotransformer ratings.

The base admittance of the two-winding transformer referenced to the source side is given by

$$Ybase_{xfm} = \frac{kVA_{xfm} \cdot 1000}{V_{1-rated}^2} \qquad (7.53)$$

The actual shunt admittance referred to the source side of the two-winding transformer is

$$Yt_{actual} = Ypu_{xfm} \cdot Ybase_{xfm} = Ypu_{xfm} \cdot \frac{kVA_{xfm} \cdot 1000}{V_{1-rated}^2} \qquad (7.54)$$

The base admittance reference to the source side of the autotransformer is given by

$$Ybase_{auto} = \frac{kVA_{auto} \cdot 1000}{V_{1-rated}^2} = \frac{\left(\frac{1 \pm n_t}{n_t}\right) \cdot kVA_{xfm} \cdot 1000}{V_{1-rated}^2} = \left(\frac{1 \pm n_t}{n_t}\right) \cdot Ybase_{xfm} \qquad (7.55)$$

Voltage Regulation

The per-unit admittance of the autotransformer is

$$Ypu_{auto} = \frac{Ypu_{xfm} \cdot Ybase_{xfm}}{Ybase_{auto}} = \frac{Ypu_{xfm} \cdot Ybase_{xfm}}{\left(\frac{1 \pm n_t}{n_t}\right) \cdot Ybase_{xfm}},$$

$$Ypu_{auto} = \frac{n_t}{(1 \pm n_t)} \cdot Ypu_{xfm}. \tag{7.56}$$

Equation 7.61 shows that the per-unit admittance based upon the autotransformer ratings is much smaller than the per-unit impedance of the two-winding transformer. For an autotransformer in the raise connection with $n_t = 0.1$, Equation 7.63 becomes

$$Ypu_{auto} = \left(\frac{0.1}{1+0.1}\right) \cdot Ypu_{xfm} = 0.0909 \cdot Ypu_{xfm}$$

It has been shown that the per-unit impedance and admittance values based upon the autotransformer kVA rating and nominal voltage are approximately one-tenth that of the values for the two-winding transformer.

7.4 STEP-VOLTAGE REGULATORS

A step-voltage regulator consists of an autotransformer and a load tap changing mechanism. The voltage change is obtained by changing the taps of the series winding of the autotransformer. The position of the tap is determined by a control circuit (line-drop compensator). Standard step regulators contain a reversing switch enabling a regulator range of ± 10%, usually in 32 steps. This amounts to a 5/8% change per step or 0.75 volts change per step on a 120-volt base. Step regulators can be connected in a "Type A" or "Type B" connection according to the ANSI/IEEE C57.15-1986 standard [2]. The more common Type B connection is shown in Figure 7.5.

The step-voltage regulator control circuit is shown in block form in Figure 7.6. The step-voltage regulator control circuit requires the following settings:

1. Voltage Level: The desired voltage (on 120-volt base) to be held at the "load center". The load center may be the output terminal of the regulator or a remote node on the feeder.
2. Bandwidth: The allowed variance of the load center voltage from the set voltage level. The voltage held at the load center will be ± one-half the bandwidth. For example, if the voltage level is set to 122 volts and the bandwidth set to 2 volts, the regulator will change taps until the load center voltage lies between 121 volts and 123 volts.
3. Time Delay: Length of time that a raise or lower operation is called for before the actual execution of the command. This prevents taps changing during a transient or short time change in current.

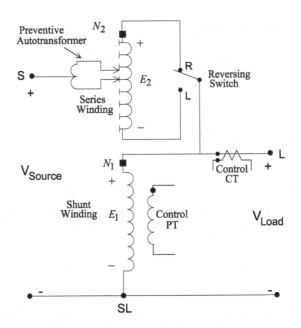

FIGURE 7.5 Type "B" step-voltage regulator

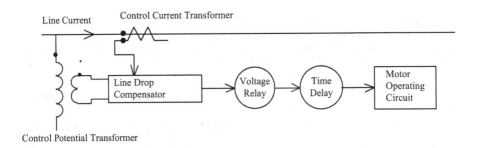

FIGURE 7.6 Step-voltage regulator control circuit

4. Line-Drop Compensator:
   a. Analog Compensator: Analog circuit set to compensate for the voltage drop (line drop) between the regulator and the load center. The settings consist of R and X settings in volts corresponding to the **equivalent** impedance between the regulator and the "load center". This setting may be zero if the regulator output terminals are the "load center".
   b. Digital Compensator: The same as the analog compensator, only the output voltage of the compensator is computed similar to a computer program. Based upon the computed output compensator, voltage taps will change to hold the "load center" voltage within specified limits.
   c. Smart Meters: With the advent of the "smart grid" it is possible for the actual voltage at the "load center" to be transmitted back to the regulator.

Voltage Regulation

Taps are then changed to hold the "load center" voltage within the prescribed limits.

The required rating of a step-regulator is based upon the kVA transformed, not the kVA rating of the line. In general, this will be 10% of the line rating since rated current flows through the series winding, which represents the ± 10% voltage change. The kVA rating of the step-voltage regulator is determined in the same manner as that of the previously discussed autotransformer.

7.4.1 Single-Phase, Step-Voltage Regulators

Because the series impedance and shunt admittance values of step-voltage regulators are so small, they will be neglected in the following equivalent circuits. It should be pointed out, however, that if it is desired to include the impedance and admittance, they can be incorporated into the equivalent circuits in the same way they were originally modeled in the autotransformer equivalent circuit.

7.4.1.1 Type A Step-Voltage Regulator

The detailed equivalent circuit and abbreviated equivalent circuit of a Type A step-voltage regulator in the "raise" position is shown in Figure 7.7.

As shown in Figure 7.7, the primary circuit of the system is connected directly to the shunt winding of the Type A regulator. The series winding is connected to the shunt winding and, in turn, via taps, to the regulated circuit. In this connection, the core excitation varies because the shunt winding is connected directly across the primary circuit.

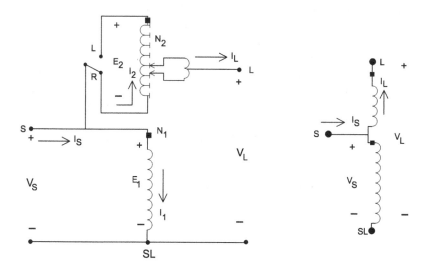

FIGURE 7.7 Type A step-voltage regulator in the raise position.

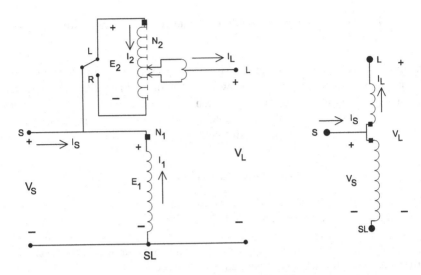

FIGURE 7.8 Type A step-voltage regulator in the lower position.

When the Type A connection is in the "lower" position, the reversing switch is connected to the "L" terminal. The effect of this reversal is to reverse the direction of the currents in the series and shunt windings. Figure 7.8 shows the equivalent circuit and abbreviated circuit of the Type A regulator in the lower position.

7.4.1.2 Type B Step-Voltage Regulator

The more common connection for step-voltage regulators is the Type B. Since this is the more common connection, the defining voltage and current equations for the voltage regulator will be developed only for the Type B connection.

The detailed and abbreviated equivalent circuits of a Type B step-voltage regulator in the "raise" position are shown in Figure 7.9.

The primary circuit of the system is connected, via taps, to the series winding of the regulator in the Type B connection. The series winding is connected to the shunt winding, which is connected directly to the regulated circuit. In a Type B regulator, the core excitation is constant because the shunt winding is connected across the regulated circuit.

The defining voltage and current equations for the regulator in the raise position are as follows:

Voltage Equations	Current Equations	
$\dfrac{E_1}{N_1} = \dfrac{E_2}{N_2}$	$N_1 \cdot I_1 = N_2 \cdot I_2$	(7.57)
$V_S = E_1 - E_2$	$I_L = I_s - I_1$	(7.58)
$V_L = E_1$	$I_2 = I_s$	(7.59)

Voltage Regulation

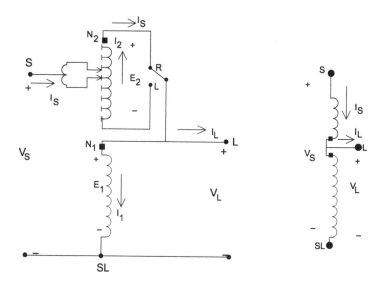

FIGURE 7.9 Type B step-voltage regulator in the raise position

Voltage Equations **Current Equations**

$$E_2 = \frac{N_2}{N_1} \cdot E_1 = \frac{N_2}{N_1} \cdot V_L \qquad I_1 = \frac{N_2}{N_1} \cdot I_2 = \frac{N_2}{N_1} \cdot I_S \qquad (7.60)$$

$$V_S = \left(1 - \frac{N_2}{N_1}\right) \cdot V_L \qquad I_L = \left(1 - \frac{N_2}{N_1}\right) I_S \qquad (7.61)$$

$$V_S = a_R \cdot V_L \qquad I_L = a_R \cdot I_S \qquad (7.62)$$

$$a_R = 1 - \frac{N_2}{N_1} \qquad (7.63)$$

Equations 7.62 and 7.63 are the necessary defining equations for modeling a Type B regulator in the raise position.

The Type B step-voltage connection in the "lower" position is shown in Figure 7.10. As in the Type A connection, note that the direction of the currents through the series and shunt windings change but the voltage polarity of the two windings remains the same.

The defining voltage and current equations for the Type B step-voltage regulator in the lower position are as follows:

Voltage Equations **Current Equations**

$$\frac{E_1}{N_1} = \frac{E_2}{N_2} \qquad N_1 \cdot I_1 = N_2 \cdot I_2 \qquad (7.64)$$

$$V_S = E_1 + E_2 \qquad I_L = I_S - I_1 \qquad (7.65)$$

$$V_L = E_1 \qquad I_2 = -I_S \qquad (7.66)$$

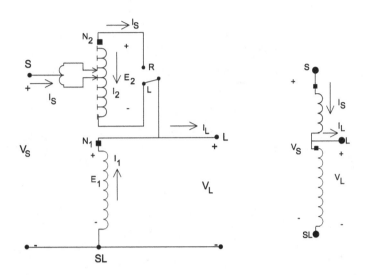

FIGURE 7.10 Type B step-voltage regulator in the lower position.

Voltage Equations   Current Equations

$$E_2 = \frac{N_2}{N_1} \cdot E_1 = \frac{N_2}{N_1} \cdot V_L \qquad I_1 = \frac{N_2}{N_1} \cdot I_2 = \frac{N_2}{N_1} \cdot (-I_S) \tag{7.67}$$

$$V_S = \left(1 + \frac{N_2}{N_1}\right) \cdot V_L \qquad I_L = \left(1 + \frac{N_2}{N_1}\right) \cdot I_S \tag{7.68}$$

$$V_S = a_R \cdot V_L \qquad I_L = a_R \cdot I_S \tag{7.69}$$

$$a_R = 1 + \frac{N_2}{N_1} \tag{7.70}$$

Equations 7.69 and 7.70 give the value of the effective regulator ratio as a function of the ratio of the number of turns on the series winding (N_2) to the number of turns on the shunt winding (N_1).

In the final analysis, the only difference between the voltage and current equations for the Type B regulator in the raise and lower positions is the sign of the ratio $\left(\frac{N_2}{N_1}\right)$. The actual ratio of the windings is not known. However, the tap position will be known. Equations 7.69 and 7.70 can be modified to give the effective regulator ratio as a function of the tap position. Each tap changes the voltage by 5/8 %volts or 0.00625 per unit. Therefore, the effective regulator ratio can be given by

$$a_R = 1 \mp 0.00625 \cdot Tap \tag{7.71}$$

In Equation 7.71, the minus sign applies to the "raise" position and the positive sign to the "lower" position.

Voltage Regulation

7.4.1.3 Generalized Constants

In previous chapters and sections of this text, generalized **abcd** constants have been developed for various devices. It can now be shown that the generalized **abcd** constants can also be applied to the step-voltage regulator. For both the Type A and Type B regulators, the relationship between the source voltage and current to the load voltage and current are of the following forms:

Type A:
$$V_S = \frac{1}{a_R} \cdot V_L; \quad I_S = a_R \cdot I_L \tag{7.72}$$

Type B: $V_S = a_R \cdot V_L;$
$$I_S = \frac{1}{a_R} \cdot I_L \tag{7.73}$$

Therefore, the generalized constants for a single-phase step-voltage regulator become the following:

Type A:
$$a = \frac{1}{a_R}; b = 0; c = 0; d = a_R \quad A = a_R B = 0 \tag{7.74}$$

Type B:
$$a = a_R; b = 0; c = 0; d = \frac{1}{a_R} \quad A = \frac{1}{a_R} B = 0, \tag{7.75}$$

where a_R is given by Equation 7.71, and the sign convention is given in Table 7.1:

7.4.1.4 The Line-Drop Compensator

The changing of taps on a regulator is controlled by the "line-drop compensator". Figure 7.11 shows an analog circuit of the compensator circuit and how it is connected to the distribution line through a potential transformer and a current transformer. Older regulators are controlled by an analog compensator circuit. Modern regulators are controlled by a digital compensator. The digital compensators require the same settings as the analog. Because it is easy to visualize, the analog circuit will be used in this section. However, understand that the modern digital compensators perform the same function for changing the taps on the regulators. Also, as the smart grid becomes popular, it will be possible to transmit the load center voltage directly to the regulator so there will not be a need for the compensator circuit.

TABLE 7.1
Sign Convention Table for a_R

	Type A	Type B
Raise	+	−
Lower	−	+

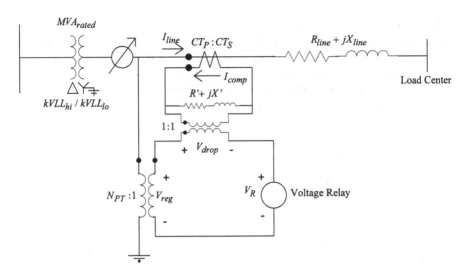

FIGURE 7.11 Line-drop compensator circuit

The purpose of the line-drop compensator is to model the voltage drop of the distribution line from the regulator to the "load center". The compensator is an analog circuit that is a scale model of the line circuit. The compensator input voltage is typically 120 volts, which requires the potential transformer in Figure 7.11 to reduce the rated voltage down to 120 volts. For a regulator connected line to ground, the rated voltage is the nominal line-to-neutral voltage, while a regulator connected line to line, the rated voltage is the line-to-line voltage. The current transformer turns ratio is specified as $CT_p:CT_s$, where the primary rating (CT_p) will typically be the rated current of the feeder. The setting that is most critical is that of R' and X' calibrated in volts. These values must represent the **equivalent** impedance from the regulator to the load center. The basic requirement is to force the per-unit line impedance to be equal to the per-unit compensator impedance. To cause this to happen, it is essential that a consistent set of base values be developed wherein the per-unit voltage and currents in the line and in the compensator are equal. The consistent set of base values is determined by selecting a base voltage and current for the line circuit and then computing the base voltage and current in the compensator by dividing the system base values by the potential transformer (PT) ratio and current transformer (CT) ratio respectively. For regulators connected line to ground, the base system voltage is selected as the rated line-to-neutral voltage (V_{LN}), and the base system current is selected as the rating of the primary winding of the current transformer (CT_p). Table 7.2, "Table of Base Values", employs these rules for a regulator connected line to ground:

With the table of base values developed, the compensator R and X settings in ohms can be computed by first computing the per-unit line impedance.

$$R_{pu} + jXpu = \frac{Rline_\Omega + jXline_\Omega}{Zbase_{line}}$$

TABLE 7.2
Table of Base Values

Base	Line Circuit	Compensator Circuit
Voltage	V_{LN}	$\dfrac{V_{LN}}{N_{PT}}$
Current	CT_P	CT_S
Impedance	$Zbase_{line} = \dfrac{V_{LN}}{CT_P}$	$Zbase_{comp} = \dfrac{V_{LN}}{N_{PT} \cdot CT_S}$

$$R_{pu} + jX_{pu} = \left(Rline_\Omega + jXline_\Omega\right) \cdot \frac{CT_P}{V_{LN}} \tag{7.76}$$

The per-unit impedance of Equation 7.76 must be the same in the line and in the compensator. The compensator impedance in ohms is computed by multiplying the per-unit impedance by the compensator base impedance.

$$Rcomp_\Omega + jXcomp_\Omega = \left(R_{pu} + jX_{pu}\right) \cdot Zbase_{comp}$$

$$Rcomp_\Omega + jXcomp_\Omega = \left(Rline_\Omega + jXline_\Omega\right) \cdot \frac{CT_P}{V_{LN}} \cdot \frac{V_{LN}}{N_{PT} \cdot CT_S}$$

$$Rcomp_\Omega + jXcomp_\Omega = \left(Rline_\Omega + jXline_\Omega\right) \cdot \frac{CT_P}{N_{PT} \cdot CT_S} \, \Omega \tag{7.77}$$

Equation 7.77 gives the value of the compensator R and X settings in ohms. The compensator R and X settings in volts are determined by multiplying the compensator R and X in ohms times the rated secondary current (CT_S) of the current transformer.

$$R' + jX' = \left(Rcomp_\Omega + jXcomp_\Omega\right) \cdot CT_S$$

$$R' + jX' = \left(Rline_\Omega + jXline_\Omega\right) \cdot \frac{CT_P}{N_{PT} \cdot CT_S} \cdot CT_S$$

$$R' + jX' = \left(Rline_\Omega + jXline_\Omega\right) \cdot \frac{CT_P}{N_{PT}} \, V \tag{7.78}$$

Knowing the equivalent impedance in ohms from the regulator to the load center, the required value for the compensator settings in volts is determined by using Equation 7.78. This is demonstrated in Example 7.4.

Example 7.4: Refer to Figure 7.11.

The substation transformer is rated 5,000 kVA, 115 kV delta – 4.16 kV grounded wye and the equivalent line impedance from the three single-phase regulators to the load center is 0.3 + j0.9 ohms. The settings for each phase regulator will be the same.

(1) Determine the potential transformer and current transformer ratings for the compensator circuit.
The rated line-to-ground voltage of each regulator is

$$V_S = 2400$$

To provide 120 volts to the compensator, the potential transformer ratio is

$$N_{PT} = \frac{2400}{120} = 20$$

The rated current of the substation transformer is

$$I_{rated} = \frac{5,000}{\sqrt{3} \cdot 4.16} = 693.9$$

The primary rating of the CT is selected as 700 amps, and if the compensator current is reduced to 5 amps, the CT ratio is

$$CT = \frac{CT_P}{CT_S} = \frac{700}{5} = 140$$

(2) Determine the R and X settings of the compensator in ohms and volts.

Applying Equation 7.78 to determine the settings in volts,

$$R' + jX' = (0.3 + j0.9) \cdot \frac{700}{20} = 10.5 + j31.5\,V$$

The R and X settings in ohms are determined by dividing the settings in volts by the rated secondary current of the current transformer.

$$R_{ohms} + jX_{ohms} = \frac{10.5 + j31.5}{5} = 2.1 + j6.3\,\Omega$$

Understand that the R and X settings on the compensator control board are calibrated in volts.

Voltage Regulation

Example 7.5: The substation transformer in Example 7.4 is supplying a three-phase load of 2,500 kVA at 4.16 kV and 0.9 power factor lag. The regulator has been set so that

$R' + jX' = 10.5 + j31.5$ V,
Voltage Level = 120 volts (desired voltage to be held at the load center),
Bandwidth = 2 volts.

Determine the tap position of the regulator that will hold the load center voltage at the desired voltage level and within the bandwidth. This means that the tap on the regulator needs to be set so that the voltage at the load center lies between 119 and 121 volts.

The first step is to calculate the actual line current.

$$I_{line} = \frac{2500}{\sqrt{3} \cdot 4.16} / -\cos^{-1}(0.9)$$

$$I_{line} = \frac{2500}{\sqrt{3} \cdot 4.16} / -a\cos(0.9) = 346.97 / -25.84 \text{ A}$$

The current in the compensator is then

$$I_{comp} = \frac{I_{line}}{CT} = \frac{346.97 / -25.84}{140} = 2.4783 / -25.84 \text{ A}$$

The input voltage to the compensator is

$$V_{reg} = \frac{VLN_{rated}}{N_{PT}} = \frac{\frac{4160}{\sqrt{3}}}{20} = \frac{2401.78 / 0}{20} = 120.09 / \underline{0} \text{ V}$$

The voltage drop in the compensator circuit is equal to the compensator current times the compensator R and X values in ohms.

$$V_{drop} = (2.1 + j6.3) \cdot 2.4783 / -25.84 = 16.458 / 45.7 \text{ V}$$

The voltage across the voltage relay is

$$V_R = V_{reg} - V_{drop} = 120.09 / \underline{0} - 16.458 / 45.7 = 109.24 / -6.19 \text{ V}$$

The voltage across the voltage relay represents the voltage at the load center. Since this is well below the minimum voltage level of 119, the voltage regulator will have to change taps in the raise position to bring the load center voltage up to the required level. Recall that on a 120-volt base, one step change on the regulator changes the voltage 0.75 volts. The number of required tap changes can then be approximated by

$$Tap = \frac{119 - 109.24}{.75} = 13.02$$

This shows that the final tap position of the regulator will be "raise 13". With the tap set at +13, the effective regulator ratio assuming a Type B regulator is

$$a_R = 1 - 0.00625 \cdot 13 = 0.9188$$

The generalized constants for modeling the regulator for this operating condition are as follows:

$$a = a_R = 0.9188$$

$$b = 0$$

$$c = 0$$

$$d = \frac{1}{0.9188} = 1.0884$$

Note that the voltage relay matches the voltage at the load center. The + 13 tap was an approximation and has resulted in a load center voltage within the bandwidth. However, since the regulator started in the neutral position, the taps will be changed one at a time until the load center voltage is inside the 119 lower bandwidth. Remember that each step changes the voltage by 0.75 volts. Since the load center voltage has been computed to be 120.6 volts, it would appear that the regulator went one step more than necessary. Table 7.3 shows what the compensator relay voltage will be as the taps change one at a time from zero to the final value.

Table 7.3 shows that when the regulator is modeled to change one tap at a time starting from the neutral position, when it reaches tap 12, the relay voltage is inside the bandwidth. For the same load, it may be that the taps will change to lower the voltage due to a previous larger load. In this case, the taps will reduce one at a time until the relay voltage is inside the 121 upper bandwidth voltage. The point is that there can be different taps for the same load depending upon whether the voltage needed to be raised or lowered from an existing tap position.

It is important to understand that the value of equivalent line impedance is not the actual impedance of the line between the regulator and the load center. Typically, the load center is located down the primary main feeder after several laterals have been tapped. As a result, the current measured by the *CT* of the regulator is not the current that flows all the way from the regulator to the load center. The only way to determine the equivalent line impedance value is to run a power-flow program of the feeder without the regulator operating. From the output of the program, the voltages at the regulator output and the load center are known. Now the "equivalent" line impedance can be computed as

$$Rline_\Omega + jXline_\Omega = \frac{V_{regulator-output} - V_{load-center}}{I_{line}} \Omega \quad (7.79)$$

Voltage Regulation

Example 7.6: Using the results of Examples 7.5, calculate the actual voltage at the load center with the tap set at +13, assuming the 2,500 kVA at 4.16 kV measured at the substation transformer low-voltage terminals.

The actual line-to-ground voltage and line current at the load-side terminals of the regulator are

$$V_L = \frac{V_S}{a} = \frac{2401.78/\underline{0}}{0.9188} = 2614.2/\underline{0}\,\text{V},$$

$$I_L = \frac{I_S}{d} = \frac{346.97/-25.84}{1.0884} = 318.77/-25.84\,\text{A}$$

The actual line-to-ground voltage at the load center is

$$V_{LC} = V_L - Z_{line} \cdot I_L = 2614.2/\underline{0} - (0.3 + j0.9) \cdot 318.77/-25.84 = 2412.8/-5.15\,\text{V}$$

On a 120-volt base, the load center voltage is

$$VLC_{120} = \frac{V_{LC}}{N_{pt}} = \frac{2412.8/-5.15}{20} = 120.6/-5.15\,\text{V}$$

In the compensator circuit,

$$V_{reg} = \frac{V_L}{N_{PT}} = \frac{2614.2/\underline{0}}{20} = 130.7/\underline{0}\,\text{V},$$

$$I_{comp} = \frac{I_{Load}}{CT} = \frac{318.8/-25.8}{140} = 2.277/-25.8\,\text{A},$$

$$V_R = V_{reg} - (R' + jX') \cdot I_{comp} = 120.6/-5.15\,\text{V}$$

TABLE 7.3
Tap Changing

Tap	Voltage
0	109.2
1	110.1
2	110.9
3	111.7
....	....
10	117.8
11	118.8
12	119.7

In Equation 7.79, the voltages must be specified in system volts and the current in system amperes.

This section has developed the model and generalized constants for Type A and Type B single-phase, step-voltage regulators. The compensator control circuit has been developed and demonstrated how this circuit controls the tap changing of the regulator. The next section will discuss the various three-phase, step-type voltage regulators.

7.4.2 Three-Phase, Step-Voltage Regulators

Three single-phase step-voltage regulators can be connected externally to form a three-phase regulator. When three single-phase regulators are connected, each regulator has its own compensator circuit, and therefore the taps on each regulator are changed separately. Typical connections for single-phase step- regulators are as follows:

1. Single-phase
2. Two regulators connected in "open wye" (sometimes referred to as "V" phase)
3. Three regulators connected in grounded wye
4. Two regulators connected in open delta
5. Three regulators connected in closed delta

A three-phase regulator has the connections between the single-phase windings internal to the regulator housing. The three-phase regulator is "gang" operated so that the taps on all windings change the same and as a result, only one compensator circuit is required. For this case, it is up to the engineer to determine which phase current and voltage will be sampled by the compensator circuit. Three-phase regulators will only be connected in a three-phase wye or closed delta.

Many times, the substation transformer will have LTC windings on the secondary. The LTC will be controlled in the same way as a gang-operated, three-phase regulator.

In the regulator models to be developed in the next sections, the phasing on the source side of the regulator will use capital letters A, B, and C. The load-side phasing will use lowercase letters a, b and c.

7.4.2.1 Wye-Connected Regulators

Three Type B single-phase regulators connected in wye are shown in Figure 7.12.

In Figure 7.12, the polarities of the windings are shown in the "raise" position. When the regulator is in the "lower" position, a reversing switch will have reconnected the series winding so that the polarity on the series winding is now at the output terminal. Regardless of whether the regulator is raising or lowering the voltage, the following equations apply:

Voltage Regulation

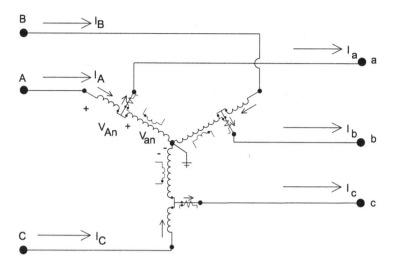

FIGURE 7.12 Wye-connected, Type B regulators.

Voltage Equations

$$\begin{bmatrix} V_{An} \\ V_{Bn} \\ V_{Cn} \end{bmatrix} = \begin{bmatrix} a_{R\_a} & 0 & 0 \\ 0 & a_{R\_b} & 0 \\ 0 & 0 & a_{R\_c} \end{bmatrix} \cdot \begin{bmatrix} V_{an} \\ V_{bn} \\ V_{cn} \end{bmatrix}, \quad (7.80)$$

where a_{R_a}, a_{R_a}, a_{R_b}, and a_{R_c} represent the effective turns ratios for the three single-phase regulators.
Equation 7.81 is of the form

$$\begin{bmatrix} VLN_{ABC} \end{bmatrix} = \begin{bmatrix} a \end{bmatrix} \cdot \begin{bmatrix} VLN_{abc} \end{bmatrix} + \begin{bmatrix} b \end{bmatrix} \cdot \begin{bmatrix} I_{abc} \end{bmatrix} \quad (7.81)$$

In Equation 7.81, the matrix [b] will be zero when the regulator impedance is neglected.
Current Equations

$$\begin{bmatrix} I_A \\ I_B \\ I_C \end{bmatrix} = \begin{bmatrix} \dfrac{1}{a_{R\_a}} & 0 & 0 \\ 0 & \dfrac{1}{a_{R\_b}} & 0 \\ 0 & 0 & \dfrac{1}{a_{R\_c}} \end{bmatrix} \cdot \begin{bmatrix} I_a \\ I_b \\ I_c \end{bmatrix} \quad (7.82)$$

$$\text{Or}\left[I_{ABC}\right] = \left[c\right]\cdot\left[VLG_{abc}\right] + \left[d\right]\left[I_{abc}\right] \qquad (7.83)$$

In Equation 7.83, the matrix [c] will be zero since the regulator shunt admittance is neglected.

Equations 7.81 and 7.83 are of the same form as the generalized equations that were developed for the three-phase line segment in Chapter 6. For a three-phase- wye-connected, step-voltage regulator neglecting the series impedance and shunt admittance, the forward and backward sweep matrices are therefore defined as follows:

$$[a] = \begin{bmatrix} a_{R\_a} & 0 & 0 \\ 0 & a_{R\_b} & 0 \\ 0 & 0 & a_{R\_c} \end{bmatrix} \qquad (7.84)$$

$$[b] = \begin{bmatrix} 0 & 0 & 0 \\ 0 & 0 & 0 \\ 0 & 0 & 0 \end{bmatrix} \qquad (7.85)$$

$$[c] = \begin{bmatrix} 0 & 0 & 0 \\ 0 & 0 & 0 \\ 0 & 0 & 0 \end{bmatrix} \qquad (7.86)$$

$$[d] = \begin{bmatrix} \dfrac{1}{a_{R\_a}} & 0 & 0 \\ 0 & \dfrac{1}{a_{R\_b}} & 0 \\ 0 & 0 & \dfrac{1}{a_{R\_c}} \end{bmatrix} \qquad (7.87)$$

$$[A] = \begin{bmatrix} \dfrac{1}{a_{R\_a}} & 0 & 0 \\ 0 & \dfrac{1}{a_{R\_b}} & 0 \\ 0 & 0 & \dfrac{1}{a_{R\_c}} \end{bmatrix} \qquad (7.88)$$

Voltage Regulation

$$[B] = \begin{bmatrix} 0 & 0 & 0 \\ 0 & 0 & 0 \\ 0 & 0 & 0 \end{bmatrix} \qquad (7.89)$$

In Equations 7.84, 7.87, and 7.88, the effective turns ratio for each regulator must satisfy

$$0.9 \le a_{R\_abc} \le 1.1 \text{ in } 32 \text{ steps of } 0.625\% \text{ per step} (0.75 \text{ volts / step on } 120 \text{ volt base}).$$

The effective turn ratios (a_{R_a}, a_{R_b}, and a_{R_c}) can take on different values when three single-phase regulators are connected in wye. It is also possible to have a three-phase regulator connected in wye where the voltage and current are sampled on only one phase and then all three phases are changed by the same number of taps.

Example 7.7: An unbalanced three-phase load is served at the end of a 10,000-foot, 12.47 kV distribution line segment. The phase generalized matrices for the line segment were computed in Example 6.2 and used in Example 6.4. The computed matrices follow:

$$[a_{line}] = \begin{bmatrix} 1 & 0 & 0 \\ 0 & 1 & 0 \\ 0 & 0 & 1 \end{bmatrix}$$

$$[b_{line}] = \begin{bmatrix} 0.8667 + j2.0417 & 0.2955 + j0.9502 & 0.2907 + j0.7290 \\ 0.2955 + j0.9502 & 0.8837 + j1.9852 & 0.2992 + j0.8023 \\ 0.2907 + j0.7290 & 0.2992 + j0.8023 & 0.8741 + j2.0172 \end{bmatrix}$$

For this line, the A and B matrices are defined as follows:

$$[A_{line}] = [a_{line}]^{-1}$$

$$[A] = \begin{bmatrix} 1 & 0 & 0 \\ 0 & 1 & 0 \\ 0 & 0 & 1 \end{bmatrix}$$

$$[B_{line}] = [a_{line}]^{-1} \cdot [b_{line}] = [Z_{abc}]$$

$$[B] = \begin{bmatrix} 0.8667 + j1.0417 & 0.2955 + j0.9502 & 0.2907 + j0.7290 \\ 0.2955 + j0.9502 & 0.8837 + j1.9852 & 0.2992 + j0.8023 \\ 0.2907 + j0.7290 & 0.2992 + j0.8023 & 0.8741 + j2.0172 \end{bmatrix}$$

In Example 6.4, the substation line-to-line voltages are balanced three phases. The line-to-neutral voltages at the substation are balanced three phases:

$$[VLN_{ABC}] = \begin{bmatrix} 7199.6/\underline{0} \\ 7199.6/\underline{-120} \\ 7199.6/\underline{120} \end{bmatrix} V$$

In Example 6.5, the unbalanced three-phase loads were

$$[kVA] = \begin{bmatrix} 2500 \\ 2000 \\ 1500 \end{bmatrix} [PF] = \begin{bmatrix} 0.9 \\ 0.85 \\ 0.95 \end{bmatrix} kVA$$

In Example 6.5, the LIT was used, and the load currents and the currents leaving the substation were

$$[I_{ABC}] = \begin{bmatrix} 374.4/\underline{-28.2} \\ 286.8/\underline{-153.9} \\ 212.6/\underline{100.5} \end{bmatrix}$$

The load voltages were calculated to be

$$[Vload_{abc}] = \begin{bmatrix} 6678.2/\underline{-2.3} \\ 6972.8/\underline{-122.1} \\ 7055.5/\underline{118.7} \end{bmatrix}.$$

The load voltages on a 120-volt base were computed to be

$$[V_{120}] = \begin{bmatrix} 111.3/\underline{-2.3} \\ 116.2/\underline{-122.1} \\ 117.6/\underline{118.7} \end{bmatrix}$$

It is obvious that the load voltages are not within the ANSI standard. To correct this problem, three single-phase Type B step-voltage regulators will be connected in wye and installed in the substation. The regulators are to be set so that each line-to-neutral load voltage on a 120-volt base will lie between 119 and 121 volts.

The potential and current transformers of the regulators are rated.

Voltage Regulation

$$N_{PT} = \frac{7200}{120} = 60$$

$$CT = \frac{600}{5} = \frac{CT_P}{CT_S} = 120$$

The voltage level and bandwidth are

Voltage Level = 120 volts,
Bandwidth = 2 volts.

The equivalent line impedance for each phase can be determined by applying Equation 7.79.

$$Zline_a = \frac{7199.6\underline{/0} - 6678.2\underline{/-2.3}}{374.4\underline{/-28.2}} = 0.8989 + j1.3024$$

$$Zline_b = \frac{7199.6\underline{/-120} - 6972.8\underline{/-122.1}}{286.8\underline{/-153.9}} = 0.1655 + j1.2007$$

$$Zline_c = \frac{7199.6\underline{/120} - 7055.5\underline{/118.7}}{212.6\underline{/100.5}} = 0.4044 + j0.9141\,\Omega$$

Even though the three regulators will change taps independently, it is the usual practice to set the R and X settings of the three regulators the same. The average value of the three-line impedances noted earlier can be used for this purpose.

$$Zline_{average} = 0.4896 + j1.1391\,\Omega$$

The compensator R and X settings are computed according to Equation 7.78.

$$R' + jX' = \left(Rline_\Omega + jXline_\Omega\right) \cdot \frac{CT_P}{N_{PT}} = (0.4896 + j1.1391) \cdot \frac{600}{60}$$

$$R' + jX' = 4.8964 + j11.3908 \text{ V}$$

The compensator controls are not calibrated to that many significant figures. The values are rounded off and set at

$$R' + jX' = 5 + j11 \text{ V}$$

For the same unbalanced loading and with the three-phase, wye-connected regulators in service, the approximate tap settings are as follows:

$$Tap_a = \frac{|119-|Vload_a||}{0.75} = \frac{|119-111.3|}{0.75} = 10.2615$$

$$Tap_b = \frac{|119-|Vload_b||}{0.75} = \frac{|119-116.2|}{0.75} = 3.7154$$

$$Tap_a = \frac{|119-|Vload_c||}{0.75} = \frac{|119-117.6|}{0.75} = 1.8787$$

Since the taps must be integers, the actual tap settings will be

$$Tap_a = +10,$$

$$Tap_b = +4,$$

$$Tap_c = +2$$

With the taps set at 10, 4, and 2, the effective turns ratio for the three regulators and the resulting generalized matrices are determined by applying Equations 7.84, 7.87, 7.88, and 7.89 for each phase:

$$[a_{reg}] = \begin{bmatrix} 1-0.00625 \cdot 10 & 0 & 0 \\ 0 & 1-0.00625 \cdot 4 & 0 \\ 0 & 0 & 1-0.00625 \cdot 2 \end{bmatrix}$$

$$= \begin{bmatrix} 0.9375 & 0 & 0 \\ 0 & 0.975 & 0 \\ 0 & 0 & 0.9875 \end{bmatrix}$$

$$[d_{reg}] = [a_{reg}]^{-1} = \begin{bmatrix} 1.0667 & 0 & 0 \\ 0 & 1.0256 & 0 \\ 0 & 0 & 1.0127 \end{bmatrix}$$

$$[A_{reg}] = [a_{reg}]^{-1} = \begin{bmatrix} 1.0667 & 0 & 0 \\ 0 & 1.0256 & 0 \\ 0 & 0 & 1.0127 \end{bmatrix}$$

Voltage Regulation

FIGURE 7.13 Simple system with a regulator and line.

$$[B_{reg}] = \begin{bmatrix} 0 & 0 & 0 \\ 0 & 0 & 0 \\ 0 & 0 & 0 \end{bmatrix}$$

With the voltage regulators connected to the source, the one-line diagram of the simple system is shown in Figure 7.13.

A ladder technique program is written following the flow chart in Figure 6.6. The program is used to compute the load voltages and currents after the regulator taps and resulting matrices have been computed. Before the program can be run, the various matrices for the regulator must be run based upon the tap positions. The following are the various matrices based upon the tap positions. Initially, the taps are all on zero.

$$i = 1, 2, 3$$

$$[Tap] = \begin{bmatrix} 0 \\ 0 \\ 0 \end{bmatrix}$$

$$[a_{reg_{i,i}}] = 1 - 0.00625 \cdot Tap_i = \begin{bmatrix} 1 & 0 & 0 \\ 0 & 1 & 0 \\ 0 & 0 & 1 \end{bmatrix}$$

$$[A_{reg}] = [a_{reg}]^{-1} = \begin{bmatrix} 1 & 0 & 0 \\ 0 & 1 & 0 \\ 0 & 0 & 1 \end{bmatrix}$$

$$[B_{reg}] = \begin{bmatrix} 0 & 0 & 0 \\ 0 & 0 & 0 \\ 0 & 0 & 0 \end{bmatrix}$$

The program is shown in Figure 7.14. The desired voltage level at the load is 120 volts. After the program has run, new tap values are calculated. The same procedure is followed as shown previously, but the new tap values are used. With the new values of $[A_{reg}]$ computed, the program is run again, and based upon the results, the taps will either change or the new taps will satisfy all the requirements.

$$\text{Tol} := .0001 \qquad \text{start} := \begin{pmatrix} 0 \\ 0 \\ 0 \end{pmatrix}$$

$$Y := \begin{vmatrix} I_{abc} \leftarrow \text{start} \\ V_{old} \leftarrow \text{start} \\ \text{for } n \in 1..30 \\ \quad \begin{pmatrix} VL_{abc} \leftarrow E_S - Z \cdot I_{abc} \\ \text{for } i \in 1..3 \\ \quad \text{Error}_i \leftarrow \left| |VL_{abc_i}| - |V_{old_i}| \right| \\ \text{Err} \leftarrow \max(\text{Error}) \\ \text{break if Err} < \text{Tol} \\ V_{old} \leftarrow VL_{abc} \\ \text{for } i \in 1..3 \\ \quad I_{abc_i} \leftarrow \dfrac{\overline{SL_i \cdot 1000}}{VL_{abc_i}} \\ \text{for } i \in 1..3 \\ \quad Vdrop_i \leftarrow \dfrac{\left| |E_{S_i}| - |VL_{abc_i}| \right|}{|E_{S_i}|} \cdot 100 \end{pmatrix} \\ Out_1 \leftarrow n \\ Out_2 \leftarrow VL_{abc} \\ Out_3 \leftarrow I_{abc} \\ Out_4 \leftarrow Vdrop \\ Out \end{vmatrix}$$

FIGURE 7.14 Ladder technique program.

Voltage Regulation

After six iterations, the results of the analysis are as follows:

$$[Vload_{abc}] = \begin{bmatrix} 7205.6/-1.9 \\ 7145.9/-122.0 \\ 7147.2/118.7 \end{bmatrix}$$

$$[V_{120}] = \begin{bmatrix} 120.1/-1.9 \\ 119.1/-122.0 \\ 119.1/118.7 \end{bmatrix}$$

$$[I_{abc}] = \begin{bmatrix} 347.0/-27.8 \\ 279.9/-153.8 \\ 209.9/100.5 \end{bmatrix}$$

$$[I_{ABC}] = \begin{bmatrix} 370.1-27.8 \\ 287.1/-153.8 \\ 212.5/100.5 \end{bmatrix}$$

The final tap positions are

$$[Tap] = \begin{bmatrix} 10 \\ 4 \\ 2 \end{bmatrix}$$

In this example, the tap positions have been determined by the analysis of the actual load voltages. Note that now all of the load voltages on the 120-volt base are within ANSI standards, assuming that the taps were actually set at +10, +4, and +2. This procedure will only work when the regulator is being fed the actual load voltages. Unfortunately, the way it really works is the compensator circuit will adjust the taps based upon the compensator relay voltage. Recall that the compensator input voltage V_{reg} is a measure of the output voltage of the regulator, and the compensator current I_{comp} is a measure of the line current out of the regulator. For this example, the compensator R and X in ohms are

$$Z_c = \frac{R + jX}{CT_s} = \frac{5 + j11}{5} = 1 + j2.2$$

Initially, the taps are set to zero with the voltages and currents at the output terminals of the regulator as follows:

$$[V_{out}] = [A_{reg}] \cdot [E_{ABC}] = \begin{bmatrix} 7199.6/\underline{0} \\ 7199.6/\underline{-120} \\ 7199.6/\underline{120} \end{bmatrix}$$

$$[I_{out}] = [d_{reg}]^{-1} \cdot [I_{ABC}] = \begin{bmatrix} 374.4/\underline{-28.2} \\ 286.8/\underline{-153.9} \\ 212.6/\underline{100.5} \end{bmatrix}$$

The voltages and currents into the compensator circuits are

$$[V_{reg}] = \frac{[V_{out}]}{N_{pt}} = \begin{bmatrix} 120.0/\underline{0} \\ 120.0/\underline{-120} \\ 120.0/\underline{120} \end{bmatrix},$$

$$[I_{comp}] = \frac{[I_{out}]}{CT} = \begin{bmatrix} 3.12/\underline{-28.2} \\ 2.39/\underline{-153.9} \\ 1.77/\underline{100.6} \end{bmatrix}$$

The compensator impedance matrix is

$$[Z_{comp}] = \begin{bmatrix} Z_c & 0 & 0 \\ 0 & Z_c & 0 \\ 0 & 0 & Z_c \end{bmatrix} = \begin{bmatrix} 1+j2.2 & 0 & 0 \\ 0 & 1+j2.2 & 0 \\ 0 & 0 & 1+j2.2 \end{bmatrix}$$

The compensator relay voltages are

$$[V_{relay}] = [V_{reg}] - [Z_{comp}] \cdot [I_{comp}] = \begin{bmatrix} 114.1/\underline{-2.3} \\ 115.1/\underline{-121.5} \\ 117.1/\underline{118.5} \end{bmatrix}$$

Since the relay voltages are not within the bandwidth, the taps will change one step at a time until the voltages are within the bandwidth. As each regulator gets the voltage within the bandwidth, it will stop while the others continue to change taps until their voltages are within the bandwidth. A quick way to determine the final taps is to recognize that each tap change will change the relay voltage by 0.75 volts.

Voltage Regulation

To determine the final tap settings, compute the difference between the low end of the voltage level and the value of the present relay voltage.

Voltage Level = 120 volts

$$V_{low} = \text{Voltage Level} - 1 = 119 \text{ volts}$$

$$i = 1 \text{ to } 3$$

$$Tap_i = \frac{V_{low} - |V_{relay\,i}|}{0.75} = \begin{bmatrix} 6.54 \\ 5.18 \\ 2.58 \end{bmatrix}$$

$$\text{Set}: Tap = \begin{bmatrix} 7 \\ 5 \\ 3 \end{bmatrix}$$

With these tap settings, the regulator stops changing taps and the relay voltages are

$$[V_{relay}] = \begin{bmatrix} 119.9 \\ 119.1 \\ 119.4 \end{bmatrix}$$

When these taps are applied to the analysis of the system, the resulting load voltages on a 120-volt base are

$$[V_{load}] = \begin{bmatrix} 117.3 \\ 120.1 \\ 119.9 \end{bmatrix}$$

The phase a voltage is not within the bandwidth. The problem is that when the example was first analyzed with the original taps, the taps had been determined by using the actual line voltage drops with the regulators in the neutral position. However, when the compensator R and X values were computed, the average of the equivalent line impedances was used for each regulator. Since the three-line currents are all different, that means the heavily loaded phase a voltage will not represent what is happening on the system. Once again, this is a problem that occurs because of the unbalanced loading.

One way to raise the load voltages is to specify a higher voltage level. Increase the voltage level to 122 volts. With the regulator changing taps one at a time until all voltage relays have a voltage just greater than 121 volts (lower bandwidth voltage), the results are

$$[Taps] = \begin{bmatrix} 9 \\ 8 \\ 5 \end{bmatrix},$$

$$[Vload_{120}] = \begin{bmatrix} 119.1 \\ 122.6 \\ 121.5 \end{bmatrix},$$

$$[V_{relay}] = \begin{bmatrix} 121.6 \\ 121.7 \\ 121.0 \end{bmatrix}$$

Example 7.7 is a long example intended to demonstrate how the engineer can determine the correct compensator R and X settings knowing the substation and load voltages and the currents leaving the substation. Generally, it will be necessary to run a power-flow study to determine these values. A simple ladder technique routine demonstrates that with the regulator tap settings, the load voltages are within the desired limits. The regulator has automatically set the taps for this load condition, and as the load changes, the taps will continue to change to hold the load voltages within the desired limits.

7.4.2.2 Closed Delta-Connected Regulators

Three single-phase Type B regulators can be connected in a closed delta, as shown in Figure 7.15. In the figure, the regulators are shown in the "raise" position.

The closed delta connection is typically used in three-wire delta feeders. Note that the potential transformers for this connection are monitoring the load-side, line-to-line voltages, and the current transformers are **not** monitoring the load-side line currents.

The relationships between the source side and currents and the voltages are needed. Equations 7.60 through 7.63 define the relationships between the series and shunt winding voltages and currents for a step-voltage regulator that must be satisfied no matter how the regulators are connected.

KVL is first applied around a closed loop, starting with the line-to-line voltage between phases A and C on the source side. Refer to Figure 7.15, which defines the various voltages.

$$V_{AB} = V_{Aa} + V_{ab} - V_{Bb} \tag{7.90}$$

Voltage Regulation

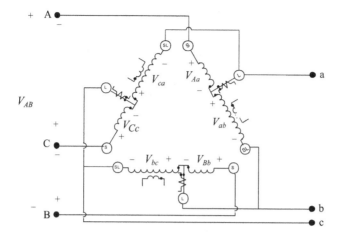

FIGURE 7.15 Closed delta-connected regulators with voltages.

But:

$$V_{Bb} = -\frac{N_2}{N_1} \cdot V_{bc} \tag{7.91}$$

$$V_{Aa} = -\frac{N_2}{N_1} \cdot V_{ab} \tag{7.92}$$

Substitute Equations 7.91 and 7.92 into Equation 7.90 and simplify.

$$V_{AB} = \left(1 - \frac{N_2}{N_1}\right) \cdot V_{ab} + \frac{N_2}{N_1} \cdot V_{bc} = a_{R\_ab} \cdot V_{ab} + \left(1 - a_{R\_bc}\right) \cdot V_{bc} \tag{7.93}$$

The same procedure can be followed to determine the relationships between the other line-to-line voltages. The final three-phase equation is

$$\begin{bmatrix} V_{AB} \\ V_{BC} \\ V_{CA} \end{bmatrix} = \begin{bmatrix} a_{R\_ab} & 1-a_{R\_bc} & 0 \\ 0 & a_{R\_bc} & 1-a_{R\_ca} \\ 1-a_{R\_ab} & 0 & a_{R\_ca} \end{bmatrix} \cdot \begin{bmatrix} V_{ab} \\ V_{bc} \\ V_{ca} \end{bmatrix} \tag{7.94}$$

Equation 7.94 is of the generalized form

$$[VLL_{ABC}] = [a] \cdot [VLL_{abc}] + [b] \cdot [I_{abc}] \tag{7.95}$$

Figure 7.16 shows the closed delta-delta connection with the defining currents.

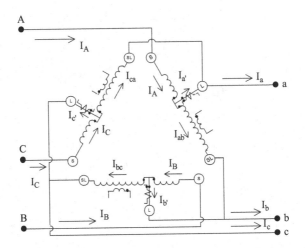

FIGURE 7.16 Closed delta-connected regulators with currents.

The relationship between source and load-line currents starts with applying KCL at the load-side terminal a.

$$I_a = I'_a + I_{ca} = I_A - I_{ab} + I_{ca} \qquad (7.96)$$

But

$$I_{ab} = \frac{N_2}{N_1} \cdot I_A \qquad (7.97),$$

$$I_{ca} = \frac{N_2}{N_1} \cdot I_C \qquad (7.98)$$

Substitute Equations 7.97 and 7.98 into Equation 7.96 and simplify.

$$I_a = \left(1 - \frac{N_2}{N_1}\right) \cdot I_A + \frac{N_2}{N_1} I_C = a_{R\_ab} \cdot I_A + \left(1 - a_{R\_ca}\right) \cdot I_C \qquad (7.99)$$

The same procedure can be followed at the other two load-side terminals. The resulting three-phase equation is

$$\begin{bmatrix} I_a \\ I_b \\ I_c \end{bmatrix} = \begin{bmatrix} a_{R\_ab} & 0 & 1 - a_{R\_ca} \\ 1 - a_{R\_ab} & a_{R\_bc} & 0 \\ 0 & 1 - a_{R\_bc} & a_{R\_ca} \end{bmatrix} \cdot \begin{bmatrix} I_A \\ I_B \\ I_C \end{bmatrix} \qquad (7.100)$$

Voltage Regulation

Equation 7.100 is of the general form

$$[I_{abc}] = [D] \cdot [I_{ABC}], \quad (7.101)$$

where: $[D] = \begin{bmatrix} a_{R\_ab} & 0 & 1-a_{R\_ca} \\ 1-a_{R\_ab} & a_{R\_bc} & 0 \\ 0 & 1-a_{R\_bc} & a_{R\_ca} \end{bmatrix}$

The general form needed for the standard model is

$$[I_{ABC}] = [c] \cdot [VLL_{ABC}] + [d] \cdot [I_{abc}],$$

where $[d] = [D]^{-1}$ \quad (7.102)

As with the wye-connected regulators, the matrices [b] and [c] are equal to zero if the series impedance and shunt admittance of each regulator is neglected.

The closed delta connection can be difficult to apply. Note in both the voltage and current equations that a change of the tap position in one regulator will affect voltages and currents in two phases. As a result, increasing the tap in one regulator will affect the tap position of the second regulator. In most cases, the bandwidth setting for the closed delta connection will have to be wider than that for wye-connected regulators.

7.4.2.3 Open Delta-Connected Regulators

Two Type B single-phase regulators can be connected in the "open" delta connection. Shown in Figure 7.17 is an open-delta connection where two single-phase regulators have been connected between phases AB and CB.

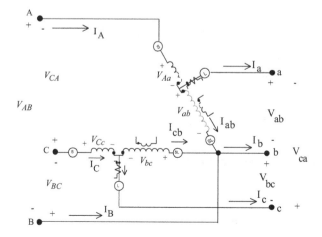

FIGURE 7.17 Open-delta connection.

Two additional open connections can be made by connecting the single-phase regulators between phases BC and AC and between phases CA and BA.

The open-delta connection is typically applied to three-wire delta feeders. Note that the potential transformers monitor the line-to-line voltages, and the current transformers monitor the line currents. Once again, the basic voltage and current relations of the individual regulators are used to determine the relationships between the source side and load-side voltages and currents. The connection shown in Figure 7.17 will be used to derive the relationships and then the relationships of the other two possible connections can follow the same procedure.

The voltage V_{AB} across the first regulator consists of the voltage across the series winding plus the voltage across the shunt winding.

$$V_{AB} = V_{Aa} + V_{ab} \tag{7.103}$$

Paying attention to the polarity marks on the series and shunt windings, the voltage across the series winding is

$$V_{Aa} = -\frac{N_2}{N_1} \cdot V_{ab} \tag{7.104}$$

Substituting Equation 7.109 into Equation 7.108 yields

$$V_{AB} = \left(1 - \frac{N_2}{N_1}\right) \cdot V_{ab} = a_{R\_ab} \cdot V_{ab} \tag{7.105}$$

Following the same procedure for the regulator connected across V_{BC}, the voltage equation is

$$V_{BC} = \left(1 - \frac{N_2}{N_1}\right) \cdot V_{bc} = a_{R\_cb} \cdot V_{bc} \tag{7.106}$$

KVL must be satisfied so that

$$V_{CA} = -\left(V_{AB} + V_{BC}\right) = -a_{R\_ab} \cdot V_{ab} - a_{R\_cb} \cdot V_{bc} \tag{7.107}$$

Equations 7.108, 7.109, and 7.110 can be put into matrix form.

$$\begin{bmatrix} V_{AB} \\ V_{BC} \\ V_{CA} \end{bmatrix} = \begin{bmatrix} a_{R\_ab} & 0 & 0 \\ 0 & a_{R\_cb} & 0 \\ -a_{R\_ab} & -a_{R\_cb} & 0 \end{bmatrix} \cdot \begin{bmatrix} V_{ab} \\ V_{bc} \\ V_{ca} \end{bmatrix} \tag{7.108}$$

Voltage Regulation

Equation 7.113 in generalized form is

$$[VLL_{ABC}] = [a_{LL}] \cdot [VLL_{abc}] + [b_{LL}] \cdot [I_{abc}], \qquad (7.109)$$

$$\text{where}: [a_{LL}] = \begin{bmatrix} a_{R\_ab} & 0 & 0 \\ 0 & a_{R\_cb} & 0 \\ -a_{R\_ab} & -a_{R\_cb} & 0 \end{bmatrix} \qquad (7.110)$$

The effective turns ratio of each regulator is given by Equation 7.76. Again, if the series impedance and shunt admittance of the regulators are neglected, $[b_{LL}]$ is zero. Equation 7.110 gives the line-to-line voltages on the source side as a function of the line-to-line voltages on the load-side of the open delta using the generalized matrices. Up to this point, the relationships between the voltages have been in terms of line-to-neutral voltages. In Chapter 8, the $[W]$ matrix is derived. This matrix is used to convert line-to-line voltages to equivalent line-to-neutral voltages.

$$[VLN_{ABC}] = [W] \cdot [VLL_{ABC}],$$

$$\text{where}\,[W] = \frac{1}{3} \cdot \begin{bmatrix} 2 & 1 & 0 \\ 0 & 2 & 1 \\ 1 & 0 & 2 \end{bmatrix} \qquad (7.111)$$

The line-to-neutral voltages are converted to line-to-line voltages by

$$[VLL_{ABC}] = [Dv] \cdot [VLN_{ABC}],$$

$$\text{where}\,[Dv] = \begin{bmatrix} 1 & -1 & 0 \\ 0 & 1 & -1 \\ -1 & 0 & 1 \end{bmatrix} \qquad (7.112)$$

Convert Equation 7.109 to line-to-neutral form.

$$[VLL_{ABC}] = [a_{LL}] \cdot [VLL_{abc}],$$

$$[VLN_{ABC}] = [W] \cdot [VLL_{ABC}] = [W] \cdot [a_{LL}] \cdot [Dv] \cdot [VLN_{abc}],$$

$$[VLN_{ABC}] = [a_{reg}] \cdot [VLN_{abc}],$$

where $\left[a_{reg} \right] = \left[W \right] \cdot \left[a_{LL} \right] \cdot \left[Dv \right]$ (7.113)

When the load-side, line-to-line voltages are needed as a function of the source side line-to-line voltages, the necessary equations are

$$\begin{bmatrix} V_{ab} \\ V_{bc} \\ V_{ca} \end{bmatrix} = \begin{bmatrix} \dfrac{1}{a_{R\_ab}} & 0 & 0 \\ 0 & \dfrac{1}{a_{R\_cb}} & 0 \\ -\dfrac{1}{a_{R\_ab}} & -\dfrac{1}{a_{R\_cb}} & 0 \end{bmatrix} \cdot \begin{bmatrix} V_{AB} \\ V_{BC} \\ V_{CA} \end{bmatrix},$$ (7.114)

$$\left[VLL_{abc} \right] = \left[A_{LL} \right] \cdot \left[VLL_{ABC} \right],$$ (7.115)

where $\left[A_{LL} \right] = \begin{bmatrix} \dfrac{1}{a_{R\_ab}} & 0 & 0 \\ 0 & \dfrac{1}{a_{R\_cb}} & 0 \\ -\dfrac{1}{a_{R\_ab}} & -\dfrac{1}{a_{R\_cb}} & 0 \end{bmatrix}$ (7.116)

Equation 7.114 is converted to line-to-neutral voltages by

$$\left[VLN_{abc} \right] = \left[W \right] \cdot \left[VLL_{abc} \right] = \left[W \right] \cdot \left[A_{LL} \right] \cdot \left[VLL_{ABC} \right],$$

$$\left[VLN_{abc} \right] = \left[W \right] \cdot \left[A_{LL} \right] \cdot \left[Dv \right] \cdot \left[VLN_{ABC} \right],$$

$$\left[VLN_{abc} \right] = \left[A_{reg} \right] \cdot \left[VLN_{ABC} \right],$$

where $\left[A_{reg} \right] = \left[W \right] \cdot \left[A_{LL} \right] \cdot \left[Dv \right]$ (7.117)

There is not a general equation for each of the elements of $[A_{reg}]$. The matrix $[A_{reg}]$ must be computed according to Equation 7.117.

Referring to Figure 7.17, the current equations are derived by applying KCL at the L node of each regulator.

$$I_A = I_a + I_{ab}$$ (7.118)

Voltage Regulation

But

$$I_{ab} = \frac{N_2}{N_1} \cdot I_A$$

Therefore, Equation 7.118 becomes

$$\left(1 - \frac{N_2}{N_1}\right) I_A = I_a \tag{7.119}$$

Therefore,

$$I_A = \frac{1}{a_{R\_ab}} \cdot I_a \tag{7.120}$$

In a similar manner, the current equation for the second regulator is given by

$$I_C = \frac{1}{a_{R\_cb}} \cdot I_c \tag{7.121}$$

Because this is a three-wire delta line, then

$$I_B = -(I_A + I_C) = -\frac{1}{a_{R\_ab}} \cdot I_a - \frac{1}{a_{R\_cb}} \cdot I_c \tag{7.122}$$

In matrix form, the current equations become

$$\begin{bmatrix} I_A \\ I_B \\ I_C \end{bmatrix} = \begin{bmatrix} \frac{1}{a_{R\_ab}} & 0 & 0 \\ -\frac{1}{a_{R\_ab}} & 0 & -\frac{1}{a_{R\_cb}} \\ 0 & 0 & \frac{1}{a_{R\_cb}} \end{bmatrix} \cdot \begin{bmatrix} I_a \\ I_b \\ I_c \end{bmatrix} \tag{7.123}$$

In generalized form, Equation 7.123 becomes

$$[I_{ABC}] = [c_{reg}] \cdot [VLN_{ABC}] + [d_{reg}] \cdot [I_{abc}], \tag{7.124}$$

where:
$$[d_{reg}] = \begin{bmatrix} \dfrac{1}{a_{R\_ab}} & 0 & 0 \\ -\dfrac{1}{a_{R\_ab}} & 0 & -\dfrac{1}{a_{R\_cb}} \\ 0 & 0 & \dfrac{1}{a_{R\_cb}} \end{bmatrix},$$

$$[c_{reg}] = \begin{bmatrix} 0 & 0 & 0 \\ 0 & 0 & 0 \\ 0 & 0 & 0 \end{bmatrix} \quad (7.125)$$

When the series impedance and shunt admittance, the constant matrix $[c_{reg}]$ will be zero.

Therefore, the source side-line currents as a function of the load-line currents are given by

$$\begin{bmatrix} I_A \\ I_B \\ I_C \end{bmatrix} = \begin{bmatrix} \dfrac{1}{a_{R1}} & 0 & 0 \\ \dfrac{1}{a_{R1}} & 0 & \dfrac{1}{a_{R2}} \\ 0 & 0 & \dfrac{1}{a_{R2}} \end{bmatrix} \cdot \begin{bmatrix} I_a \\ I_b \\ I_c \end{bmatrix}, \quad (7.126)$$

$$[I_{ABC}] = [d_{reg}] \cdot [I_{abc}], \quad (7.127)$$

where

$$[d_{reg}] = \begin{bmatrix} \dfrac{1}{a_{R1}} & 0 & 0 \\ -\dfrac{1}{a_{R1}} & 0 & -\dfrac{1}{a_{R2}} \\ 0 & 0 & \dfrac{1}{a_{R2}} \end{bmatrix} \quad (7.128)$$

The determination of the R and X compensator settings for the open delta follows the same procedure as that of the wye-connected regulators. However, care must be taken to recognize that in the open-delta connection the voltages applied to the compensator are line-to-line and the currents are line currents. The open-delta-connected regulators will maintain only two of the line-to-line voltages at the load center within defined limits (Figure 7.18). The third line-to-line voltage will be dictated by the

Voltage Regulation

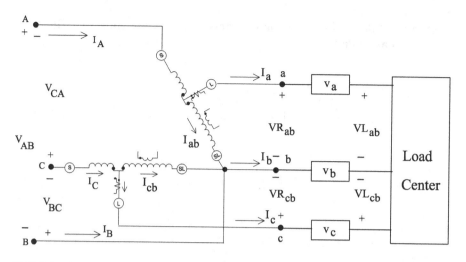

FIGURE 7.18 Open delta connected to a load center.

other two (KVL). Therefore, it is possible that the third voltage may not be within the defined limits.

With reference to Figure 7.18, an equivalent impedance between the regulators and the load center must be computed. Since each regulator is sampling line-to-line voltages and a line current, the equivalent impedance is computed by taking the appropriate line-to-line voltage drop and dividing by the sampled line current. For the open delta connection shown in Figure 7.18, the equivalent impedances are computed as

$$Zeq_a = \frac{VR_{ab} - VL_{ab}}{I_a}, \qquad (7.129)$$

$$Zeq_c = \frac{VR_{cb} - VL_{cb}}{I_c} \qquad (7.130)$$

The units of these impedances will be in system ohms. They must be converted to compensator volts by applying Equation 7.78. For the open delta connection, the potential transformer will transform the system line-to-line rated voltage down to 120 volts. Example 7.8 demonstrates how the compensator R and X settings are determined by knowing the line-to-line voltages at the regulator and at the load center.

The load is delta connected with values of

$$[kVA] = \begin{bmatrix} 2500 \\ 2000 \\ 1500 \end{bmatrix} [PF] = \begin{bmatrix} 0.90 \\ 0.85 \\ 0.95 \end{bmatrix}$$

Example 7.8: A three-wire delta system is shown in Figure 7.19. The voltages at Node S are

$$[VLL_{ABC}] = \begin{bmatrix} 12,470\underline{/0} \\ 12,470\underline{/-120} \\ 12,470\underline{/120} \end{bmatrix},$$

$$[VLN_{ABC}] = [W] \cdot [VLL_{ABC}] = \begin{bmatrix} 7199.6\underline{/-30} \\ 7199.6\underline{/-150} \\ 7199.6\underline{/90} \end{bmatrix}$$

The three-wire delta line conductor is 336,400 26/7 ACSR with spacings, as shown in Figure 7.20.

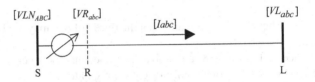

FIGURE 7.19 Circuit for Example 7.8.

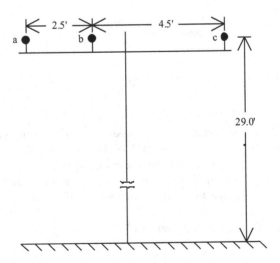

FIGURE 7.20 Three-wire delta line spacings.

Voltage Regulation

The line is 10,000-feet long, and the total phase impedance matrix is

$$[Z_{abc}] = \begin{bmatrix} 0.7600 + j2.6766 & 0.1805 + j1.1627 & 0.1805 + j1.3761 \\ 0.1805 + j1.1627 & 0.7600 + j2.6766 & 0.1805 + j1.4777 \\ 0.1805 + j1.3761 & 0.1805 + j1.4777 & 0.7600 + j2.6766 \end{bmatrix}$$

For this connection, the potential transformer ratio and current transformer ratios are selected to be

$$N_{pt} = \frac{12,470}{120} = 103.92,$$

$$CT = \frac{500}{5} = 100$$

The program of Example 7.7 (Figure 7.14) is modified so that the line-to-line voltages at the load are computed and used to compute the load currents flowing in the line. In the program, the regulator taps are initially set in the neutral position. With the regulators set in neutral on a 120-volt base, the load center voltages are computed to be

$$[VLL_{abc}] = [Dv] \cdot [VL_{abc}],$$

$$\begin{bmatrix} V120_{ab} \\ V120_{bc} \\ V120_{ca} \end{bmatrix} = \frac{[VLL_{abc}]}{N_{pt}} = \frac{1}{103.92} \cdot \begin{bmatrix} 11,883.0/-2.0 \\ 11,943.7/-121.4 \\ 12022.1/\underline{118.0} \end{bmatrix} = \begin{bmatrix} 114.4/-2.0 \\ 114.9/-121.4 \\ 115.7/\underline{118.0} \end{bmatrix} V$$

The line currents are

$$[I_{abc}] = \begin{bmatrix} 303.2/-46.9 \\ 336.2/\underline{76.1} \\ 236.2/\underline{57.1} \end{bmatrix}$$

Two single-phase Type B regulators are to be installed in an open delta connection. The regulators are to be connected between phases A-B and B-C, as shown in Figure 7.18. The voltage level will be set at 120 volts with a bandwidth of 2 volts. As computed earlier in the chapter, the load center voltages are not within the desired limits of 120 ± 1 volts.

The compensator R and X settings for each regulator must first be determined using the results of the power-flow study. The first regulator monitors the voltage V_{ab} and the line current I_a. The equivalent line impedance for this regulator is

$$Zeq_a = \frac{VR_{ab} - VL_{ab}}{I_a} = 0.3224 + j2.3844 \, \Omega$$

The second regulator monitors the voltage V_{cb} and the line current I_c. In the computation of the equivalent line impedance, it is necessary to use the c-b voltages, which are the negative of the given b-c voltages.

$$Zeq_c = \frac{VR_{cb} - VL_{cb}}{I_c} = \frac{-VR_{bc} + VL_{bc}}{I_c} = 2.1776 + j1.3772 \, \Omega$$

Unlike the wye-connected regulators, the compensator settings for the two regulators will be different. The settings calibrated in volts are

$$R'_{ab} + jX'_{ab} = Z_a \cdot \frac{CT_P}{N_{PT}} = (0.3224 + j2.3844) \cdot \frac{500}{103.92} = 1.5511 + j11.4726 \, \text{V},$$

$$R'_{cb} + jX'_{cb} = Z_c \cdot \frac{CT_P}{N_{PT}} = (2.1776 + j1.3772) \cdot \frac{500}{103.92} = 10.4776 + j6.6263 \, \text{V}$$

The compensator settings will be set to

$$R'_{ab} + X'_{ab} = 1.6 + j11.5,$$

$$R'_{cb} + jX'_{cb} = 10.5 + j6.6 \, \text{V}$$

With regulators installed, in the neutral position, and with the same loading, the currents and voltages in the compensator circuits are as follows:

$$Vcomp_{ab} = \frac{VR_{ab}}{N_{pt}} = \frac{12,470/\underline{0}}{103.92} = 120/\underline{0} \, \text{V}$$

$$Vcomp_{cb} = \frac{-VR_{bc}}{N_{pt}} = \frac{12,470/60}{103.92} = 120/60 \, \text{V}$$

$$Icomp_a = \frac{I_a}{CT} = 3.0321/\underline{-46.9} \, \text{A}$$

Voltage Regulation

$$Icomp_c = \frac{I_c}{CT} = 2.3621\underline{/57.1}\,\text{A}$$

The compensator impedances in ohms are determined by dividing the settings in volts by the secondary rating of the current transformer.

$$R_{ab} + jX_{ab} = \frac{R'_{ab} + jX'_{ab}}{CT_{secondary}} = \frac{1.6 + j11.5}{5} = 0.32 + j2.3\,\Omega$$

$$R_{cb} + jX_{cb} = \frac{R'_{cb} + jX'_{cb}}{CT_{secondary}} = \frac{10.5 + j6.6}{5} = 2.1 + j1.32\,\Omega$$

The voltages across the voltage relays in the two compensator circuits are

$$Vrelay_{ab} = Vcomp_{ab} - (R_{ab} + jX_{ab}) \cdot Icomp_a = 114.3\underline{/-2.0}\,\text{V},$$

$$Vrelay_{cb} = Vcomp_{cb} - (R_{cb} + jX_{cb}) \cdot Icomp_c = 114.9\underline{/58.6}\,\text{V}$$

Since the voltages are below the lower limit of 119, the control circuit will send "raise" commands to change the taps one at a time on both regulators. For analysis purposes, the approximate number of tap changes necessary to bring the load center voltage into the lower limit of the bandwidth for each regulator will be

$$Tap_{ab} = \frac{|119 - 114.3|}{0.75} = 6.2422 \approx 6,$$

$$Tap_{cb} = \frac{|119 - 114.9|}{0.75} = 5.43 \approx 5$$

With the taps set at 6 and 5, a check can be made to determine if the voltages at the load center are now within the limits.

With the taps adjusted, the regulator ratios are

$$a_{R\_ab} = 1.0 - 0.00625 \cdot Tap_{ab} = 0.9625,$$

$$a_{R\_cb} = 1.0 = 0.00625 \cdot Tap_{cb} = 0.9688$$

To determine the load-side regulator voltages and currents, the matrix $[A_{LL}]$ (Equation 7.115) is then converted to the equivalent $[A_{reg}]$ matrix where the system line-to-neutral voltages are used.

$$[A_{LL}] = \begin{bmatrix} \dfrac{1}{0.9625} & 0 & 0 \\ 0 & \dfrac{1}{0.9688} & 0 \\ -\dfrac{1}{0.9625} & -\dfrac{1}{0.9688} & 0 \end{bmatrix} = \begin{bmatrix} 1.039 & 0 & 0 \\ 0 & 1.0323 & 0 \\ -1.039 & -1.0323 & 0 \end{bmatrix}$$

$$[A_{reg}] = [W] \cdot [A_{LL}] \cdot [D] = \begin{bmatrix} 0.6926 & -0.3486 & -0.3441 \\ -0.3463 & 0.6904 & -0.3441 \\ -0.3463 & -0.3419 & 0.6882 \end{bmatrix}$$

Using Equation 7.128, the current matrix $[d_{reg}]$ is computed to be

$$[d_{reg}] = \begin{bmatrix} 1.039 & 0 & 0 \\ -1.039 & 0 & -1.0323 \\ 0 & 0 & 1.0323 \end{bmatrix}$$

With the taps set at +6 and +5, the output line-to-neutral voltages from the regulators are

$$[VR_{abc}] = [A_{reg}] \cdot [VLN_{ABC}] = \begin{bmatrix} 7480.1/\underline{-29.8} \\ 7455.9/\underline{-150.1} \\ 7431.9/\underline{90.2} \end{bmatrix} V$$

The line-to-line voltages are

$$[VRLL_{abc}] = [Dv] \cdot [VR_{abc}] = \begin{bmatrix} 12{,}955.8/\underline{0} \\ 12{,}872.3/\underline{-120} \\ 12{,}914.3/\underline{120} \end{bmatrix} V$$

The load-line side currents from the regulators are

$$[I_{abc}] = [D_{reg}] \cdot [I_{ABC}] = \begin{bmatrix} 291.6/\underline{-46.7} \\ 323.5/\underline{176.3} \\ 227.4/\underline{57.3} \end{bmatrix} A$$

Voltage Regulation

The source currents are

$$[I_{ABC}] = \begin{bmatrix} 302.9/-46.7 \\ 335.4/176.1 \\ 234.7/57.3 \end{bmatrix} \text{A}$$

There are two ways to test if the voltages at the load center are within the limits. The first method is to compute the relay voltages in the compensator circuits. The procedure is the same as was done initially to determine the load center voltages. First, the voltages and currents in the compensator circuits are computed.

$$Vcomp_{ab} = \frac{VR_{ab}}{N_{pt}} = \frac{12,955.8/0}{103.92} = 124.7/0 \text{ V}$$

$$Vcomp_{cb} = \frac{-VR_{bc}}{N_{pt}} = \frac{12,872.3/60}{103.92} = 123.9/60 \text{ V}$$

$$Icomp_a = \frac{I_a}{CT} = \frac{291.6/-46.7}{100} = 2.916/-58.0 \text{ A}$$

$$Icomp_c = \frac{I_c}{CT} = \frac{227.4/57.3}{100} = 2.274/57.3 \text{ A}$$

The voltages across the voltage relays are computed to be

$$Vrelay_{ab} = Vcomp_{ab} - (R_{ab} + jX_{ab}) \cdot Icomp_a = 119.2/-1.9 \text{ V},$$

$$Vrelay_{cb} = Vcomp_{cb} - (R_{cb} + jX_{cb}) \cdot Icomp_c = 119.0/58.7 \text{ V}$$

Since both voltages are within the bandwidth, no further tap changing will be necessary.

The actual voltages at the load center can be computed using the output voltages and currents from the regulator and then computing the voltage drop to the load center.

With reference to Figure 7.19, the equivalent line-to-neutral and actual line-to-line voltages at the load are

$$[VL_{abc}] = [VR_{abc}] - [Z_{abc}] \cdot [I_{abc}] = \begin{bmatrix} 7178.8/-31.8 \\ 7126.6/-151.8 \\ 7195.6/88.8 \end{bmatrix} \text{V},$$

$$[VLLL_{abc}] = [Dv] \cdot [VL_{abc}] = \begin{bmatrix} 12{,}392.4 / -1.9 \\ 12{,}366.5 / -121.3 \\ 12{,}481.8 / 118.5 \end{bmatrix} V$$

Dividing the load center line-to-line voltages by the potential transformer ratio gives the load line-to-line voltages on the 120-volt base as

$$V120_{ab} = 119.25 / -1.9,$$

$$V120_{bc} = 119.0 / -121.3,$$

$$V120_{ca} = 120.1 / 118.5 \text{ V}$$

Note how the actual load voltages on the 120-volt base match very closely the values computed across the compensator relays. It is also noted that, in this case, the third line-to-line voltage is also within the bandwidth. That will not always be the case.

This example is very long but has been included to demonstrate how the compensator circuit is set and then how it will adjust taps so that the voltages at a remote load center node will be held within the set limits. In actual practice, the only responsibilities of the engineer will be to correctly determine the R and X settings of the compensator circuit and to determine the desired voltage level and bandwidth.

The open delta regulator connection using phases A-B and C-B has been presented. There are two other possible open delta connections using phase B-C and A-C and then C-A and B-A. Generalized matrices for these additional two connections can be developed using the procedures presented in this section.

7.5 SUMMARY

It has been shown that all possible connections for Type B step-voltage regulators can be modeled using the generalized matrices. The derivations in this chapter were limited to three-phase connections. If a single-phase regulator is connected line-to-neutral or two regulators connected in open wye, then the [a] and [d] matrices will be of the same form as that of the wye-connected regulators, only the terms in the rows and columns associated with the missing phases would be zero. The same can be said for a single-phase regulator connected line-to-line. Again the rows and columns associated with the missing phases would be set to zero in the matrices developed for the open delta connection.

The generalized matrices developed in this chapter are of exactly the same form as those developed for the three-phase line segments. In the next chapter, the generalized matrices for all three-phase transformers will be developed.

Voltage Regulation

PROBLEMS

7.1 A single-phase transformer is rated 100 kVA, 2,400–240 volts. The impedances and shunt admittance of the transformer are

$Z_1 = 0.65 + j0.95\ \Omega$ (high-voltage winding impedance),
$Z_2 = 0.0052 + j0.0078\ \Omega$ (low-voltage winding impedance),
$Y_m = 2.56 \times 10^{-4} - j11.37 \times 10^{-4}$ S (referred to the high-voltage winding).

Modify the MATLAB script M0701.m to
(a) determine the a, b, c, d constants, and the A and B constants;
(b) determine the transformer is serving an 80 kW, 0.85 lagging power factor load at 230 volts;
(c) determine the primary voltage, current, and complex power; and
(d) determine the per-unit transformer impedance and shunt admittance based upon the transformer ratings.

7.2 The single-phase transformer of Problem 7.1 is to be connected as a step-down autotransformer to transform the voltage from 2,400 volts down to 2,160 volts.
(a) Draw the connection diagram, including the series impedance and shunt admittance.
(b) Determine the autotransformer kVA rating.
(c) Determine the a, b, c, d, A, and B generalized constants.
(d) The autotransformer is serving a load of 80 kVA, 0.95 lagging power factor at a voltage of 2,000 volts. Including the impedance and shunt admittance, determine the input voltage, current and complex power.
(e) Determine the per-unit impedance and shunt admittance based upon the autotransformer rating. How do these values compare to the per-unit values of Problem 7.1?

7.3 A "Type B", step-voltage regulator is installed to regulate the voltage on a 7,200-volt, single-phase lateral. The potential transformer and current transformer ratios connected to the compensator circuit are
Potential transformer: 7,200–120 V
Current transformer: 500:5 A
The R and X settings in the compensator circuit are $R = 5$ V and $X = 10$ V.
The regulator tap is set on the +10 position when the voltage and current on the source side of the regulator are
$V_{source} = 7200$ V and $I_{source} = 375$ at a power factor of 0.866 lagging power factor. Write a MATLAB script to
(a) determine the voltage at the load center,
(b) determine the equivalent line impedance between the regulator and the load center, and
(c) assuming that the voltage level on the regulator has been set at 120 volts with a bandwidth of 2 V, determine what tap the regulator will move to.

7.4 Refer to Figure 7.11. The substation transformer is rated 24 MVA, 230 kV delta-13.8 kV wye. Three single-phase, Type B regulators are connected in wye. The equivalent line impedance between the regulators and the load center node is

$$Z_{line} = 0.264 + j0.58 \, \Omega/\text{mile}$$

The distance to the load center node is 10,000 feet. Modify the MATLAB script M0707.m to
(a) determine the appropriate PT and CT ratios;
(b) determine the R' and X' settings in ohms and volts for the compensator circuit;
(c) the substation is serving a balanced three-phase load of 16 MVA, 0.9 lagging power factor when the output line-to-line voltages of the substation are balanced 13.8 kV and the regulators are set in the neutral position, so assume the voltage level is set at 121 volts and a bandwidth of 2 volts and determine the final tap position for each regulator (they will be the same). The regulators have 32%–5/8% taps (16 raise and 16 lower);
(d) determine what the regulator tap settings for a load of 24 MVA, 0.9 lagging power factor would be with the output voltages of the substation transformer balanced at three-phase 13.8 kV;
(e) determine what the load center voltages for the load of part d would be.

7.5 Three Type B, step-voltage regulators are connected in wye and located on the secondary bus of a 12.47 kV substation. The feeder is serving an unbalanced load. A power-flow study has been run, and the voltages at the substation and the load center node are

$$[Vsub_{abc}] = \begin{bmatrix} 7200/\underline{0} \\ 7200/\underline{-120} \\ 7200/\underline{120} \end{bmatrix} V,$$

$$[VLC_{abc}] = \begin{bmatrix} 6890.6/-1.49 \\ 6825.9/-122.90 \\ 6990.5/117.05 \end{bmatrix} V$$

The currents at the substation are

$$[I_{abc}] = \begin{bmatrix} 362.8/\underline{-27.3} \\ 395.4/\underline{-154.7} \\ 329.0/\underline{98.9} \end{bmatrix} A$$

Voltage Regulation

The regulator potential transformer ratio is 7,200–120, and the current transformer ratio is 500:5. The voltage level of the regulators is set at 121 volts and the bandwidth at 2 volts.

(a) Determine the equivalent line impedance per phase between the regulator and the load center.

(b) The compensators on each regulator are to be set with the same R and X values. Specify these values in volts and in ohms.

7.6 The impedance compensator settings for the three-step regulators of Problem 7.5 have been set as

$$R' = 3.0 \text{ V}, \quad X' = 9.3 \text{ V}$$

The voltages and currents at the substation bus are

$$[Vsub_{abc}] = \begin{bmatrix} 7200/\underline{0} \\ 7200/\underline{-120} \\ 7200/\underline{120} \end{bmatrix} \text{V},$$

$$[I_{abc}] = \begin{bmatrix} 320.6/\underline{-27.4} \\ 409.0/\underline{-155.1} \\ 331.5/\underline{98.2} \end{bmatrix} \text{A}$$

Determine the final tap settings for each regulator.

7.7 For a different load condition for the system of Problem 7.5, the taps on the regulators have been automatically set by the compensator circuit to

$$Tap_a = +8, \quad Tap_b = +11, \quad Tap_c = +6.$$

The load reduces so that the voltages and currents at the substation bus are

$$[Vsub_{abc}] = \begin{bmatrix} 7200/\underline{0} \\ 7200/\underline{-120} \\ 7200/\underline{120} \end{bmatrix} \text{V},$$

$$[I_{abc}] = \begin{bmatrix} 177.1/\underline{-28.5} \\ 213.4/\underline{-156.4} \\ 146.8/\underline{98.3} \end{bmatrix} \text{A}$$

Determine the new final tap settings for each regulator.

7.8 The load center node for the regulators described in Problem 7.5 is located 1.5 miles from the substation. There are no lateral taps between the substation and the load center. The phase impedance matrix of the line segment (Problem 4.1) is

$$[\bar{z}_{abc}] = \begin{bmatrix} 0.3375 + j1.0478 & 0.1535 + j0.3849 & 0.1559 + j0.5017 \\ 0.1535 + j0.3849 & 0.3414 + j1.0348 & 0.1580 + j0.4236 \\ 0.1559 + j0.5017 & 0.1580 + j0.4236 & 0.3465 + j1.0179 \end{bmatrix} \Omega/\text{mile}$$

A wye-connected, unbalanced constant impedance load is located at the load center node. The load impedances are

$$ZL_a = 19 + j11 \ \Omega, \quad ZL_b = 22 + j12 \ \Omega, \quad ZL_c = 18 + j10 \ \Omega$$

The voltages at the substation are a balanced three phase of 7,200 volts line-to-to neutral. The regulators are set on neutral.
(a) Determine the line-to-neutral voltages at the load center.
(b) Determine the R and X settings in volts for the compensator.
(c) Determine the required tap settings to hold the load center voltages within the desired limits.

7.9 The R and X settings for the line in Problem 7.8 have been set to $2.3 + j7.4$ volts. For this problem, the loads are wye connected and modeled such that the per-phase load kVA and power factor (constant PQ loads) are

$$[kVA] = \begin{bmatrix} 1200 \\ 1600 \\ 1000 \end{bmatrix} \quad [PF] = \begin{bmatrix} 0.90 \\ 0.85 \\ 0.95 \end{bmatrix}$$

Determine
(a) the final regulator tap positions,
(b) the compensator relay voltages, and
(c) the load line-to-neutral voltages on a 120-volt base.

7.10 The phase impedance matrix for a three-wire line segment is:

$$[\bar{z}_{abc}] = \begin{bmatrix} 0.4013 + j1.4133 & 0.0953 + j0.8515 & 0.0953 + j0.7802 \\ 0.0953 + j0.8515 & 0.4013 + j1.4133 & 0.0953 + j0.7266 \\ 0.0953 + j0.7802 & 0.0953 + j0.7266 & 0.4013 + j1.4133 \end{bmatrix} \Omega/\text{mile}$$

The line is serving an unbalanced load so that at the substation transformer line-to-line voltages and output currents are

$$[VLL_{abc}] = \begin{bmatrix} 12{,}470\underline{/0} \\ 12{,}470\underline{/-120} \\ 12{,}470\underline{/120} \end{bmatrix} V,$$

Voltage Regulation

$$[I_{abc}] = \begin{bmatrix} 307.9/-54.6 \\ 290.6/178.6 \\ 268.2/65.3 \end{bmatrix} A$$

Two Type B, step-voltage regulators are connected in an open delta at the substation using phases A-B and C-B. The potential transformer ratios are 12,470/120 and the current transformer ratios are 500:5. The voltage level is set at 121 volts with a 2-volt bandwidth.
(a) Determine the line-to-line voltages at the load center.
(b) Determine the R and X compensator settings in volts. For the open-delta connection, the R and X settings will be different on each regulator.
(c) Determine the final tap positions of the two voltage regulators.

7.11 The regulators in Problem 7.10 have gone to the +9 tap on both regulators for a particular load. The load is reduced so that the currents leaving the substation transformer with the regulators in the +9 position are

$$[I_{abc}] = \begin{bmatrix} 144.3/-53.5 \\ 136.3/179.6 \\ 125.7/66.3 \end{bmatrix} A$$

Determine the final tap settings on each regulator for this new load condition.

7.12 Use the system of Example 7.8 with the delta-connected loads changed to

$$[kVA] = \begin{bmatrix} 1800 \\ 1500 \\ 2000 \end{bmatrix} [PF] = \begin{bmatrix} 0.90 \\ 0.95 \\ 0.92 \end{bmatrix}$$

The source voltages, potential transformer, and current transformer ratings are those in the example. The desired voltage level is set at 122 volts with a bandwidth of 2 volts. For this load condition,
(a) use the R and X compensator values from Example 7.8,
(b) use the required tap positions,
(c) determine the final relay voltages, and
(d) determine the final load line-to-line load voltages and the line currents.

WINDMIL ASSIGNMENT:

Use System 3 and add a step-voltage regulator connected between the source and the three-phase OH line. Call this "System 4". The regulator is to be set with a specified voltage level of 122 volts at Node 2. The potential transformer ratio is 7,200–20 and the CT ratio is 200:5. Call the regulator Reg-1. Follow these steps in the user's manual on how to install the three-phase, wye-connected regulators.

1. Follow the steps outlined in the user's manual to have Windmil determine the R and X settings to hold the specified voltage level at Node 2.
2. Run "Voltage Drop". Check the node voltages and in particular the voltage at Node 2.
3. What taps did the regulators go to?

In Example 7.7, MATLAB script M0707.m was used to calculate the compensator R and X setting. Follow that procedure to compute the R and X settings and compare them to the Windmil settings.

REFERENCES

1. *American Nation Standard for Electric Power – Systems and Equipment Voltage Ratings (60 Hertz)*, ANSI C84.1-1995, National Electrical Manufacturers Association, Rosslyn, Virginia, 1996.
2. *IEEE Standard Requirements, Terminology, and Test Code for Step-Voltage and Induction-Voltage Regulators*, ANSI/IEEE C57.15-1986, Institute of Electrical and Electronic Engineers, New York, NY, 1988.

8 Three-Phase Transformer Models

Three-phase transformer banks are found in the distribution substation where the voltage is transformed from the transmission or sub-transmission level to the distribution feeder level. In most cases, the substation transformer will be a three-phase unit, perhaps with high voltage no-load taps and, perhaps, low voltage LTC. For a four-wire wye feeder, the most common substation transformer connection is the delta – grounded wye. A three-wire delta feeder will typically have a delta-delta transformer connection in the substation. Three-phase transformer banks out on the feeder will provide the final voltage transformation to the customer's load. A variety of transformer connections can be applied. The load can be a pure three-phase or a combination of single-phase lighting load, split-phase electric vehicle chargers, and a three-phase load such as an induction motor. In the analysis of a distribution feeder, it is important that the various three-phase transformer connections be modeled correctly.

Unique models of three-phase transformer banks applicable to radial distribution feeders will be developed in this chapter. Models for the following three-phase connections are included:

- Delta – Grounded Wye
- Ungrounded Wye – Delta
- Grounded Wye – Delta
- Open Wye – Open Delta
- Grounded Wye – Grounded Wye
- Delta – Delta
- Open Delta – Open Delta

8.1 INTRODUCTION

Figure 8.1 defines the various voltages and currents for all three-phase transformer banks connected between the source-side Node-**n** and the load-side Node-**m**.

In Figure 8.1, the models can represent a step-down (source side to load side) or a step-up (source side to load side) transformer bank. The notation is such that the capital letters **A**, **B**, **C**, **N** will always refer to the **source** side (Node-**n**) of the bank and the lowercase letters **a**, **b**, **c**, **n** will always refer to the **load** side (Node-**m**) of the bank. It is assumed that all variations of the wye-delta connections are connected in the "American standard 30°" connection. The described phase notation and the standard phase shifts for positive sequence voltages and currents are as follows:

Step-Down Connection

$$V_{AB} \text{ leads } V_{ab} \text{ by } 30° \qquad (8.1)$$

$$I_A \text{ leads } I_a \text{ by } 30° \qquad (8.2)$$

Step-Up Connection

$$V_{ab} \text{ leads } V_{AB} \text{ by } 30° \qquad (8.3)$$

$$I_a \text{ leads } I_A \text{ by } 30° \qquad (8.4)$$

8.2 GENERALIZED MATRICES

The transformer bank models to be used in power-flow and short-circuit studies are shown in Figure 8.1. In the "forward sweep" of the ladder technique, the voltages at the line side are defined as a function of the voltages at the source side and the currents at the line side. In the forward sweep, only the downstream voltages are computed. The required equation is

$$[VLN_{abc}] = [A_t] \cdot [VLN_{ABC}] - [B_t] \cdot [I_{abc}] \qquad (8.5)$$

In the "backward sweep" of the ladder technique, only the upstream currents are computed. The matrix equation is

$$[I_{ABC}] = [c_t] \cdot [VLN_{ABC}] + [d_t] \cdot [I_{abc}] \qquad (8.6)$$

In Equations 8.5 and 8.6, the matrices [VLN_{ABC}] and [VLN_{abc}] represent the line-to-neutral voltages for an ungrounded wye connection or the line-to-ground voltages

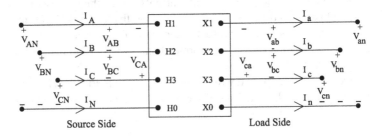

FIGURE 8.1 General three-phase transformer bank.

Three-Phase Transformer Models

for a grounded wye connection. For a delta connection, the voltage matrices represent "equivalent" line-to-neutral voltages, as given in Equation 8.7.

$$[VLN_{abc}] = [W] \cdot [VLL_{abc}], \qquad (8.7)$$

where

$$[W] = \frac{1}{3} \cdot \begin{bmatrix} 2 & 1 & 0 \\ 0 & 2 & 1 \\ 1 & 0 & 2 \end{bmatrix}$$

The current matrices represent the line currents regardless of the transformer winding connection.

In the modified ladder technique, Equation 8.5 is used to compute new node voltages downstream from the source using the most recent line currents. In the backward sweep, only Equation 8.6 is used to compute the source-side line currents using the newly computed load-side line currents.

8.3 THE DELTA-GROUNDED WYE STEP-DOWN CONNECTION

The delta-grounded wye step-down connection is a popular connection that is typically used in a distribution substation serving a four-wire feeder system. Another application of the connection is to provide service to a load that is primarily single phase. Because of the wye connection, three single-phase circuits are available thereby making it possible to balance the single-phase loading on the transformer bank.

Three single-phase transformers can be connected delta – grounded wye in a "standard 30° step-down connection", as shown in Figure 8.2.

Note in Figure 8.2 that at no load, the line-to-ground voltages are the negative of the "ideal" secondary voltages. That is at no-load $V_{ag} = - Vt_a$.

8.3.1 Voltages

The positive sequence phasor diagrams of the voltages in Figure 8.2 show the relationships between the various positive sequence voltages. Note that the voltage from a to ground is the negative of the ideal transformer voltage Vt_a. The same is true for phases b and c. Shown in Figure 8.2, the primary line-to-line voltage from A to B leads the secondary line-to-line voltage from a to b by 30°.

Care must be taken to observe the polarity marks on the individual transformer windings. KVL for all loads gives the line-to-line voltage between phase **a** and **b** as

$$V_{ab} = V_{ag} - V_{bg} \qquad (8.8)$$

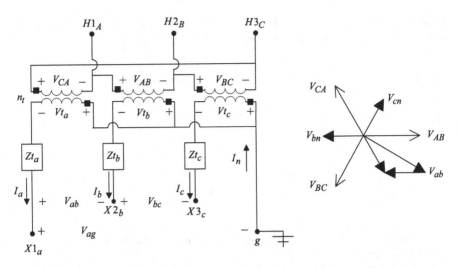

FIGURE 8.2 Standard delta – grounded wye connection with voltages.

The phasors of the positive sequence voltages of Equation 8.8 are shown in Figure 8.2.

The magnitude changes between the voltages can be defined in terms of the actual winding turns ratio (n_t). With reference to Figure 8.2, these ratios are defined as follows

$$n_t = \frac{kVLL_{\text{Rated Primary}}}{kVLN_{\text{Rated Secondary}}} \quad (8.9)$$

With reference to Figure 8.2, the line-to-line voltages on the primary side of the transformer connection as a function of the ideal secondary side voltages are given by

$$\begin{bmatrix} V_{AB} \\ V_{BC} \\ V_{CA} \end{bmatrix} = n_t \cdot \begin{bmatrix} 0 & 1 & 0 \\ 0 & 0 & 1 \\ 1 & 0 & 0 \end{bmatrix} \cdot \begin{bmatrix} Vt_a \\ Vt_b \\ Vt_c \end{bmatrix} \quad [VLL_{ABC}] = [AV] \cdot [Vt_{abc}],$$

$$\text{where } [AV] = n_t \cdot \begin{bmatrix} 0 & 1 & 0 \\ 0 & 0 & 1 \\ 1 & 0 & 0 \end{bmatrix} \quad (8.10)$$

Equation 8.10 gives the primary line-to-line voltages at Node-**n** as a function of the ideal secondary voltages. However, what is needed is a relationship between "equivalent" line-to-neutral voltages at Node-**n** and the ideal secondary voltages. The question is how is the equivalent line-to-neutral voltages determined knowing the

Three-Phase Transformer Models 221

line-to-line voltages? One approach is to apply the theory of symmetrical components.

The known line-to-line voltages are transformed to their sequence voltages by

$$[VLL_{012}] = [A_s]^{-1} \cdot [VLL_{ABC}], \qquad (8.11)$$

where

$$[A_s] = \begin{bmatrix} 1 & 1 & 1 \\ 1 & a_s^2 & a_s \\ 1 & a_s & a_s^2 \end{bmatrix}, \qquad (8.12)$$

$$a_s = 1.0\underline{/120}.$$

By definition, the zero-sequence, line-to-line voltage is always zero. The relationship between the positive and negative sequence line-to-neutral and line-to-line voltages is known. These relationships in matrix form are given by

$$\begin{bmatrix} VLN_0 \\ VLN_1 \\ VLN_2 \end{bmatrix} = \begin{bmatrix} 1 & 0 & 0 \\ 0 & t_s^* & 0 \\ 0 & 0 & t_s \end{bmatrix} \cdot \begin{bmatrix} VLL_0 \\ VLL_1 \\ VLL_2 \end{bmatrix} \quad [VLN_{012}] = [T] \cdot [VLL_{012}],$$

$$\text{where} : t = \frac{1}{\sqrt{3}}\underline{/30} \qquad (8.13)$$

Since the zero-sequence, line-to-line voltage is zero, the (1,1) term of the matrix [T] can be any value. For the purposes here, the (1,1) term is chosen to have a value of 1.0. Knowing the sequence line-to-neutral voltages, the equivalent line-to-neutral voltages can be determined.

The equivalent line-to-neutral voltages as a function of the sequence line-to-neutral voltages are

$$[VLN_{ABC}] = [A_s] \cdot [VLN_{012}] \qquad (8.14)$$

Substitute Equation 8.13 into Equation 8.14.

$$[VLN_{ABC}] = [A_s] \cdot [T] \cdot [VLL_{012}] \qquad (8.15)$$

Substitute Equation 8.11 into Equation 8.15.

$$[VLN_{ABC}] = [A_s] \cdot [T] \cdot [A_s]^{-1} \cdot [VLL_{ABC}] = [W] \cdot [VLL_{ABC}], \qquad (8.16)$$

where

$$[W] = [A_s] \cdot [T] \cdot [A_s]^{-1} = \frac{1}{3} \cdot \begin{bmatrix} 2 & 1 & 0 \\ 0 & 2 & 1 \\ 1 & 0 & 2 \end{bmatrix} \quad (8.17)$$

Equation 8.17 provides a method of computing equivalent line-to-neutral voltages from a knowledge of the line-to-line voltages. This is an important relationship that will be used in a variety of ways as other three-phase transformer connections are studied.

Equation 8.16 can be substituted into Equation 8.10.

$$[VLN_{ABC}] = [W] \cdot [VLL] = [W] \cdot [AV] \cdot [Vt_{abc}] = [a_t] \cdot [Vt_{abc}], \quad (8.18)$$

where

$$[a_t] = [W] \cdot [AV] = \frac{n_t}{3} \cdot \begin{bmatrix} 0 & 2 & 1 \\ 1 & 0 & 2 \\ 2 & 1 & 0 \end{bmatrix} \quad (8.19)$$

Equation 8.19 defines the generalized $[a_t]$ matrix for the delta-grounded wye step-down connection.

The ideal secondary voltages as a function of the secondary line-to-ground voltages and the secondary line currents are

$$[Vt_{abc}] = [VLG_{abc}] + [Zt_{abc}] \cdot [I_{abc}], \quad (8.20)$$

where

$$[Zt_{abc}] = \begin{bmatrix} Zt_a & 0 & 0 \\ 0 & Zt_b & 0 \\ 0 & 0 & Zt_c \end{bmatrix} \quad (8.21)$$

Notice in Equation 8.21 that there is no restriction that the impedances of the three transformers be equal.

Substitute Equation 8.20 into Equation 8.18.

$$[VLN_{ABC}] = [a_t] \cdot ([VLG_{abc}] + [Zt_{abc}] \cdot [I_{abc}]),$$

$$[VLN_{ABC}] = [a_t] \cdot [VLG_{abc}] + [b_t] \cdot [I_{abc}], \quad (8.22)$$

Three-Phase Transformer Models

where

$$[a_t] = [W] \cdot [AV] = \frac{n_t}{3} \cdot \begin{bmatrix} 0 & 2 & 1 \\ 1 & 0 & 2 \\ 2 & 1 & 0 \end{bmatrix},$$

$$[b_t] = [a_t] \cdot [Zt_{abc}] = \frac{n_t}{3} \cdot \begin{bmatrix} 0 & 2 \cdot Zt_b & Zt_c \\ Zt_a & 0 & 2 \cdot Zt_b \\ 2 \cdot Zt_a & Zt_b & 0 \end{bmatrix} \quad (8.23)$$

The generalized matrices $[a_t]$ and $[b_t]$ have now been defined. The derivation of the generalized matrices $[A_t]$ and $[B_t]$ begins with solving Equation 8.10 for the ideal secondary voltages.

$$[Vt_{abc}] = [AV]^{-1} \cdot [VLL_{ABC}] \quad (8.24)$$

The line-to-line voltages as a function of the equivalent line-to-neutral voltages are

$$[VLL_{ABC}] = [Dv] \cdot [VLN_{ABC}], \quad (8.25)$$

where

$$[Dv] = \begin{bmatrix} 1 & -1 & 0 \\ 0 & 1 & -1 \\ -1 & 0 & 1 \end{bmatrix}. \quad (8.26)$$

Substitute Equation 8.25 into Equation 8.24.

$$[Vt_{abc}] = [AV]^{-1} \cdot [Dv] \cdot [VLN_{ABC}] = [A_t] \cdot [VLN_{ABC}], \quad (8.27)$$

where

$$[A_t] = [AV]^{-1} \cdot [Dv] = \frac{1}{n_t} \cdot \begin{bmatrix} -1 & 0 & 1 \\ 1 & -1 & 0 \\ 0 & 1 & -1 \end{bmatrix} \quad (8.28)$$

Substitute Equations 8.20 into Equation 8.27.

$$[VLG_{abc}] + [Zt_{abc}] \cdot [I_{abc}] = [A_t] \cdot [VLN_{ABC}] \quad (8.29)$$

Rearrange Equation 8.29.

$$[VLG_{abc}] = [A_t] \cdot [VLN_{ABC}] - [B_t] \cdot [I_{abc}], \qquad (8.30)$$

where

$$[B_t] = [Zt_{abc}] = \begin{bmatrix} Zt_a & 0 & 0 \\ 0 & Zt_b & 0 \\ 0 & 0 & Zt_c \end{bmatrix} \qquad (8.31)$$

Equation 8.22 is referred to as the "backward sweep voltage equation", and Equation 8.30 is referred to as the "forward sweep voltage equation". Equations 8.22 and 8.30 apply only for the step-down delta-grounded wye transformer. Note that these equations are in the same form as those derived in earlier chapters for line segments and step-voltage regulators.

8.3.2 Currents

The 30° connection specifies that the positive sequence current entering the H1 terminal will lead the positive sequence current, leaving the X1 terminal by 30°. Figure 8.3 shows the same connection as Figure 8.2 but with the currents instead of the voltages displayed.

As with the voltages, the polarity marks on the transformer windings must be observed for the currents. For example, in Figure 8.3, the current I_a is entering the polarity mark on the low voltage winding so the current I_{AC} flowing out of the polarity mark on the high voltage winding will be in phase with I_a. This relationship is shown in the phasor diagrams for positive sequence currents in Figure 8.3. Note that the primary line current on phase A leads the secondary phase a current by 30°.

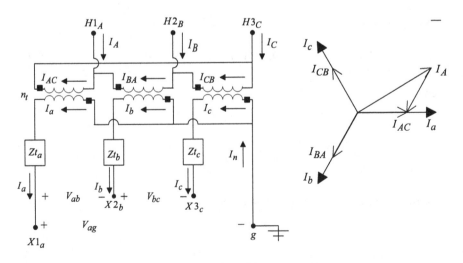

FIGURE 8.3 Delta-grounded wye connection with currents.

Three-Phase Transformer Models

The line currents can be determined as a function of the delta currents by applying KCL.

$$\begin{bmatrix} I_A \\ I_B \\ I_C \end{bmatrix} = \begin{bmatrix} 1 & -1 & 0 \\ 0 & 1 & -1 \\ -1 & 0 & 1 \end{bmatrix} \cdot \begin{bmatrix} I_{AC} \\ I_{BA} \\ I_{CB} \end{bmatrix} \qquad (8.32)$$

In condensed form, Equation 8.32 is

$$[I_{ABC}] = [Di] \cdot [ID_{ABC}],$$

$$\text{where } [Di] = \begin{bmatrix} 1 & -1 & 0 \\ 0 & 1 & -1 \\ -1 & 0 & 1 \end{bmatrix} \qquad (8.33)$$

The matrix equation relating the delta primary currents to the secondary line currents is given by

$$\begin{bmatrix} I_{AC} \\ I_{BA} \\ I_{CB} \end{bmatrix} = \frac{1}{n_t} \cdot \begin{bmatrix} 1 & 0 & 0 \\ 0 & 1 & 0 \\ 0 & 0 & 1 \end{bmatrix} \cdot \begin{bmatrix} I_a \\ I_b \\ I_c \end{bmatrix}, \qquad (8.34)$$

$$[ID_{ABC}] = [AI] \cdot [I_{abc}]$$

$$\text{where } [AI] = \frac{1}{n_t} \cdot \begin{bmatrix} 1 & 0 & 0 \\ 0 & 1 & 0 \\ 0 & 0 & 1 \end{bmatrix} \qquad (8.35)$$

Substitute Equation 8.35 into Equation 8.33.

$$[I_{ABC}] = [Di] \cdot [AI] \cdot [I_{abc}] = [c_t] \cdot [VLG_{abc}] + [d_t] \cdot [I_{abc}], \qquad (8.36)$$

where

$$[d_t] = [Di] \cdot [AI] = \frac{1}{n_t} \cdot \begin{bmatrix} 1 & -1 & 0 \\ 0 & 1 & -1 \\ -1 & 0 & 1 \end{bmatrix}, \qquad (8.37)$$

$$[c_t] = \begin{bmatrix} 0 & 0 & 0 \\ 0 & 0 & 0 \\ 0 & 0 & 0 \end{bmatrix} \qquad (8.38)$$

Equation 8.36 (referred to as the "backward sweep current equations") provides a direct method of computing the phase line currents at Node-**n** knowing the phase line currents at Node-**m**. Again, this equation is in the same form as that previously derived for three-phase line segments and three-phase step-voltage regulators.

The equations derived in this section are for the step-down connection. The next Section 8.4 will summarize the matrices for the delta-grounded wye step-up connection.

Example 8.1: An example system with three buses. Bus 1 is connected to a voltage source, Bus 1 and Bus 2 are connected by a delta – grounded wye step-down transformer, Bus 2 and Bus 3 are connected by a 10,000-foot line.

In the example system of Figure 8.4, an unbalanced constant impedance load is being served at the end of a 10,000-foot section of a three-phase line. The 10,000-foot-long line is being fed from a substation transformer rated 5,000 kVA, 115 kV delta – 12.47 kV grounded wye with a per-unit impedance of 0.085/85. The phase conductors of the line are 336,400 26/7 ACSR with a neutral conductor 4/0 ACSR. The configuration and computation of the phase impedance matrix are given in Example 4.1. From that example, the phase impedance matrix was computed to be

$$[z_{line}] = \begin{bmatrix} 0.4576 + j1.0780 & 0.1560 + j0.5017 & 0.1535 + j0.3849 \\ 0.1560 + j0.5017 & 0.4666 + j1.0482 & 0.1580 + j0.4236 \\ 0.1535 + j0.3849 & 0.1580 + j0.4236 & 0.4615 + j1.0651 \end{bmatrix} \text{ohms/mile,}$$

$$dist = \frac{10,000}{5,280} \text{ mile,}$$

$$[Zline_{abc}] = dist \cdot [z_{line}] = \begin{bmatrix} 0.8667 + j2.0417 & 0.2955 + j0.9502 & 0.2907 + j0.7290 \\ 0.2955 + j0.9502 & 0.8837 + j1.9852 & 0.2992 + j0.8023 \\ 0.2907 + j0.7290 & 0.2992 + j0.8023 & 0.8741 + j2.0172 \end{bmatrix}$$

The general matrices for the line are

$$[A_{line}] = \begin{bmatrix} 1 & 0 & 0 \\ 0 & 1 & 0 \\ 0 & 0 & 1 \end{bmatrix} [B_{line}] = [Zline_{abc}][d_{line}] = \begin{bmatrix} 1 & 0 & 0 \\ 0 & 1 & 0 \\ 0 & 0 & 1 \end{bmatrix}.$$

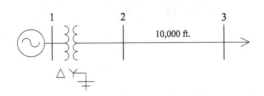

FIGURE 8.4 Example system.

Three-Phase Transformer Models

The transformer impedance needs to be converted to ohms referenced to the low voltage side of the transformer. The low-side base impedance is

$$Z_{base} = \frac{12.47^2}{5000 \cdot 1000} = 31.1.$$

The transformer impedance referenced to the low voltage side is

$$Zt = (0.085\underline{/85}) \cdot 31.3 = 0.2304 + j2.6335 \, \Omega$$

The transformer phase impedance matrix is

$$[Zt_{abc}] = \begin{bmatrix} 0.2304 + j2.6335 & 0 & 0 \\ 0 & 0.2304 + j2.6335 & 0 \\ 0 & 0 & 0.2304 + j2.6335 \end{bmatrix} \Omega$$

The unbalanced constant impedance load is connected in grounded wye. The load impedance matrix is specified to be

$$[Zload_{abc}] = \begin{bmatrix} 12 + j6 & 0 & 0 \\ 0 & 13 + j4 & 0 \\ 0 & 0 & 14 + j5 \end{bmatrix} \Omega$$

The unbalanced line-to-line voltages at Node 1 serving the substation transformer are given as

$$[VLL_{ABC}] = \begin{bmatrix} 115,000\underline{/0} \\ 116,500\underline{/-115.5} \\ 123,538\underline{/121.7} \end{bmatrix} V$$

(a) Determine the generalized matrices for the transformer:
The transformer turns ratio is

$$n_t = \frac{kVLL_{high}}{kVLN_{low}} = \frac{115}{\frac{12.47}{\sqrt{3}}} = 15.9732$$

From Equation 8.19,

$$[a_t] = \frac{n_t}{3} \cdot \begin{bmatrix} 0 & 2 & 1 \\ 1 & 0 & 2 \\ 2 & 1 & 0 \end{bmatrix} = \begin{bmatrix} 0 & 10.6488 & 5.3244 \\ 5.3244 & 0 & 10.6488 \\ 10.6488 & 5.3244 & 0 \end{bmatrix}$$

From Equation 8.23,

$$b_t = a_t \cdot Zt_{abc} = \frac{n_t}{3} \cdot \begin{bmatrix} 0 & 2 \cdot Zt_b & Zt_c \\ Zt_a & 0 & 2 \cdot Zt_c \\ 2 \cdot Zt_a & Zt_b & 0 \end{bmatrix},$$

$$b_t = \begin{bmatrix} 0 & 2.4535 + j28.0432 & 1.2267 + j14.0216 \\ 1.2267 + j14.0216 & 0 & 2.4535 + j28.0432 \\ 2.4535 + j28.0432 & 2.4535 + j28.0432 & 0 \end{bmatrix}$$

From Equation 8.37,

$$[d_t] = \frac{1}{n_t} \cdot \begin{bmatrix} 1 & -1 & 0 \\ 0 & 1 & -1 \\ -1 & 0 & 1 \end{bmatrix} = \begin{bmatrix} 0.0626 & -0.0626 & 0 \\ 0 & 0.0626 & -0.0626 \\ -0.0626 & 0 & 0.0626 \end{bmatrix}$$

From Equation 8.28,

$$A_t = \frac{1}{n_t} \cdot \begin{bmatrix} -1 & 0 & 1 \\ 1 & -1 & 0 \\ 0 & 1 & -1 \end{bmatrix} = \begin{bmatrix} -0.0626 & 0 & 0.0626 \\ 0.0626 & -0.0626 & 0 \\ 0 & 0.0626 & -0.0626 \end{bmatrix}$$

From Equation 8.31,

$$[B_t] = [Zt_{abc}] = \begin{bmatrix} 0.2304 + j2.6335 & 0 & 0 \\ 0 & 0.2304 + j2.6335 & 0 \\ 0 & 0 & 0.2304 + j2.6335 \end{bmatrix}$$

(b) Given the line-to-line voltages at Node 1, determine the "ideal" transformer voltages:
From Equation 8.10,

$$AV = n_t \cdot \begin{bmatrix} 0 & 1 & 0 \\ 0 & 0 & 1 \\ 1 & 0 & 0 \end{bmatrix} = \begin{bmatrix} 0 & 15.9732 & 0 \\ 0 & 0 & 15.9732 \\ 15.9732 & 0 & 0 \end{bmatrix},$$

$$Vt_{abc} = AV^{-1} \cdot VLL_{ABC} = \begin{bmatrix} 7734.0952/\underline{121.7} \\ 7199.5579/\underline{0} \\ 7293.4651/\underline{-115.5} \end{bmatrix} V$$

(c) Determine the load currents.
Since the load is modeled as constant impedances, the system is linear and the analysis can combine all of the impedances (transformer, line, and load) to an equivalent impedance matrix.
KVL gives

$$[Vt_{abc}] = ([Zt_{abc}] + [Zline_{abc}] + [Zload_{abc}]) \cdot [I_{abc}] = [Zeq_{abc}] \cdot [I_{abc}],$$

$$[Zeq_{abc}] = \begin{bmatrix} 13.0971 + j10.6751 & 0.2955 + j0.9502 & 0.2907 + j.7290 \\ 0.2955 + j0.9502 & 14.1141 + j8.6187 & 0.2992 + j0.8023 \\ 0.2907 + j.7290 & 0.2992 + j0.8023 & 15.1045 + j9.6507 \end{bmatrix} \Omega$$

Three-Phase Transformer Models

The line currents can now be computed.

$$[I_{abc}] = [Zeq_{abc}]^{-1} \cdot [Vt_{abc}] = \begin{bmatrix} 471.7/\underline{84.9} \\ 456.7/\underline{-30.1} \\ 427.3/\underline{-146.5} \end{bmatrix} A$$

(d) Determine the line-to-ground voltages at the load in volts and on a 120-volt base.

$$[Vload_{abc}] = [Zload_{abc}] \cdot [I_{abc}] = \begin{bmatrix} 6328.1/\underline{111.4} \\ 6212.2/\underline{-13.0} \\ 6352.6/\underline{-126.9} \end{bmatrix} V$$

The load voltages on a 120-volt base are

$$[Vload_{120}] = \begin{bmatrix} 105.5/\underline{111.4} \\ 103.5/\underline{-13.0} \\ 105.9/\underline{-126.9} \end{bmatrix}.$$

The line-to-ground voltages at Node 2 are

$$[VLG_{abc}] = [a_{line}] \cdot [Vload_{abc}] + [b_{line}] \cdot [I_{abc}] = \begin{bmatrix} 6965.4/\underline{114.0} \\ 6580.6/\underline{-8.6} \\ 6691.4/\underline{-123.3} \end{bmatrix} V$$

(e) Using the backward sweep voltage equation, determine the equivalent line-to-neutral voltages and the line-to-line voltages at Node 1.

$$[VLN_{ABC}] = [a_t] \cdot [VLG_{abc}] + [b_t] \cdot [I_{abc}] = \begin{bmatrix} 69.443/\underline{-30.3} \\ 65,263/\underline{-147.5} \\ 70,272/\underline{94.0} \end{bmatrix} V$$

$$[VLL_{ABC}] = [Dv] \cdot [VLN_{ABC}] = \begin{bmatrix} 115,000/\underline{0} \\ 116,500/\underline{-115.5} \\ 123.538/\underline{121.7} \end{bmatrix} V$$

It is always comforting to be able to work back and compute what was initially given. In this case, the line-to-line voltages at Node 1 have been computed, and the same values result that were given at the start of the problem.

(f) Use the forward sweep voltage equation to verify that the line-to-ground voltages at Node 2 can be computed knowing the equivalent line-to-neutral voltages at Node 1 and the currents leaving Node 2.

$$[VLG_{abc}] = [A_t] \cdot [VLN_{ABC}] - [B_t] \cdot [I_{abc}] = \begin{bmatrix} 6965.4\underline{/114.0} \\ 6580.6\underline{/-8.6} \\ 6691.4\underline{/-123.3} \end{bmatrix} V$$

These are the same values of the line-to-ground voltages at Node 2 that were determined working from the load toward the source.

Example 8.1 has demonstrated the application of the forward and backward sweep equations. The example also provides verification that the same voltages and currents result working from the load toward the source or from the source toward the load.

In Example 8.2, the system of Example 8.1 is used, only this time the source voltages at Node 1 are specified and the three-phase load is specified as constant PQ. Because this makes the system non-linear, the LIT must be used to solve for the system voltages and currents.

Example 8.2: Use the system of Example 8.1. The source voltages at Node 1 are

$$[ELL_{ABC}] = \begin{bmatrix} 115{,}000\underline{/0} \\ 115{,}000\underline{/-120} \\ 115{,}000\underline{/120} \end{bmatrix}$$

The wye connected loads are

$$[kVA] = \begin{bmatrix} 1700 \\ 1200 \\ 1500 \end{bmatrix} [PF] = \begin{bmatrix} 0.90 \\ 0.85 \\ 0.95 \end{bmatrix}$$

The complex powers of the loads are computed to be

$$SL_i = kVA_i \cdot e^{j \cdot acos(PF_i)} = \begin{bmatrix} 1530 + j741.0 \\ 1020 + j632.1 \\ 1425 + j468.4 \end{bmatrix} kW + jkvar$$

The LIT must be used to analyze the system. A simple Mathcad program is initialized with

$$[I_{start}] = \begin{bmatrix} 0 \\ 0 \\ 0 \end{bmatrix} \quad Tol = 0.000001 \quad VM = \frac{12{,}470}{\sqrt{3}} = 7199.5579$$

The Mathcad program is Figure 8.5:

Three-Phase Transformer Models

$$X := \begin{vmatrix} I_{abc} \leftarrow \text{Start} \\ Iload_{abc} \leftarrow \text{Start} \\ V_{old} \leftarrow \text{Start} \\ ELN_{ABC} \leftarrow W \cdot ELL_{ABC} \\ \text{for } n \in 1..200 \\ \quad \begin{vmatrix} V2LN_{abc} \leftarrow A_t \cdot ELN_{ABC} - B_t \cdot I_{abc} \\ V3LN_{abc} \leftarrow A_{line} \cdot V2LN_{abc} - B_{line} \cdot Iload_{abc} \\ \text{for } j \in 1..3 \\ \quad \begin{vmatrix} Iload_{abc_j} \leftarrow \dfrac{\overline{SL_j \cdot 1000}}{V3LN_{abc_j}} \end{vmatrix} \\ \text{for } k \in 1..3 \\ \quad \begin{vmatrix} Error_k \leftarrow \dfrac{\left| V3LN_{abc_k} - V_{old_k} \right|}{VM} \end{vmatrix} \\ Error_{max} \leftarrow \max(Error) \\ \text{break if } Error_{max} < Tol \\ V_{old} \leftarrow V3LN_{abc} \\ I_{abc} \leftarrow d_{line} \cdot Iload_{abc} \\ I_{ABC} \leftarrow d_t \cdot I_{abc} \end{vmatrix} \\ Out_1 \leftarrow V3LN_{abc} \\ Out_2 \leftarrow V2LN_{abc} \\ Out_3 \leftarrow I_{abc} \\ Out_4 \leftarrow I_{ABC} \\ Out_5 \leftarrow n \\ Out \end{vmatrix}$$

FIGURE 8.5 Example 8.2 Mathcad program.

Note in this routine that in the forward sweep, the secondary transformer voltages are first computed, and then those are used to compute the voltages at the loads. At the end of the routine, the newly calculated line currents are taken back to the top of the routine and used to compute the new voltages. This continues until the error in the difference between the two most recently calculated load voltages is less than the tolerance. As a last step, after conversion, the primary currents of the transformer are computed.

After nine iterations, the load voltages and currents are

$$[VLN_{load}] = \begin{bmatrix} 6490.1/-66.7 \\ 6772.4/\underline{176.2} \\ 6699.4/\underline{53.9} \end{bmatrix},$$

$$[I_{abc}] = \begin{bmatrix} 261.9/-92.5 \\ 177.2/\underline{144.4} \\ 223.9/\underline{35.7} \end{bmatrix}$$

The primary currents are

$$[I_{ABC}] = \begin{bmatrix} 24.3/-70.0 \\ 20.5/-175.2 \\ 27.4/\underline{63.8} \end{bmatrix}$$

The magnitude of the load voltages on a 120-volt base are

$$[Vload_{120}] = \begin{bmatrix} 108.2/-66.6 \\ 112.9/\underline{176.2} \\ 111.7/\underline{53.9} \end{bmatrix}$$

Needless to say, these voltages are not acceptable. To correct this problem three step-voltage regulators connected in wye are installed at the secondary terminals of the substation transformer as shown in Figure 8.6. The voltage level set on the regulator is 120 volts with a bandwidth of 2 volts.

Using the method as outlined in Chapter 7, the initial steps for the three regulators are

$$Tap_i = \frac{|119 - |Vload_{120_i}||}{0.75} = \begin{bmatrix} 14.44 \\ 8.17 \\ 9.79 \end{bmatrix}$$

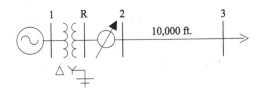

FIGURE 8.6 Voltage regulators installed.

Three-Phase Transformer Models

$$\text{Round off tap}: Tap = \begin{bmatrix} 14 \\ 8 \\ 10 \end{bmatrix}$$

With these tap positions, the load voltages rounded to the nearest volt are

$$Vload_{120} = \begin{bmatrix} 119.8/\underline{-66.2} \\ 118.9/\underline{176.3} \\ 119.7/\underline{54.1} \end{bmatrix} \text{volts}$$

Because the phase b voltage is low, the phase b tap is changed to 9.

$$[Tap] = \begin{bmatrix} 14 \\ 9 \\ 10 \end{bmatrix}$$

The regulator turns ratios are

$$aR_i = 1 - .00625 \cdot Tap_i = \begin{bmatrix} 0.9125 \\ 0.9437 \\ 0.9375 \end{bmatrix}$$

The regulator matrices are

$$[A_{reg}] = [d_{reg}] = \begin{bmatrix} \dfrac{1}{aR_1} & 0 & 0 \\ 0 & \dfrac{1}{aR_2} & 0 \\ 0 & 0 & \dfrac{1}{aR_3} \end{bmatrix} = \begin{bmatrix} 1.0959 & 0 & 0 \\ 0 & 1.0596 & 0 \\ 0 & 0 & 1.0667 \end{bmatrix},$$

$$[B_{reg}] = \begin{bmatrix} 0 & 0 & 0 \\ 0 & 0 & 0 \\ 0 & 0 & 0 \end{bmatrix}$$

At the start of the Mathcad routine, the following equation is added:

$$I_{reg} \leftarrow Start$$

In the Mathcad routine, the first three equations inside the n loop are as follows:

$$VR_{abc} \leftarrow A_t \cdot ELN_{ABC} - B_t \cdot I_{reg}$$

$$V2LN_{abc} \leftarrow A_{reg} \cdot VR_{abc} - B_{reg} \cdot I_{abc}$$

$$V3LN_{abc} \leftarrow A_{line} \cdot V2LN_{abc} - B_{line} \cdot I_{abc}$$

At the end of the loop, the following equations are added:

$$I_{reg} \leftarrow d_{reg} \cdot I_{abc}$$

$$I_{ABC} \leftarrow d_t \cdot I_{reg}$$

With the three regulators installed the load voltages on a 120-volt base are

$$[Vload_{abc}] = \begin{bmatrix} 119.8/\underline{-66.2} \\ 119.7/\underline{176.3} \\ 119.7/\underline{54.1} \end{bmatrix} \text{volts}$$

As has been seen from this example, as more elements of a system are added, there will be one equation for each of the system elements for the forward sweep and backward sweeps. This concept will be further developed in later chapters.

8.4 THE DELTA-GROUNDED WYE STEP-UP CONNECTION

Figure 8.7 shows the connection diagram for the delta-grounded wye step-up connection.

The no-load phasor diagrams for the voltages and currents are also shown in Figure 8.7. Note that the high-side (primary), line-to-line voltage from A to B lags the low-side (secondary), line-to-line voltage from a to b by 30°, and the same can be said for the high- and low-side line currents.

The development of the generalized matrices follows the same procedure as was used for the step-down connection. Only two matrices differ between the two connections.

The primary (low-side) line-to-line voltages are given by

$$\begin{bmatrix} V_{AB} \\ V_{BC} \\ V_{CA} \end{bmatrix} = n_t \cdot \begin{bmatrix} 1 & 0 & 0 \\ 0 & 1 & 0 \\ 0 & 0 & 1 \end{bmatrix} \cdot \begin{bmatrix} Vt_a \\ Vt_b \\ Vt_c \end{bmatrix} \quad [VLL_{ABC}] = [AV] \cdot [Vt_{abc}],$$

Three-Phase Transformer Models

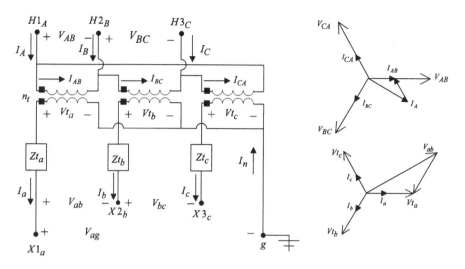

FIGURE 8.7 Delta-grounded wye step-up connection.

$$\text{where}\,[AV] = n_t \cdot \begin{bmatrix} 1 & 0 & 0 \\ 0 & 1 & 0 \\ 0 & 0 & 1 \end{bmatrix},$$

$$n_t = \frac{kVLL_{Rated\ Primary}}{kVLN_{Rated\ Secondary}}. \tag{8.39}$$

The primary delta currents are given by

$$\begin{bmatrix} I_{AB} \\ I_{BC} \\ I_{CA} \end{bmatrix} = \frac{1}{n_t} \cdot \begin{bmatrix} 1 & 0 & 0 \\ 0 & 1 & 0 \\ 0 & 0 & 1 \end{bmatrix} \cdot \begin{bmatrix} I_a \\ I_b \\ I_c \end{bmatrix} \quad [ID_{ABC}] = [AI] \cdot [I_{abc}],$$

$$\text{where}\,[AI] = \frac{1}{n_t} \cdot \begin{bmatrix} 1 & 0 & 0 \\ 0 & 1 & 0 \\ 0 & 0 & 1 \end{bmatrix}. \tag{8.40}$$

The primary line currents are given by

$$\begin{bmatrix} I_A \\ I_B \\ I_C \end{bmatrix} = \begin{bmatrix} 1 & 0 & -1 \\ -1 & 1 & 0 \\ 0 & -1 & 1 \end{bmatrix} \cdot \begin{bmatrix} I_{AB} \\ I_{BC} \\ I_{CA} \end{bmatrix} \quad [I_{ABC}] = [Di] \cdot [ID_{ABC}],$$

where $[Di] = \begin{bmatrix} 1 & 0 & -1 \\ -1 & 1 & 0 \\ 0 & -1 & 1 \end{bmatrix}$ (8.41)

The forward sweep matrices are as follows:
Applying Equation 8.28,

$$[A_t] = AV^{-1} \cdot Di = \frac{1}{n_t} \cdot \begin{bmatrix} 1 & 0 & -1 \\ 0 & 1 & -1 \\ -1 & 0 & 1 \end{bmatrix}$$ (8.42)

Applying Equation 8.31,

$$[B_t] = [Zt_{abc}] = \begin{bmatrix} Zt_{ab} & 0 & 0 \\ 0 & Zt_{bc} & 0 \\ 0 & 0 & Zt_{ca} \end{bmatrix}$$ (8.43)

The backward sweep matrices are as follows:
Applying Equation 8.19,

$$[a_t] = [W] \cdot [AV] = \frac{n_t}{3} \cdot \begin{bmatrix} 2 & 1 & 0 \\ 0 & 2 & 1 \\ 1 & 0 & 2 \end{bmatrix}$$ (8.44)

Applying Equation 8.23,

$$[b_t] = [a_t] \cdot [Zt_{abc}] = \frac{n_t}{3} \cdot \begin{bmatrix} 2 \cdot Zt_{ab} & Zt_{bc} & 0 \\ 0 & 2 \cdot Zt_{bc} & Zt_{ca} \\ Zt_{ab} & 0 & 2 \cdot Zt_{ca} \end{bmatrix}$$ (8.45)

Applying Equation 8.37,

$$[d_t] = [Di] \cdot [AI] = \frac{1}{n_t} \cdot \begin{bmatrix} 1 & 0 & -1 \\ 0 & 1 & -1 \\ -1 & 0 & 1 \end{bmatrix}$$ (8.46)

8.5 THE UNGROUNDED WYE-DELTA STEP-DOWN CONNECTION

Three single-phase transformers can be connected in a wye-delta connection. The neutral of the wye can be grounded or ungrounded.

Three-Phase Transformer Models

The grounded wye connection is characterized by the following:

- The grounded wye provides a path for zero-sequence currents for line-to-ground faults upstream from the transformer bank. This causes the transformers to be susceptible to burnouts on the upstream faults.
- If one phase of the primary circuit is opened, the transformer bank will continue to provide three-phase service by operating as an open wye–open delta bank. However, the two remaining transformers may be subject to an overload condition leading to burnout.

The most common connection is the ungrounded wye-delta. This connection is typically used to provide service to a combination single-phase "lighting" load and a three-phase "power" load such as an induction motor. The generalized constants for the ungrounded wye-delta transformer connection will be developed following the same procedure as was used for the delta – grounded wye.

Three single-phase transformers can be connected in an ungrounded wye "standard 30° step-down connection", as shown in Figure 8.8.

The voltage phasor diagrams in Figure 8.7 illustrate that the high-side positive sequence line-to-line voltage leads the low-side positive sequence line-to-line voltage by 30°. Also, the same phase shift occurs between the high-side, line-to-neutral voltage and the low-side "equivalent" line-to-neutral voltage. The negative sequence phase shift is such that the high-side negative sequence voltage will lag the low-side negative sequence voltage by 30°.

Figure 8.8 illustrates that the positive sequence line current on the high side of the transformer (Node-**n**) leads the low-side line current (Node-**m**) by 30°. It can also be shown that the negative sequence high-side line current will lag the negative sequence low-side line current by 30°.

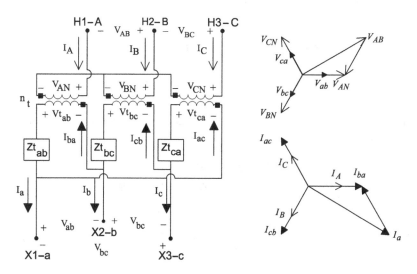

FIGURE 8.8 Standard ungrounded wye-delta connection step-down.

The definition for the "turns ratio n_t" will be the same as Equation 8.9 with the exception that the numerator will be the line-to-neutral voltage, and the denominator will be the line-to-line voltage. It should be noted in Figure 8.7 that the "ideal" low-side transformer voltages for this connection will be line-to-line voltages. Also, the "ideal" low-side currents are the currents flowing inside the delta.

The basic "ideal" transformer voltage and current equations as a function of the "turns ratio" are

$$\begin{bmatrix} V_{AN} \\ V_{BN} \\ V_{CN} \end{bmatrix} = \begin{bmatrix} n_t & 0 & 0 \\ 0 & n_t & 0 \\ 0 & 0 & n_t \end{bmatrix} \cdot \begin{bmatrix} Vt_{ab} \\ Vt_{bc} \\ Vt_{ca} \end{bmatrix}, \qquad (8.47)$$

where

$$n_t = \frac{kVLN_{\text{Rated Primary}}}{kVLL_{\text{Rated Secondary}}},$$

$$[VLN_{ABC}] = [AV] \cdot [Vt_{abc}], \qquad (8.48)$$

$$\begin{bmatrix} I_A \\ I_B \\ I_C \end{bmatrix} = \frac{1}{n_t} \cdot \begin{bmatrix} 1 & 0 & 0 \\ 0 & 1 & 0 \\ 0 & 0 & 1 \end{bmatrix} \cdot \begin{bmatrix} I_{ba} \\ I_{cb} \\ I_{ac} \end{bmatrix}, \qquad (8.49)$$

$$[I_{ABC}] = [AI] \cdot [ID_{abc}],$$

$$[ID_{abc}] = [AI]^{-1} \cdot [I_{ABC}] \qquad (8.50)$$

Solving Equation 8.48 for the "ideal" delta transformer voltages,

$$[Vt_{abc}] = [AV]^{-1} \cdot [VLN_{ABC}] \qquad (8.51)$$

The line-to-line voltages at Node-**m** as a function of the "ideal" transformer voltages and the delta currents are given by

$$\begin{bmatrix} V_{ab} \\ V_{bc} \\ V_{ca} \end{bmatrix} = \begin{bmatrix} Vt_{ab} \\ Vt_{bc} \\ Vt_{ca} \end{bmatrix} - \begin{bmatrix} Zt_{ab} & 0 & 0 \\ 0 & Zt_{bc} & 0 \\ 0 & 0 & Zt_{ca} \end{bmatrix} \cdot \begin{bmatrix} ID_{ba} \\ ID_{cb} \\ ID_{ac} \end{bmatrix}, \qquad (8.52)$$

$$[VLL_{abc}] = [Vt_{abc}] - [Zt_{abc}] \cdot [ID_{abc}] \qquad (8.53)$$

Substitute Equations 8.50 and 8.51 into Equation 8.53.

Three-Phase Transformer Models

$$[VLL_{abc}] = [AV]^{-1} \cdot [VLN_{ABC}] - [ZNt_{abc}] \cdot [I_{ABC}], \qquad (8.54)$$

where

$$[ZNt_{abc}] = [Zt_{abc}] \cdot [AI]^{-1} = \begin{bmatrix} n_t \cdot Zt_{ab} & 0 & 0 \\ 0 & n_t \cdot Zt_{bc} & 0 \\ 0 & 0 & n_t \cdot Zt_{ca} \end{bmatrix} \qquad (8.55)$$

The line currents on the delta side of the transformer bank as a function of the wye-transformer currents are given by

$$[I_{abc}] = [Di] \cdot [ID_{abc}], \qquad (8.56)$$

where

$$[Di] = \begin{bmatrix} 1 & 0 & -1 \\ -1 & 1 & 0 \\ 0 & -1 & 1 \end{bmatrix} \qquad (8.57)$$

Substitute Equation 8.50 into Equation 8.56.

$$[I_{abc}] = [Di] \cdot [AI]^{-1} \cdot [I_{ABC}] = [DY] \cdot [I_{ABC}], \qquad (8.58)$$

where

$$[DY] = [Di] \cdot [AI]^{-1} = \begin{bmatrix} n_t & 0 & -n_t \\ -n_t & n_t & 0 \\ 0 & -n_t & n_t \end{bmatrix} \qquad (8.59)$$

Because the matrix [Di] is singular, it is not possible to use Equation 8.56 to develop an equation relating the wye-side line currents at Node-**n** to the delta-side line currents at Node-**m**. To develop the necessary matrix equation, three independent equations must be written. Two independent KCL equations at the vertices of the delta can be used. Because there is no path for the high-side currents to flow to ground, they must sum to zero and, therefore, so must the delta currents in the transformer secondary sum to zero. This provides the third independent equation. The resulting three independent equations in matrix form are given by

$$\begin{bmatrix} I_a \\ I_b \\ 0 \end{bmatrix} = \begin{bmatrix} 1 & 0 & -1 \\ -1 & 1 & 0 \\ 1 & 1 & 1 \end{bmatrix} \cdot \begin{bmatrix} I_{ba} \\ I_{cb} \\ I_{ac} \end{bmatrix} \qquad (8.60)$$

Solving Equation 8.60 for the delta currents.

$$\begin{bmatrix} I_{ba} \\ I_{cb} \\ I_{ac} \end{bmatrix} = \begin{bmatrix} 1 & 0 & -1 \\ -1 & 1 & 0 \\ 1 & 1 & 1 \end{bmatrix}^{-1} \cdot \begin{bmatrix} I_a \\ I_b \\ 0 \end{bmatrix} = \frac{1}{3} \cdot \begin{bmatrix} 1 & -1 & 1 \\ 1 & 2 & 1 \\ -2 & -1 & 1 \end{bmatrix} \cdot \begin{bmatrix} I_a \\ I_b \\ 0 \end{bmatrix} \quad (8.61)$$

$$[ID_{abc}] = [L0] \cdot [I_{ab0}] \quad (8.62)$$

Equation 8.62 can be modified to include the phase c current by setting the third column of the [L0] matrix to zero.

$$\begin{bmatrix} I_{ba} \\ I_{cb} \\ I_{ac} \end{bmatrix} = \frac{1}{3} \cdot \begin{bmatrix} 1 & -1 & 0 \\ 1 & 2 & 0 \\ -2 & -1 & 0 \end{bmatrix} \cdot \begin{bmatrix} I_a \\ I_b \\ I_c \end{bmatrix} \quad (8.63)$$

$$[ID_{abc}] = [L] \cdot [I_{abc}] \quad (8.64)$$

Solve Equation 8.50 for $[I_{ABC}]$ and substitute into Equation 8.64.

$$[I_{ABC}] = [AI] \cdot [L] \cdot [I_{abc}] = [d_t] \cdot [I_{abc}], \quad (8.65)$$

where

$$[d_t] = [AI] \cdot [L] = \frac{1}{3 \cdot n_t} \cdot \begin{bmatrix} 1 & -1 & 0 \\ 1 & 2 & 0 \\ -2 & -1 & 0 \end{bmatrix} \quad (8.66)$$

Equation 8.66 defines the generalized constant matrix $[d_t]$ for the ungrounded wye-delta step-down transformer connection. In the process of the derivation, a very convenient equation (8.63) evolved that can be used anytime the currents in a delta need to be determined knowing the line currents. However, it must be understood that this equation will only work when the delta currents sum to zero, which means an ungrounded neutral on the primary.

The generalized matrices $[a_t]$ and $[b_t]$ can now be developed. Solve Equation 8.54 for [VLN_{ABC}].

$$[VLN_{ABC}] = [AV] \cdot [VLL_{abc}] + [AV] \cdot [ZNt_{abc}] \cdot [I_{ABC}] \quad (8.67)$$

Substitute Equation 8.65 into Equation 8.67.

$$[VLN_{ABC}] = [AV] \cdot [VLL_{abc}] + [AV] \cdot [ZNt_{abc}] \cdot [d_t] \cdot [I_{abc}],$$

Three-Phase Transformer Models

$$[VLL_{abc}] = [Dv] \cdot [VLN_{abc}],$$

where $[Dv] = \begin{bmatrix} 1 & -1 & 0 \\ 0 & 1 & -1 \\ -1 & 0 & 1 \end{bmatrix},$

$$[VLN_{ABC}] = [AV] \cdot [Dv] \cdot [VLN_{abc}] + [AV] \cdot [ZNt_{abc}] \cdot [d_t] \cdot [I_{abc}],$$

$$[VLN_{ABC}] = [a_t] \cdot [VLN_{abc}] + [b_t] \cdot [I_{abc}], \tag{8.68}$$

where

$$[a_t] = [AV] \cdot [Dv] = n_t \cdot \begin{bmatrix} 1 & -1 & 0 \\ 0 & 1 & -1 \\ -1 & 0 & 1 \end{bmatrix}, \tag{8.69}$$

$$[b_t] = [AV] \cdot [ZNt_{abc}] \cdot [d_t] = \frac{n_t}{3} \cdot \begin{bmatrix} Zt_{ab} & -Zt_{ab} & 0 \\ Zt_{bc} & 2 \cdot Zt_{bc} & 0 \\ -2 \cdot Zt_{ca} & -Zt_{ca} & 0 \end{bmatrix} \tag{8.70}$$

The generalized constant matrices have been developed for computing voltages and currents from the load toward the source (backward sweep). The forward sweep matrices can be developed by referring back to Equation 8.54, which is repeated here for convenience.

$$[VLL_{abc}] = [AV]^{-1} \cdot [VLN_{ABC}] - [ZNt_{abc}] \cdot [I_{ABC}] \tag{8.71}$$

Equation 8.16 is used to compute the equivalent line-to-neutral voltages as a function of the line-to-line voltages.

$$[VLN_{abc}] = [W] \cdot [VLL_{abc}] \tag{8.72}$$

Substitute Equation 8.71 into Equation 8.72.

$$[VLN_{abc}] = [W] \cdot [AV]^{-1} \cdot [VLN_{ABC}] - [W] \cdot [ZNt_{abc}] \cdot [d_t] \cdot [I_{abc}],$$

$$[VLN_{abc}] = [A_t] \cdot [VLN_{ABC}] - [B_t] \cdot [I_{abc}], \tag{8.73}$$

where

$$[A_t] = [W] \cdot [AV]^{-1} = \frac{1}{3 \cdot n_t} \cdot \begin{bmatrix} 2 & 1 & 0 \\ 0 & 2 & 1 \\ 1 & 0 & 2 \end{bmatrix}, \quad (8.74)$$

$$[B_t] = [W] \cdot [ZNt_{abc}] \cdot [d_t] = \frac{1}{9} \cdot \begin{bmatrix} 2 \cdot Zt_{ab} + Zt_{bc} & 2 \cdot Zt_{bc} - 2 \cdot Zt_{ab} & 0 \\ 2 \cdot Zt_{bc} - 2 \cdot Zt_{ca} & 4 \cdot Zt_{bc} - Zt_{ca} & 0 \\ Zt_{ab} - 4 \cdot Zt_{ca} & -Zt_{ab} - 2 \cdot Zt_{ca} & 0 \end{bmatrix} \quad (8.75)$$

The generalized matrices have been developed for the ungrounded wye-delta transformer connection. The derivation has applied basic circuit theory and the basic theories of transformers. The result of the derivations is to provide an easy method of analyzing the operating characteristics of the transformer connection. Example 8.3 will demonstrate the application of the generalized matrices for this transformer connection.

Example 8.3: An ungrounded wye-delta step-down transformer feeding a three-phase, delta-connected load.

Figure 8.9 shows three single-phase transformers in an ungrounded wye-delta step-down connection serving a combination single-phase and three-phase load in a delta connection. The voltages at the load are a balanced three phase of 240 volts line-to-line. The net loading by phase is

S_{ab} = 100 kVA at 0.9 lagging power factor,
$S_{bc} = S_{ca}$ = 50 kVA at 0.8 lagging power factor.

The transformers are rated as follows:

Phase A-N: 100 kVA, 7200 – 240 volts, Z = 0.01 + j0.04 per-unit
Phases B-N and C-N: 50 kVA, 7200 – 240 volts, Z = 0.015 + j0.035 per-unit

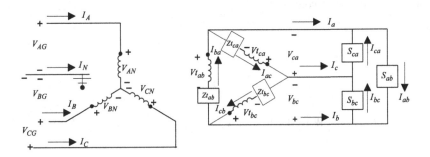

FIGURE 8.9 Ungrounded wye-delta step-down with unbalanced load.

Three-Phase Transformer Models

Determine the following:

1. the currents in the load,
2. the secondary line currents,
3. the equivalent line-to-neutral secondary voltages,
4. the primary line-to-neutral and line-to-line voltages, and
5. the primary line currents.

Before the analysis can start, the transformer impedances must be converted to actual values in ohms and located inside the delta-connected secondary windings.

"Lighting" transformer:

$$Z_{base} = \frac{0.24^2 \cdot 1000}{100} = 0.576,$$

$$Zt_{ab} = (0.01 + j0.4) \cdot 0.576 = 0.0058 + j.023 \, \Omega.$$

"Power" transformers:

$$Z_{base} = \frac{0.24^2 \cdot 1000}{50} = 1.152,$$

$$Zt_{bc} = Zt_{ca} = (0.015 + j0.35) \cdot 1.152 = 0.0173 + j0.0403 \, \Omega.$$

The transformer impedance matrix can now be defined:

$$[Zt_{abc}] = \begin{bmatrix} .0058 + j.023 & 0 & 0 \\ 0 & .0173 + j.0403 & 0 \\ 0 & 0 & .0173 + j.0403 \end{bmatrix} \Omega$$

The turns ratio of the transformers is $n_t = \frac{7200}{240} = 30$. Define all of the matrices.

$$[W] = \frac{1}{3} \cdot \begin{bmatrix} 2 & 1 & 0 \\ 0 & 2 & 1 \\ 1 & 0 & 2 \end{bmatrix} \quad [Dv] = \begin{bmatrix} 1 & -1 & 0 \\ 0 & 1 & -1 \\ -1 & 0 & 1 \end{bmatrix} \quad [Di] = \begin{bmatrix} 1 & 0 & -1 \\ -1 & 1 & 0 \\ 0 & -1 & 1 \end{bmatrix}$$

$$[a_t] = n_t \cdot \begin{bmatrix} 1 & -1 & 0 \\ 0 & 1 & -1 \\ -1 & 0 & 1 \end{bmatrix} = \begin{bmatrix} 30 & -30 & 0 \\ 0 & 30 & -30 \\ -30 & 0 & 30 \end{bmatrix}$$

$$[b_t] = \frac{n_t}{3} \cdot \begin{bmatrix} Zt_{ab} & -Zt_{ab} & 0 \\ Zt_{bc} & 2 \cdot Zt_{bc} & 0 \\ -2 \cdot Zt_{ca} & -Zt_{ca} & 0 \end{bmatrix} = \begin{bmatrix} 0.0576 + j.2304 & -0.576 - j.2304 & 0 \\ 0.1728 + j.4032 & 0.3456 + j.8064 & 0 \\ -0.3456 - j.8064 & -0.1728 - j.4032 & 0 \end{bmatrix}$$

$$[c_t] = \begin{bmatrix} 0 & 0 & 0 \\ 0 & 0 & 0 \\ 0 & 0 & 0 \end{bmatrix}$$

$$[d_t] = \frac{1}{3 \cdot n_T} \cdot \begin{bmatrix} 1 & -1 & 0 \\ 1 & 2 & 0 \\ -2 & -1 & 0 \end{bmatrix} = \begin{bmatrix} 0.0111 & -0.0111 & 0 \\ 0.0111 & 0.0222 & 0 \\ -0.0222 & -0.0111 & 0 \end{bmatrix}$$

$$[A_t] = \frac{1}{3 \cdot n_t} \cdot \begin{bmatrix} 2 & 1 & 0 \\ 0 & 2 & 1 \\ 1 & 0 & 2 \end{bmatrix} = \begin{bmatrix} .0222 & .0111 & 0 \\ 0 & .0222 & .0111 \\ .0111 & 0 & .0222 \end{bmatrix}$$

$$[B_t] = \frac{1}{9} \cdot \begin{bmatrix} 2 \cdot Zt_{ab} + Zt_{bc} & 2 \cdot Zt_{bc} - 2 \cdot Zt_{ab} & 0 \\ 2 \cdot Zt_{bc} - 2 \cdot Zt_{ca} & 4 \cdot Zt_{bc} - Zt_{ca} & 0 \\ Zt_{ab} - 4 \cdot Zt_{ca} & -Zt_{ab} - 2 \cdot Zt_{ca} & 0 \end{bmatrix}$$

$$[B_t] = \begin{bmatrix} 0.0032 + j.0096 & 0.0026 + j.0038 & 0 \\ 0 & 0.0058 + j.0134 & 0 \\ -0.007 - j.0154 & -0.0045 - j.0115 & 0 \end{bmatrix}$$

Define the line-to-line load voltages.

$$[VLL_{abc}] = \begin{bmatrix} 240/\underline{0} \\ 240/\underline{-120} \\ 240/\underline{120} \end{bmatrix} V$$

Define the loads.

$$[SD_{abc}] = \begin{bmatrix} 100/\underline{acos(0.9)} \\ 50/\underline{acos(0.8)} \\ 50/\underline{acos(0.8)} \end{bmatrix} = \begin{bmatrix} 90 + j43.589 \\ 40 + j30 \\ 40 + j30 \end{bmatrix} kVA$$

Calculate the delta load currents.

$$ID_i = \left(\frac{SD_i \cdot 1000}{VLL_{abc_i}} \right)^* A$$

$$[ID_{abc}] = \begin{bmatrix} I_{ab} \\ I_{bc} \\ I_{ca} \end{bmatrix} = \begin{bmatrix} 416.7/\underline{-25.84} \\ 208.3/\underline{-156.87} \\ 208.3/\underline{83.13} \end{bmatrix} A$$

Three-Phase Transformer Models

Compute the secondary line currents.

$$[I_{abc}] = [Di] \cdot [ID_{abc}] = \begin{bmatrix} 522.9/-47.97 \\ 575.3/-119.06 \\ 360.8/53.13 \end{bmatrix} A$$

Compute the equivalent secondary line-to-neutral voltages.

$$[VLN_{abc}] = [W] \cdot [VLL_{abc}] = \begin{bmatrix} 138.56/-30 \\ 138.56/-150 \\ 138.56/90 \end{bmatrix} A$$

Use the generalized constant matrices to compute the primary line-to-neutral voltages and line-to-line voltages.

$$[VLN_{ABC}] = [a_t] \cdot [VLN_{abc}] + [b_t] \cdot [I_{abc}] = \begin{bmatrix} 7367.6/1.4 \\ 7532.3/-119.1 \\ 7406.2/121.7 \end{bmatrix} V$$

$$[VLL_{ABC}] = [Dv] \cdot [VLN_{ABC}] = \begin{bmatrix} 12,9356/31.54 \bullet \\ 12,8845/-88.95 \\ 12,8147/151.50 \end{bmatrix} kV$$

The high primary line currents are

$$[I_{ABC}] = [d_t] \cdot [I_{abc}] = \begin{bmatrix} 11.54/-28.04 \\ 8.95/-166.43 \\ 7.68/101.16 \end{bmatrix} A.$$

It is interesting to compute the operating kVA of the three transformers. Taking the product of the transformer voltage times the conjugate of the current gives the operating kVA of each transformer.

$$ST_i = \frac{VLN_{ABC_i} \cdot (I_{ABC_i})^*}{1000} = \begin{bmatrix} 85.02/29.46 \\ 67.42/47.37 \\ 56.80/20.58 \end{bmatrix} kVA$$

The operating power factors of the three transformers are

$$[PF] = \begin{bmatrix} \cos(29.46) \\ \cos(47.37) \\ \cos(20.58) \end{bmatrix} = \begin{bmatrix} 87.1 \\ 67.7 \\ 93.6 \end{bmatrix} \%.$$

Note that the operating kVAs do not match very closely the rated kVAs of the three transformers. In particular, the transformer on phase A did not serve the total load of 100 kVA that is directly connected to its terminals. That transformer is operating below rated kVA, while the other two transformers are overloaded. In fact, the transformer connected to phase B is operating 35% above rated kVA. Because of this overload, the ratings of the three transformers should be changed so that the phase B and C transformers are rated 75 kVA. Finally, the operating power factors of the three transformers bear little resemblance to the load power factors.

Example 8.3 demonstrates how the generalized constant matrices can be used to determine the operating characteristics of the transformers. In addition, the example shows that the obvious selection of transformer ratings will lead to an overload condition on the two power transformers. The beauty in this is that if the generalized constant matrices have been applied in a computer program, it is a simple task to change the transformer kVA ratings and be assured that none of the transformers will be operating in an overload condition.

Example 8.3 has demonstrated the "backward" sweep to compute the primary voltages and currents. As before when the source (primary) voltages are given along with the load PQ, the LIT must be used to analyze the transformer connection.

Example 8.4: The Mathcad program that has been used in previous examples is modified to demonstrate the LIT for computing the load voltages given the source voltages and load power and reactive powers (PQ load). In Example 8.4, the computed source voltages from Example 8.3 are specified, along with the same loads. From Example 8.3, the source voltages are

$$[VLL_{ABC}] = \begin{bmatrix} 12{,}935.6\underline{/31.5} \\ 12{,}884.5\underline{/-88.9} \\ 12{,}814.7\underline{/151.5} \end{bmatrix}$$

The initial conditions are

$$[Start] = \begin{bmatrix} 0 \\ 0 \\ 0 \end{bmatrix} \quad Tol = 0.000001 \quad VM = 240$$

The modified Mathcad program is shown in Figure 8.10.

With balanced source voltages specified, after six iterations, the load voltages are computed to be exactly what they were, as specified in Example 8.3:

$$[VLL_{abc}] = \begin{bmatrix} 240\underline{/0} \\ 240\underline{/-120} \\ 240.0\underline{/120} \end{bmatrix} V$$

Three-Phase Transformer Models

$$X := \begin{vmatrix} I_{abc} \leftarrow \text{Start} \\ V_{old} \leftarrow \text{Start} \\ VLN_{ABC} \leftarrow W \cdot VLL_{ABC} \\ \text{for } n \in 1..200 \\ \quad \begin{vmatrix} VLN_{abc} \leftarrow A_t \cdot VLN_{ABC} - B_t \cdot I_{abc} \\ VLL_{abc} \leftarrow Dv \cdot VLN_{abc} \\ \text{for } j \in 1..3 \\ \quad ID_{abc_j} \leftarrow \dfrac{\overline{SL_j \cdot 1000}}{VLL_{abc_j}} \\ \text{for } k \in 1..3 \\ \quad Error_k \leftarrow \dfrac{|VLL_{abc_k} - V_{old_k}|}{VM} \\ Error_{max} \leftarrow \max(Error) \\ \text{break if } Error_{max} < Tol \\ V_{old} \leftarrow VLL_{abc} \\ I_{abc} \leftarrow Di \cdot ID_{abc} \\ I_{ABC} \leftarrow d_t \cdot I_{abc} \end{vmatrix} \\ Out_1 \leftarrow VLN_{abc} \\ Out_2 \leftarrow VLL_{abc} \\ Out_3 \leftarrow I_{abc} \\ Out_4 \leftarrow I_{ABC} \\ Out_5 \leftarrow n \\ Out \end{vmatrix}$$

FIGURE 8.10 Example 8.4 Mathcad program.

Example 8.4 has demonstrated how the simple Mathcad program can be modified to analyze the ungrounded wye-delta step-down transformer bank connection.

8.6 THE UNGROUNDED WYE-DELTA STEP-UP CONNECTION

The connection diagram for the step-up connection is shown in Figure 8.11

The only difference in the matrices between the step-up and step-down connections are the definitions of the turns ratio n_t, [AV] and [AI]. For the step-up connection,

$$n_t = \frac{kVLN_{Rated\ Primary}}{kVLL_{Rated\ Secondary}}, \tag{8.76}$$

$$\begin{bmatrix} V_{AN} \\ V_{BN} \\ V_{CN} \end{bmatrix} = n_t \cdot \begin{bmatrix} 0 & 0 & -1 \\ -1 & 0 & 0 \\ 0 & -1 & 0 \end{bmatrix} \cdot \begin{bmatrix} Vt_{ab} \\ Vt_{bc} \\ Vt_{ca} \end{bmatrix},$$

$$[VLN_{ABC}] = [AV] \cdot [Vt_{abc}],$$

$$\text{where } [AV] = n_t \cdot \begin{bmatrix} 0 & 0 & -1 \\ -1 & 0 & 0 \\ 0 & -1 & 0 \end{bmatrix}, \tag{8.77}$$

$$\begin{bmatrix} I_A \\ I_B \\ I_C \end{bmatrix} = \frac{1}{n_t} \cdot \begin{bmatrix} 0 & 0 & -1 \\ -1 & 0 & 0 \\ 0 & -1 & 0 \end{bmatrix} \cdot \begin{bmatrix} ID_{ba} \\ ID_{cb} \\ ID_{ac} \end{bmatrix},$$

$$[I_{ABC}] = [AI] \cdot [ID_{abc}],$$

$$\text{where } [AI] = \frac{1}{n_t} \cdot \begin{bmatrix} 0 & 0 & -1 \\ -1 & 0 & 0 \\ 0 & -1 & 0 \end{bmatrix} \tag{8.78}$$

8.7 THE GROUNDED WYE-DELTA STEP-DOWN CONNECTION

The connection diagram for the standard 30° grounded wye (high) – delta (low) transformer connection grounded through an impedance of Z_g is shown in Figure 8.12. Note that the primary is grounded through an impedance of Z_g.

Basic Transformer Equations:
The turns ratio is given by

$$n_t = \frac{kVLN_{Rated\ Primary}}{kVLL_{Rated\ Secondary}} \tag{8.79}$$

The basic "ideal" transformer voltage and current equations as a function of the turns ratio are

$$\begin{bmatrix} V_{AN} \\ V_{BN} \\ V_{CN} \end{bmatrix} = n_t \cdot \begin{bmatrix} 1 & 0 & 0 \\ 0 & 1 & 0 \\ 0 & 0 & 1 \end{bmatrix} \cdot \begin{bmatrix} Vt_{ab} \\ Vt_{bc} \\ Vt_{ca} \end{bmatrix},$$

$$[VLN_{ABC}] = [AV] \cdot [Vt_{abc}],$$

Three-Phase Transformer Models

Example 8.5: The equations for the forward and backward sweep matrices, as defined in Section 8.3, can be applied using the definitions in Equations 8.76, 8.77, and 8.78. The system of Example 8.3 is modified so that the transformer connection is step-up. The transformers have the same ratings, but now the rated voltages for the primary and secondary are as follows:

$$\text{Primary: } VLL_{pri} = 240 \, VLN_{pri} = 138.6 \text{ Volts}$$

$$\text{Secondary: } VLL \text{ Volts}_{sec}$$

$$n_t = \frac{VLN_{pri}}{VLL_{sec}} = \frac{138.6}{12470} = 0.0111$$

The transformer impedances must be computed in ohms relative to the delta secondary and then used to compute the new forward and backward sweep matrices. When this is done, the new matrices are as follows:

$$[d_t] = [AI]^{-1} \cdot [L] = \begin{bmatrix} 60 & 30 & 0 \\ -30 & 30 & 0 \\ -30 & -60 & 0 \end{bmatrix}$$

$$[A_t] = [W] \cdot [AV]^{-1} = \begin{bmatrix} 0 & -60 & -30 \\ -30 & 0 & -60 \\ -60 & -30 & 0 \end{bmatrix}$$

$$[B_t] = [W] \cdot [ZNt_{abc}] \cdot [d_t] = \begin{bmatrix} 8.64 + j25.92 & 6.91 + j10.37 & 0 \\ 0 & 15.55 + j36.28 & 0 \\ -19.01 - j41.47 & -12.09 - j31.10 & 0 \end{bmatrix}$$

Using these matrices and the same loads, specify the primary line-to-line voltages to be

$$[VLL_{ABC}] = \begin{bmatrix} 240/\underline{30} \\ 240/\underline{-90} \\ 240/\underline{150} \end{bmatrix}$$

Using the Mathcad program of Figure 8.9, the LIT computes the load voltages as

$$[VLL_{abc}] = \begin{bmatrix} 12{,}055/\underline{58.2} \\ 11{,}982/\underline{-61.3} \\ 12{,}106/\underline{178.7} \end{bmatrix}$$

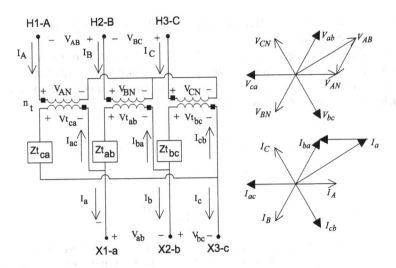

FIGURE 8.11 Ungrounded wye-delta step-up connection.

$$\text{where}\left[AV\right] = n_t \cdot \begin{bmatrix} 1 & 0 & 0 \\ 0 & 1 & 0 \\ 0 & 0 & 1 \end{bmatrix}, \tag{8.80}$$

$$\begin{bmatrix} I_A \\ I_B \\ I_C \end{bmatrix} = \frac{1}{n_t} \cdot \begin{bmatrix} 1 & 0 & 0 \\ 0 & 1 & 0 \\ 0 & 0 & 1 \end{bmatrix} \cdot \begin{bmatrix} I_{ba} \\ I_{cb} \\ I_{ac} \end{bmatrix},$$

$$\left[I_{ABC}\right] = \left[AI\right] \cdot \left[ID_{abc}\right]$$

$$\left[ID_{abc}\right] = \left[AI\right]^{-1} \cdot \left[I_{ABC}\right],$$

$$\text{where}\left[AI\right] = \frac{1}{n_t} \cdot \begin{bmatrix} 1 & 0 & 0 \\ 0 & 1 & 0 \\ 0 & 0 & 1 \end{bmatrix} \tag{8.81}$$

Solving Equation 8.80 for the "ideal" transformer voltages,

$$\left[Vt_{abc}\right] = \left[AV\right]^{-1} \cdot \left[VLN_{ABC}\right] \tag{8.82}$$

The line-to-neutral transformer primary voltages as a function of the system line-to-ground voltages are given by the following:

$$V_{AN} = V_{AG} - Z_g \cdot \left(I_A + I_B + I_C\right),$$

Three-Phase Transformer Models

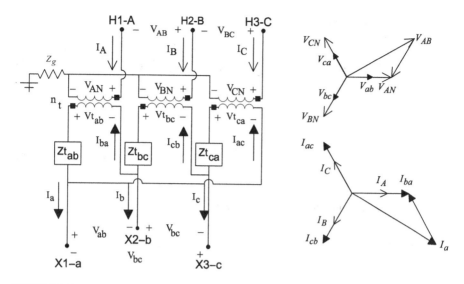

FIGURE 8.12 The grounded wye-delta step-down connection

$$V_{BN} = V_{BG} - Z_g \cdot (I_A + I_B + I_C),$$

$$V_{CN} = V_{CG} - Z_g \cdot (I_A + I_B + I_C),$$

$$\begin{bmatrix} V_{AN} \\ V_{BN} \\ V_{CN} \end{bmatrix} = \begin{bmatrix} V_{AG} \\ V_{BG} \\ V_{CG} \end{bmatrix} - \begin{bmatrix} Z_g & Z_g & Z_g \\ Z_g & Z_g & Z_g \\ Z_g & Z_g & Z_g \end{bmatrix} \cdot \begin{bmatrix} I_A \\ I_B \\ I_C \end{bmatrix},$$

$$[VLN_{ABC}] = [VLG_{ABC}] - [ZG] \cdot [I_{ABC}],$$

$$\text{where } [ZG] = \begin{bmatrix} Z_g & Z_g & Z_g \\ Z_g & Z_g & Z_g \\ Z_g & Z_g & Z_g \end{bmatrix} \quad (8.83)$$

The line-to-line voltages on the delta side are given by

$$\begin{bmatrix} V_{ab} \\ V_{bc} \\ V_{ca} \end{bmatrix} = \begin{bmatrix} Vt_{ab} \\ Vt_{bc} \\ Vt_{ca} \end{bmatrix} - \begin{bmatrix} Zt_{ab} & 0 & 0 \\ 0 & Zt_{bc} & 0 \\ 0 & 0 & Zt_{ca} \end{bmatrix} \cdot \begin{bmatrix} ID_{ba} \\ ID_{cb} \\ ID_{ac} \end{bmatrix},$$

$$[VLL_{abc}] = [Vt_{abc}] - [Zt_{abc}] \cdot [ID_{abc}] \quad (8.84)$$

Substitute Equation 8.82 into 8.84.

$$[VLL_{abc}] = [AV]^{-1} \cdot [VLN_{ABC}] - [Zt_{abc}] \cdot [ID_{abc}] \quad (8.85)$$

Substitute Equation 8.81 into 8.85.

$$[VLL_{abc}] = [AV]^{-1} \cdot [VLN_{ABC}] - ([Zt_{abc}] \cdot [AI]^{-1}) \cdot [I_{ABC}] \quad (8.86)$$

Substitute Equation 8.83 into Equation 8.86.

$$[VLL_{abc}] = [AV]^{-1} \cdot ([VLG_{ABC}] - [ZG] \cdot [I_{ABC}]) - ([Zt_{abc}] \cdot [AI]^{-1}) \cdot [I_{ABC}]$$

$$[VLL_{abc}] = [AV]^{-1} \cdot [VLG_{ABC}] - ([AV]^{-1} \cdot [ZG] + [Zt_{abc}] \cdot [AI]^{-1}) \cdot [I_{ABC}] \quad (8.87)$$

Equation 8.81 gives the delta secondary currents as a function of the primary wye-side line currents. The secondary line currents are related to the secondary delta currents by

$$\begin{bmatrix} I_a \\ I_b \\ I_c \end{bmatrix} = \begin{bmatrix} 1 & 0 & -1 \\ -1 & 1 & 0 \\ 0 & -1 & 1 \end{bmatrix} \cdot \begin{bmatrix} ID_{ba} \\ ID_{cb} \\ ID_{ac} \end{bmatrix},$$

$$[I_{abc}] = [Di] \cdot [ID_{abc}]. \quad (8.88)$$

The real problem of transforming currents from one side to the other occurs in the case when the line currents on the delta secondary side $[I_{abc}]$ are known and the transformer secondary currents $[ID_{abc}]$ and primary line currents on the wye side $[I_{ABC}]$ are needed. The only way a relationship can be developed is to recognize that the sum of the line-to-line voltages on the delta secondary of the transformer bank must add up to zero. Three independent equations can be written as follows:

$$I_a = I_{ba} - I_{ac}$$

$$I_b = I_{cb} - I_{ba} \quad (8.89)$$

KVL around the delta secondary windings gives

$$Vt_{ab} - Zt_{ab} \cdot I_{ba} + Vt_{bc} - Zt_{bc} \cdot I_{cb} + Vt_{ca} - Zt_{ca} \cdot I_{ac} = 0 \quad (8.90)$$

Three-Phase Transformer Models

Replacing the "ideal" secondary delta voltages with the primary line-to-neutral voltages,

$$\frac{V_{AN}}{n_t} + \frac{V_{BN}}{n_t} + \frac{V_{CN}}{n_t} = Zt_{ab} \cdot I_{ba} + Zt_{bc} \cdot I_{cb} + Zt_{ca} \cdot I_{ac} \qquad (8.91)$$

Multiply both sides of Equation 8.91 by the turns ratio n_t.

$$V_{AN} + V_{BN} + V_{CN} = n_t \cdot Zt_{ab} \cdot I_{ba} + n_t \cdot Zt_{bc} \cdot I_{cb} + n_t \cdot Zt_{ca} \cdot I_{ac} \qquad (8.92)$$

Determine the left side of Equation 8.92 as a function of the line-to-ground voltages using Equation 8.83.

$$V_{AN} + V_{BN} + V_{CN} = V_{AG} + V_{BG} + V_{CG} - 3 \cdot Z_g \cdot (I_A + I_B + I_C)$$

$$V_{AN} + V_{BN} + V_{CN} = V_{AG} + V_{BG} + V_{CG} - 3 \cdot \frac{1}{n_t} \cdot Z_g \cdot (I_{ba} + I_{cb} + I_{ac}) \qquad (8.93)$$

Substitute Equation 8.93 into Equation 8.92.

$$V_{AG} + V_{BG} + V_{CG} - \frac{3}{n_t} \cdot Z_g \cdot (I_{ba} + I_{cb} + I_{ac}) = n_t \cdot Zt_{ab} \cdot I_{ba} + n_t \cdot Zt_{bc} \cdot I_{cb} + n_t \cdot Zt_{ca} \cdot I_{ac},$$

$$Vsum = \left(n_t \cdot Zt_{ab} + \frac{3}{n_t} \cdot Z_g\right) \cdot I_{ba} + \left(n_t \cdot Zt_{bc} + \frac{3}{n_t} \cdot Z_g\right) \cdot I_{cb} + \left(n_t \cdot Zt_{ca} + \frac{3}{n_t} \cdot Z_g\right) \cdot I_{ac},$$

$$\text{where } V_{sum} = V_{AG} + V_{BG} + V_{CG} \qquad (8.94)$$

Equations 8.88, 8.89, and 8.94 can be put into matrix form.

$$\begin{bmatrix} I_a \\ I_b \\ V_{sum} \end{bmatrix} = \begin{bmatrix} 1 & 0 & -1 \\ -1 & 1 & 0 \\ n_t \cdot Zt_{ab} + \frac{3}{n_t} \cdot Z_g & n_t \cdot Zt_{bc} + \frac{3}{n_t} \cdot Z_g & n_t \cdot Zt_{ca} + \frac{3}{n_t} \cdot Z_g \end{bmatrix} \cdot \begin{bmatrix} I_{ba} \\ I_{cb} \\ I_{ac} \end{bmatrix} \qquad (8.95)$$

Equation 8.95 in general form is

$$[X] = [F] \cdot [ID_{abc}] \qquad (8.96)$$

Solve for $[ID_{abc}]$.

$$[ID_{abc}] = [F]^{-1} \cdot [X] = [G] \cdot [X] \qquad (8.97)$$

Equation 8.97 in full form is

$$[ID_{abc}] = \begin{bmatrix} G_{11} & G_{12} & G_{13} \\ G_{21} & G_{22} & G_{23} \\ G_{31} & G_{32} & G_{33} \end{bmatrix} \cdot \begin{bmatrix} I_a \\ I_b \\ V_{AG} + V_{BG} + V_{CG} \end{bmatrix},$$

$$[ID_{abc}] = \begin{bmatrix} G_{13} & G_{13} & G_{13} \\ G_{23} & G_{23} & G_{23} \\ G_{33} & G_{33} & G_{33} \end{bmatrix} \cdot \begin{bmatrix} V_{AG} \\ V_{BG} \\ V_{CG} \end{bmatrix} + \begin{bmatrix} G_{11} & G_{12} & 0 \\ G_{21} & G_{22} & 0 \\ G_{31} & G_{32} & 0 \end{bmatrix} \cdot \begin{bmatrix} I_a \\ I_b \\ I_c \end{bmatrix} \quad (8.98)$$

Equation 8.98 in shorten form is

$$[ID_{abc}] = [G1] \cdot [VLG_{ABC}] + [G2] \cdot [I_{abc}]. \quad (8.99)$$

Substitute Equation 8.81 into Equation 8.98.

$$[I_{ABC}] = [AI] \cdot [ID_{abc}] = [AI] \cdot ([G1] \cdot [VLG_{ABC}] + [G2] \cdot [I_{abc}]),$$

$$[I_{ABC}] = [x_t] \cdot [VLG_{ABC}] + [d_t] \cdot [I_{abc}],$$

where $[x_t] = [AI] \cdot [G1]$,

$$[d_t] = [AI] \cdot [G2] \quad (8.100)$$

Equation 8.100 is used in the "backward" sweep to compute the primary currents based upon the secondary currents and primary LG voltages.

The "forward" sweep equation is determined by substituting Equation 8.100 into Equation 8.87.

$$[VLL_{abc}] = [AV]^{-1} \cdot [VLG_{ABC}] - ([AV]^{-1} \cdot [ZG] + [Zt_{abc}] \cdot [AI]^{-1}) \cdot [I_{ABC}]$$

$$[VLL_{abc}] = [AV]^{-1} \cdot [VLG_{ABC}] -$$

$$([Zt_{abc}] \cdot [AI]^{-1} + [AV]^{-1} \cdot [ZG]) \cdot ([x_t] \cdot [VLG_{ABC}] + [d_t] \cdot [I_{abc}])$$

Define: $[X1] = [Zt_{abc}] \cdot [AI]^{-1} + [AV]^{-1} \cdot [ZG]$

$$[VLL_{abc}] = ([AV]^{-1} - [X1] \cdot [x_t]) \cdot [VLG_{ABC}] - [X1] \cdot [d_t] \cdot [I_{abc}]$$

Three-Phase Transformer Models

$$[VLN_{abc}] = [W] \cdot [VLL_{abc}]$$

$$[VLN_{abc}] = [W] \cdot \left(\left([AV]^{-1} - [X1] \cdot [x_t]\right) \cdot [VLG_{ABC}] - [X1] \cdot [d_t] \cdot [I_{abc}]\right)[VLN_{abc}] =$$

$$[VLN_{abc}] = [W] \cdot \left([AV]^{-1} - [X1] \cdot [x_t]\right) \cdot [VLG_{ABC}] - [W] \cdot [X1] \cdot [d_t] \cdot [I_{abc}] \quad (8.101)$$

The final form of Equation 8.101 gives the equation for the forward sweep.

$$[VLN_{abc}] = [A_t] \cdot [VLG_{ABC}] - [B_t] \cdot [I_{abc}],$$

$$\text{where} [A_t] = [W] \cdot \left([AV]^{-1} - [X1] \cdot [x_t]\right),$$

$$[B_t] = [W] \cdot [X1] \cdot [d_t] \quad (8.102)$$

As can be seen, the major difference between this and the ungrounded connection is in the line currents and the ground current on the primary side. Experience has shown that the value of the neutral grounding resistance should not exceed the transformer impedance relative to the primary side. If the ground resistance is too big, the program will not converge.

The question is should the neutral for the wye-delta connection be grounded or not? In Chapter 10, the short-circuit calculations for the grounded wye-delta transformer bank will be developed. In this development, it will be shown that during a grounded fault upstream from the transformer bank, there will be a back-feed current from the grounded wye-delta bank back to the grounded fault. This typically results in blowing the transformer fuses for the upstream ground fault. With that in mind, the grounded wye-delta transformer connection should not be used.

8.8 OPEN WYE – OPEN DELTA

A common load to be served on a distribution feeder is a combination of a single-phase lighting load and a three-phase power load. Many times, the three-phase power load will be an induction motor. This combination load can be served by a grounded or ungrounded wye-delta connection, as previously described, or by an "open wye – open delta" connection. When the three-phase load is small compared to the single-phase load, the open wye – open delta connection is commonly used. The open wye – open delta connection requires two transformers, but the connection will provide three-phase, line-to-line voltages to the combination load. Figure 8.13 shows the open wye – open delta connection and the primary and secondary positive sequence voltage phasors.

Example 8.6: The system of Examples 8.3 and 8.4 is changed so that the same transformers are connected in a grounded wye-delta step-down connection to serve the same load. The neutral ground resistance is 5 ohm. The computed matrices are as follows:

$$[x_t] = \begin{bmatrix} 0.0053 - j0.0061 & 0.0053 - j0.0061 & 0.0053 - j0.0061 \\ 0.0053 - j0.0061 & 0.0053 - j0.0061 & 0.0053 - j0.0061 \\ 0.0053 - j0.0061 & 0.0053 - j0.0061 & 0.0053 - j0.0061 \end{bmatrix}$$

$$[d_t] = \begin{bmatrix} 0.0128 + j0.0002 & -0.0128 - j0.0002 & 0 \\ 0.0128 + j0.0002 & 0.0206 - j0.0002 & 0 \\ -0.0206 + j0.0002 & -0.0128 - j0.0002 & 0 \end{bmatrix}$$

$$[A_t] = \begin{bmatrix} 0.0128 + j0.0002 & 0.0017 + j0.0002 & -0.0094 + j0.0002 \\ -0.0128 - j0.0002 & 0.0094 - j0.0002 & -0.0017 - j0.0002 \\ 0 & -0.0111 & 0.0111 \end{bmatrix}$$

$$[B_t] = \begin{bmatrix} 0.0043 + j0.0112 & 0.0014 + j0.0022 & 0 \\ 0.0014 + j0.0022 & 0.0043 + j0.0112 & 0 \\ -0.0058 - j0.0134 & -0.0058 - j0.0134 & 0 \end{bmatrix}$$

The only change in the program from Example 8.4 is for the equation computing the primary line currents.

$$[I_{ABC}] = [x_t] \cdot [VLG_{ABC}] + [d_t] \cdot [I_{abc}]$$

The source voltages are balanced by 12,470 volts. After five iterations, the resulting load line-to-line load voltages are

$$[VLL_{abc}] = \begin{bmatrix} 232.6/\underline{-1.7} \\ 231.0/\underline{-121.4} \\ 233.0/\underline{118.8} \end{bmatrix}$$

The voltage unbalance is computed to be

$$V_{unbalance} = 0.53\%.$$

The currents are

$$[I_{abc}] = \begin{bmatrix} 540.8/\underline{-49.5} \\ 594.1/\underline{168.5} \\ 373.0/\underline{51.7} \end{bmatrix},$$

Three-Phase Transformer Models

$$[I_{ABC}] = \begin{bmatrix} 13.7/\underline{-28.6} \\ 7.8/\underline{-160.4} \\ 7.1/\underline{87.3} \end{bmatrix},$$

$$I_g = -(I_A + I_B + I_C) = 5.4/\underline{157.8}.$$

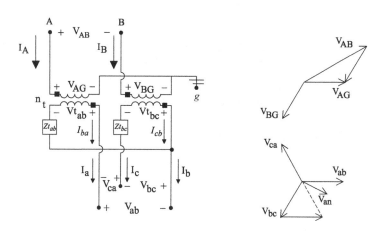

FIGURE 8.13 Open wye – open delta connection

With reference to Figure 8.13, the basic "ideal" transformer voltages as a function of the "turns ratio" are

$$\begin{bmatrix} V_{AG} \\ V_{BG} \\ V_{CG} \end{bmatrix} = n_t \cdot \begin{bmatrix} 1 & 0 & 0 \\ 0 & 1 & 0 \\ 0 & 0 & 0 \end{bmatrix} \cdot \begin{bmatrix} Vt_{ab} \\ Vt_{bc} \\ Vt_{ca} \end{bmatrix},$$

$$[VLG_{ABC}] = [AV] \cdot [Vt_{abc}] \tag{8.103}$$

The currents as a function of the turns ratio are given by

$$I_A = \frac{1}{n_t} \cdot I_{ba} = \frac{1}{n_t} \cdot I_a,$$

$$I_B = \frac{1}{n_t} \cdot I_{cb} = -\frac{1}{n_t} \cdot I_c,$$

$$I_b = -(I_a + I_c). \tag{8.104}$$

Equation 8.104 can be expressed in matrix form by

$$\begin{bmatrix} I_A \\ I_B \\ I_C \end{bmatrix} = \frac{1}{n_t} \cdot \begin{bmatrix} 1 & 0 & 0 \\ 0 & 0 & -1 \\ 0 & 0 & 0 \end{bmatrix} \cdot \begin{bmatrix} I_a \\ I_b \\ I_c \end{bmatrix},$$

$$[I_{ABC}] = [d_t] \cdot [I_{abc}],$$

$$\text{where } [d_t] = \frac{1}{n_t} \cdot \begin{bmatrix} 1 & 0 & 0 \\ 0 & 0 & -1 \\ 0 & 0 & 0 \end{bmatrix} \quad (8.105)$$

The secondary line currents as a function of the primary line currents are

$$\begin{bmatrix} I_a \\ I_b \\ I_c \end{bmatrix} = n_t \cdot \begin{bmatrix} 1 & 0 & 0 \\ -1 & 1 & 0 \\ 0 & -1 & 0 \end{bmatrix} \cdot \begin{bmatrix} I_A \\ I_B \\ I_C \end{bmatrix},$$

$$[I_{abc}] = [AI] \cdot [I_{ABC}],$$

$$\text{where } [AI] = n_t \cdot \begin{bmatrix} 1 & 0 & 0 \\ -1 & 1 & 0 \\ 0 & -1 & 0 \end{bmatrix} \cdot \begin{bmatrix} I_A \\ I_B \\ I_C \end{bmatrix} \quad (8.106)$$

The ideal voltages in the secondary can be determined by

$$Vt_{ab} = V_{ab} + Zt_{ab} \cdot I_a,$$

$$Vt_{bc} = V_{bc} - Zt_{bc} \cdot I_c \quad (8.107)$$

Substitute Equation 8.107 into Equations 8.103.

$$V_{AG} = n_t \cdot Vt_{ab} = n_t \cdot V_{ab} + n_t \cdot Zt_{ab} \cdot I_a,$$

$$V_{BG} = n_t \cdot Vt_{bc} = n_t \cdot V_{bc} - n_t \cdot Zt_{bc} \cdot I_c \quad (8.108)$$

Equation 8.108 can be put into three-phase matrix form as

$$\begin{bmatrix} V_{AG} \\ V_{BG} \\ V_{CG} \end{bmatrix} = n_t \cdot \begin{bmatrix} 1 & 0 & 0 \\ 0 & 1 & 0 \\ 0 & 0 & 0 \end{bmatrix} \cdot \begin{bmatrix} V_{ab} \\ V_{bc} \\ V_{ca} \end{bmatrix} + n_t \cdot \begin{bmatrix} Zt_{ab} & 0 & 0 \\ 0 & 0 & -Zt_{bc} \\ 0 & 0 & 0 \end{bmatrix} \cdot \begin{bmatrix} I_a \\ I_b \\ I_c \end{bmatrix},$$

Three-Phase Transformer Models

$$[VLG_{ABC}] = [AV] \cdot [VLL_{abc}] + [b_t] \cdot [I_{abc}] \qquad (8.109)$$

The secondary line-to-line voltages in Equation 8.109 can be replaced by the equivalent line-to-neutral secondary voltages.

$$[VLG_{ABC}] = [AV] \cdot [Dv] \cdot [VLN_{abc}] + [b_t] \cdot [I_{abc}],$$

$$[VLG_{ABC}] = [a_t] \cdot [VLN_{abc}] + [b_t] \cdot [I_{abc}],$$

where $[a_t] = [AV] \cdot [Dv]$,

$$[b_t] = n_t \cdot \begin{bmatrix} Zt_{ab} & 0 & 0 \\ 0 & 0 & -Zt_{bc} \\ 0 & 0 & 0 \end{bmatrix} \qquad (8.110)$$

Equations 8.109 and 8.110 give the matrix equations for the backward sweep. The forward sweep equation can be determined by solving Equation 8.108 for the two line-to-line secondary voltages.

$$V_{ab} = \frac{1}{n_t} \cdot V_{AG} - Zt_{ab} \cdot I_a$$

$$V_{bc} = \frac{1}{n_t} \cdot V_{BG} - Zt_{bc} \cdot I_c \qquad (8.111)$$

The third line-to-line voltage V_{ca} must be equal to the negative sum of the other two line-to-line voltages (KVL). In matrix form, the desired equation is:

$$\begin{bmatrix} V_{ab} \\ V_{bc} \\ V_{ca} \end{bmatrix} = \frac{1}{n_t} \cdot \begin{bmatrix} 1 & 0 & 0 \\ 0 & 1 & 0 \\ -1 & -1 & 0 \end{bmatrix} \cdot \begin{bmatrix} V_{AG} \\ V_{BG} \\ V_{CG} \end{bmatrix} - \begin{bmatrix} Zt_{ab} & 0 & 0 \\ 0 & 0 & -Zt_{bc} \\ -Zt_{ab} & 0 & Zt_{bc} \end{bmatrix} \cdot \begin{bmatrix} I_a \\ I_b \\ I_c \end{bmatrix},$$

$$[VLL_{abc}] = [BV] \cdot [VLG_{ABC}] - [Zt_{abc}] \cdot [I_{abc}] \qquad (8.112)$$

The equivalent secondary line-to-neutral voltages are then given by

$$[VLN_{abc}] = [W] \cdot [VLL_{ABC}] = [W][BV] \cdot [VLG_{ABC}] - [W] \cdot [Zt_{abc}] \cdot [I_{abc}] \qquad (8.113)$$

The forward sweep equation is given by

$$[VLN_{abc}] = [A_t] \cdot [VLG_{ABC}] - [B_t] \cdot [I_{abc}],$$

where $[A_t] = [W] \cdot [BV] = \dfrac{1}{3 \cdot n_t} \cdot \begin{bmatrix} 2 & 1 & 0 \\ -1 & 1 & 0 \\ -1 & -2 & 0 \end{bmatrix}$,

$$[B_t] = [W] \cdot [Zt_{abc}] = \dfrac{1}{3} \cdot \begin{bmatrix} 2 \cdot Zt_{ab} & 0 & -Zt_{bc} \\ -Zt_{ab} & 0 & -Zt_{bc} \\ -Zt_{ab} & 0 & 2 \cdot Zt_{bc} \end{bmatrix} \quad (8.114)$$

Example 8.7 The unbalanced load of Example 8.3 is to be served by the "leading" open wye – open delta connection using phases A and B. The primary line-to-line voltages are balanced 12.47 kV.

The "lighting" transformer is rated: 100 kVA, 7200 Wye – 240 delta, $Z = 1.0 + j4.0$ %.
The "power" transformer is rated: 50 kVA, 7200 Wye – 240 delta, $Z = 1.5 + j3.5$ %.

Use the forward/backward sweep to compute

1. the load line-to-line voltages,
2. the secondary line currents,
3. the load currents,
4. the primary line currents, and
5. load voltage unbalance.

The transformer impedances referred to the secondary are the same as in Example 8.7 since the secondary rated voltages are still 240 volts.
The required matrices for the forward and backward sweeps are as follows:

$$[A_t] = \begin{bmatrix} 0.0222 & 0.0111 & 0 \\ -0.0111 & 0.0111 & 0 \\ -0.0111 & -0.0222 & 0 \end{bmatrix}$$

$$[B_t] = \begin{bmatrix} 0.0038 + j0.0154 & 0 & -0.0058 - j0.0134 \\ -0.0019 - j0.0077 & 0 & -0.0058 - j0.0134 \\ -0.0019 - j0.0077 & 0 & 0.0115 + j0.0269 \end{bmatrix}$$

$$[d_t] = \begin{bmatrix} 0.0333 & 0 & 0 \\ 0 & 0 & -0.0333 \\ 0 & 0 & 0 \end{bmatrix}$$

The same Mathcad program from Example 8.3 can be used for this example. After seven iterations, the results are as follows:

$$[VLL_{abc}] = \begin{bmatrix} 228.3/\underline{-1.4} \\ 231.4/\underline{-123.4} \\ 222.7/\underline{116.9} \end{bmatrix}$$

$$[I_{abc}] = \begin{bmatrix} 548.2/\underline{-50.3} \\ 606.5/\underline{167.8} \\ 381.0/\underline{50.5} \end{bmatrix}$$

$$[ID_{abc}] = \begin{bmatrix} 438.0/\underline{-27.3} \\ 216.1/\underline{-160.4} \\ 224.6/\underline{80.0} \end{bmatrix}$$

$$[I_{ABC}] = \begin{bmatrix} 18.3/\underline{-50.3} \\ 12.7/\underline{-129.5} \\ 0 \end{bmatrix}$$

$$V_{unbalance} = 2.11\%$$

The open wye – open delta connection derived in this section utilized phases A and B on the primary. This is just one of three possible connections. The other two possible connections would use phases B and C and then phases C and A. The generalized matrices will be different from those just derived. The same procedure can be used to derive the matrices for the other two connections.

The terms "leading" and "lagging" connection are also associated with the open wye – open delta connection. When the lighting transformer is connected across the leading of the two phases, the connection is referred to as the "leading" connection. Similarly, when the lighting transformer is connected across the lagging of the two phases, the connection is referred to as the "lagging" connection. For example, if the bank is connected to phases A and B, and the lighting transformer is connected from phase A to ground, this would be referred to as the "leading" connection because the voltage A-G leads the voltage B-G by 120°. Reverse the connection and it would now be called the "lagging" connection. Obviously, there is a leading and lagging connection for each of the three possible open wye – open delta connections.

Note the significant difference in voltage unbalance between this and Example 8.6. While it is economical to serve the load with two rather than three transformers, it is recognized that the open connection will lead to a much higher voltage unbalance.

8.9 THE GROUNDED WYE – GROUNDED WYE CONNECTION

The grounded wye – grounded wye connection is primarily used to supply single-phase and three-phase loads on four-wire multi-grounded systems. The grounded wye – grounded wye connection is shown in Figure 8.14.

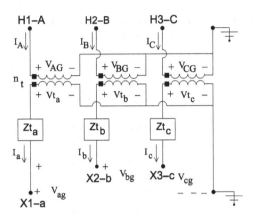

FIGURE 8.14 Grounded wye – grounded wye connection.

Unlike the delta-wye and wye-delta connections, there is no phase shift between the voltages and the currents on the two sides of the bank. This makes the derivation of the generalized constant matrices much easier. The ideal transformer equations are

$$n_t = \frac{VLN_{Rated\ Primary}}{VLN_{Rated\ Secondary}}, \tag{8.115}$$

$$\begin{bmatrix} V_{AG} \\ V_{BG} \\ V_{CG} \end{bmatrix} = n_t \cdot \begin{bmatrix} 1 & 0 & 0 \\ 0 & 1 & 0 \\ 0 & 0 & 1 \end{bmatrix} \cdot \begin{bmatrix} Vt_a \\ Vt_b \\ Vt_c \end{bmatrix},$$

$$[VLG_{ABC}] = [AV] \cdot [Vt_{abc}],$$

$$\text{where } [AV] = n_t \cdot \begin{bmatrix} 1 & 0 & 0 \\ 0 & 1 & 0 \\ 0 & 0 & 1 \end{bmatrix}, \tag{8.116}$$

$$\begin{bmatrix} I_A \\ I_B \\ I_C \end{bmatrix} = \frac{1}{n_t} \cdot \begin{bmatrix} 1 & 0 & 0 \\ 0 & 1 & 0 \\ 0 & 0 & 1 \end{bmatrix} \cdot \begin{bmatrix} I_a \\ I_b \\ I_c \end{bmatrix},$$

$$[I_{ABC}] = [AI] \cdot [I_{abc}],$$

$$\text{where } [AI] = \frac{1}{n_t} \cdot \begin{bmatrix} 1 & 0 & 0 \\ 0 & 1 & 0 \\ 0 & 0 & 1 \end{bmatrix} \tag{8.117}$$

Three-Phase Transformer Models

With reference to Figure 8.14, the ideal transformer voltages on the secondary windings can be computed by

$$\begin{bmatrix} Vt_a \\ Vt_b \\ Vt_c \end{bmatrix} = \begin{bmatrix} V_{ag} \\ V_{bg} \\ V_{cg} \end{bmatrix} + \begin{bmatrix} Zt_a & 0 & 0 \\ 0 & Zt_b & 0 \\ 0 & 0 & Zt_c \end{bmatrix} \cdot \begin{bmatrix} I_a \\ I_b \\ I_c \end{bmatrix},$$

$$[Vt_{abc}] = [VLG_{abc}] + [Zt_{abc}] \cdot [I_{abc}] \quad (8.118)$$

Substitute Equation 8.118 into Equation 8.116.

$$[VLG_{ABC}] = [AV] \cdot ([VLG_{abc}] + [Zt_{abc}] \cdot [I_{abc}]),$$

$$[VLG_{ABC}] = [a_t] \cdot [VLG_{abc}] + [b_t] \cdot [I_{abc}] \quad (8.119)$$

Equation 8.119 is the backward sweep equation with the $[a_t]$ and $[b_t]$ matrices defined by

$$[a_t] = [AV] = n_t \cdot \begin{bmatrix} 1 & 0 & 0 \\ 0 & 1 & 0 \\ 0 & 0 & 1 \end{bmatrix}, \quad (8.120)$$

$$[b_t] = [AV] \cdot [Zt_{abc}] = n_t \cdot \begin{bmatrix} Zt_a & 0 & 0 \\ 0 & Zt_b & 0 \\ 0 & 0 & Zt_c \end{bmatrix} \quad (8.121)$$

The primary line currents as a function of the secondary line currents are given by

$$[I_{ABC}] = [d_t] \cdot [I_{abc}],$$

$$\text{where} [d_t] = [AI] = \frac{1}{n_t} \cdot \begin{bmatrix} 1 & 0 & 0 \\ 0 & 1 & 0 \\ 0 & 0 & 1 \end{bmatrix} \quad (8.122)$$

The forward sweep equation is determined by solving Equation 8.119 for the secondary line-to-ground voltages.

$$[VLG_{abc}] = [AV]^{-1} \cdot [VLG_{ABC}] - [Zt_{abc}] \cdot [I_{abc}],$$

$$[VLG_{abc}] = [A_t] \cdot [VLG_{ABC}] - [B_t] \cdot [I_{abc}],$$

where $[A_t] = [AV]^{-1}$,

$$[B_t] = [Zt_{abc}] \quad (8.123)$$

The modeling and analysis of the grounded wye – grounded wye connection does not present any problems. Without the phase shift, there is a direct relationship between the primary and secondary voltages and currents as had been demonstrated in the derivation of the generalized constant matrices.

8.10 THE DELTA – DELTA CONNECTION

The delta – delta connection is primarily used on three-wire systems to provide service to a three-phase load or a combination of three-phase and single-phase loads. Three single-phase transformers connected in a delta-delta are shown in Figure 8.15.

The basic "ideal" transformer voltage and current equations as a function of the "turns ratio" are

$$n_t = \frac{VLL_{Rated\ Primary}}{VLL_{Rated\ Secondary}}, \quad (8.124)$$

$$\begin{bmatrix} VLL_{AB} \\ VLL_{BC} \\ VLL_{CA} \end{bmatrix} = n_t \cdot \begin{bmatrix} 1 & 0 & 0 \\ 0 & 1 & 0 \\ 0 & 0 & 1 \end{bmatrix} \cdot \begin{bmatrix} Vt_{ab} \\ Vt_{bc} \\ Vt_{ca} \end{bmatrix},$$

$$[VLL_{ABC}] = [AV] \cdot [Vt_{abc}],$$

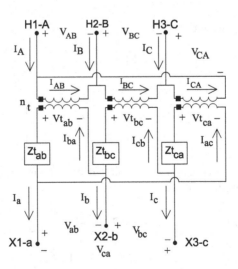

FIGURE 8.15 Delta – delta connection

$$\text{where} [AV] = n_t \cdot \begin{bmatrix} 1 & 0 & 0 \\ 0 & 1 & 0 \\ 0 & 0 & 1 \end{bmatrix}, \qquad (8.125)$$

$$\begin{bmatrix} I_{AB} \\ I_{BC} \\ I_{CA} \end{bmatrix} = \frac{1}{n_t} \cdot \begin{bmatrix} 1 & 0 & 0 \\ 0 & 1 & 0 \\ 0 & 0 & 1 \end{bmatrix} \cdot \begin{bmatrix} I_{ba} \\ I_{cb} \\ I_{ac} \end{bmatrix}$$

$$[ID_{ABC}] = [AI] \cdot [ID_{abc}],$$

$$\text{where} [AI] = \frac{1}{n_t} \cdot \begin{bmatrix} 1 & 0 & 0 \\ 0 & 1 & 0 \\ 0 & 0 & 1 \end{bmatrix} \qquad (8.126)$$

Solve Equation 8.126 for the secondary side delta currents.

$$[ID_{abc}] = [AI]^{-1} \cdot [ID_{ABC}] \qquad (8.127)$$

The line currents as a function of the delta currents on the source side are given by

$$\begin{bmatrix} I_A \\ I_B \\ I_C \end{bmatrix} = \begin{bmatrix} 1 & 0 & -1 \\ -1 & 1 & 0 \\ 0 & -1 & 1 \end{bmatrix} \cdot \begin{bmatrix} I_{AB} \\ I_{BC} \\ I_{CA} \end{bmatrix},$$

$$[I_{ABC}] = [Di] \cdot [ID_{ABC}],$$

$$\text{where} [Di] = \begin{bmatrix} 1 & 0 & -1 \\ -1 & 1 & 0 \\ 0 & -1 & 1 \end{bmatrix} \qquad (8.128)$$

Substitute Equation 8.126 into Equation 8.128.

$$[I_{ABC}] = [Di] \cdot [AI] \cdot [ID_{abc}] \qquad (8.129)$$

Since [AI] is a diagonal matrix, Equation 8.129 can be rewritten as

$$[I_{ABC}] = [AI] \cdot [Di] \cdot [ID_{abc}] \qquad (8.130)$$

The load-side line currents as a function of the load-side delta currents:

$$[I_{abc}] = [Di] \cdot [ID_{abc}],$$

$$\text{or} [ID_{abc}] = [Di]^{-1} \cdot [I_{abc}],$$

even though $[Di]^{-1}$ is singular. (8.131)

Applying Equation 8.131, Equation 8.130 becomes

$$[I_{ABC}] = [AI] \cdot [Di] \cdot [Di]^{-1} \cdot [I_{abc}],$$

$$[I_{ABC}] = [AI] \cdot [I_{abc}] \quad (8.132)$$

Turn Equation 8.132 around to solve for the load-side line currents as a function of the source side line currents.

$$[I_{abc}] = [AI]^{-1} \cdot [I_{ABC}] \quad (8.133)$$

Equations 8.132 and 8.133 merely demonstrate that the line currents on the two sides of the transformer are in phase and differ only by the turns ratio of the transformer windings. In the per-unit system, the per-unit line currents on the two sides of the transformer are exactly equal.

The ideal line-to-line voltages on the secondary side as a function of the line-to-line voltages the delta currents and the transformer impedances are given by

$$[Vt_{abc}] = [VLL_{abc}] + [Zt_{abc}] \cdot [ID_{abc}],$$

$$\text{where} [Zt_{abc}] = \begin{bmatrix} Zt_{ab} & 0 & 0 \\ 0 & Zt_{bc} & 0 \\ 0 & 0 & Zt_{ca} \end{bmatrix} \quad (8.134)$$

Substitute Equation 8.134 into Equation 8.125.

$$[VLL_{ABC}] = [AV] \cdot [VLL_{abc}] + [AV] \cdot [Zt_{abc}] \cdot [ID_{abc}] \quad (8.135)$$

Solve Equation 8.135 for the load-side, line-to-line voltages.

$$[VLL_{abc}] = [AV]^{-1} \cdot [VLL_{ABC}] - [Zt_{abc}] \cdot [ID_{abc}] \quad (8.136)$$

Three-Phase Transformer Models

The delta currents $[ID_{abc}]$ in Equations 8.135 and 8.136 need to be replaced by the secondary line currents $[I_{abc}]$. To develop the needed relationship, three independent equations are needed. The first two come from applying KCL at two vertices of the delta-connected secondary.

$$I_a = I_{ba} - I_{ac}$$

$$I_b = I_{cb} - I_{ba} \tag{8.137}$$

The third equation comes from recognizing that the sum of the primary line-to-line voltages, and therefore the secondary ideal transformer voltages must sum to zero. KVL around the delta windings gives

$$Vt_{ab} - Zt_{ab} \cdot I_{ba} + Vt_{bc} - Zt_{bc} \cdot I_{cb} + Vt_{ca} - Zt_{ca} \cdot I_{ac} = 0. \tag{8.138}$$

Replacing the "ideal" delta voltages with the source-side, line-to-line voltages.

$$\frac{V_{AB}}{n_t} + \frac{V_{BC}}{n_t} + \frac{V_{CA}}{n_t} = Zt_{ab} \cdot I_{ba} + Zt_{bc} \cdot I_{cb} + Zt_{ca} \cdot I_{ac} \tag{8.139}$$

Since the sum of the line-to-line voltages must equal zero (KVL), and the turns ratios of the three transformers are equal, Equation 8.139 is simplified to

$$0 = Zt_a \cdot I_{ba} + Zt_b \cdot I_{cb} + Zt_c \cdot I_{ac} \tag{8.140}$$

Note in Equation 8.140 that if the three transformer impedances are equal, then the sum of the delta currents will add to zero, meaning that the zero-sequence delta currents will be zero.

Equations 8.137 and 8.140 can be put into matrix form:

$$\begin{bmatrix} I_a \\ I_b \\ 0 \end{bmatrix} = \begin{bmatrix} 1 & 0 & -1 \\ -1 & 1 & 0 \\ Zt_{ab} & Zt_{bc} & Zt_{ca} \end{bmatrix} \cdot \begin{bmatrix} I_{ba} \\ I_{cb} \\ I_{ac} \end{bmatrix},$$

$$[I0_{abc}] = [F] \cdot [ID_{abc}],$$

$$\text{where} [I0_{abc}] = \begin{bmatrix} I_a \\ I_b \\ 0 \end{bmatrix},$$

$$[F] = \begin{bmatrix} 1 & 0 & -1 \\ -1 & 1 & 0 \\ Zt_{ab} & Zt_{bc} & Zt_{ca} \end{bmatrix} \tag{8.141}$$

Solve Equation 8.141 for the load-side delta currents.

$$[ID_{abc}] = [F]^{-1} \cdot [IO_{abc}] = [G] \cdot [IO_{abc}],$$

where $[G] = [F]^{-1}$. (8.142)

Writing Equation 8.142 in matrix form gives

$$\begin{bmatrix} I_{ba} \\ I_{cb} \\ I_{ac} \end{bmatrix} = \begin{bmatrix} G_{11} & G_{12} & G_{13} \\ G_{21} & G_{22} & G_{23} \\ G_{31} & G_{32} & G_{33} \end{bmatrix} \cdot \begin{bmatrix} I_a \\ I_b \\ 0 \end{bmatrix}$$ (8.143)

From Equations 8.142 and 8.143, it is seen that the delta currents are a function of the transformer impedances and just the line currents in phases a and b. Equation 8.143 can be modified to include the line current in phase c by setting the last column of the [G] matrix to zeros.

$$\begin{bmatrix} I_{ba} \\ I_{cb} \\ I_{ac} \end{bmatrix} = \begin{bmatrix} G_{11} & G_{12} & 0 \\ G_{21} & G_{22} & 0 \\ G_{31} & G_{32} & 0 \end{bmatrix} \cdot \begin{bmatrix} I_a \\ I_b \\ I_c \end{bmatrix},$$

$$[ID_{abc}] = [G1] \cdot [I_{abc}],$$

where $[G1] = \begin{bmatrix} G_{11} & G_{12} & 0 \\ G_{21} & G_{22} & 0 \\ G_{31} & G_{32} & 0 \end{bmatrix}$ (8.144)

When the impedances of the transformers are equal, the sum of the delta currents will be zero, meaning that there is no circulating zero-sequence current in the delta windings.

Substitute Equation 8.144 into Equation 8.135.

$$[VLL_{ABC}] = [AV] \cdot [VLL_{abc}] + [AV] \cdot [Zt_{abc}] \cdot [G1] \cdot [I_{abc}]$$ (8.145)

The generalized matrices are defined in terms of the line-to-neutral voltages on the two sides of the transformer bank. Equation 8.145 is modified to be in terms of equivalent line-to-neutral voltages.

$$[VLN_{ABC}] = [W] \cdot [VLL_{ABC}]$$

$$[VLN_{ABC}] = [W] \cdot [AV] \cdot [Dv] \cdot [VLN_{abc}] + [W] \cdot [AV] \cdot [Zt_{abc}] \cdot [G1] \cdot [I_{abc}]$$ (8.146)

Three-Phase Transformer Models

Equation 8.146 is in the general form.

$$[VLN_{ABC}] = [a_t] \cdot [VLN_{abc}] + [b_t] \cdot [I_{abc}],$$

where $[a_t] = [W] \cdot [AV] \cdot [Dv]$

$$[b_t] = [W] \cdot [AV] \cdot [Zt_{abc}] \cdot [G1] \qquad (8.147)$$

Equation 8.133 gives the generalized equation for currents.

$$[I_{ABC}] = [AI] \cdot [I_{abc}] = [d_t] \cdot [I_{abc}],$$

where $[d_t] = [AI]$. $\qquad (8.148)$

The forward sweep equations can be derived by modifying Equation 8.136 in terms of equivalent line-to-neutral voltages.

$$[VLN_{abc}] = [W] \cdot [VLL_{abc}]$$

$$[VLN_{abc}] = [W] \cdot [AV]^{-1} \cdot [Dv] \cdot [VLN_{ABC}] - [W] \cdot [Zt_{abc}] \cdot [G1] \cdot [I_{abc}] \quad (8.149)$$

The forward sweep equation is

$$[VLN_{abc}] = [A_t] \cdot [VLN_{ABC}] - [B_t] \cdot [I_{abc}],$$

where $[A_t] = [W] \cdot [AV]^{-1} \cdot [Dv],$

$$[B_t] = [W] \cdot [Zt_{abc}] \cdot [G1] \qquad (8.150)$$

The forward and backward sweep matrices for the delta – delta connection have been derived. Once again it has been a long process to get to the final six equations that define the matrices. The derivation provides an excellent exercise in the application of basic transformer theory and circuit theory. Once the matrices have been defined for a particular transformer connection, the analysis of the connection is a relatively simple task. Example 8.8 will demonstrate the analysis of this connection using the generalized matrices.

After six iterations, the results are as follows:

$$[VLL_{abc}] = \begin{bmatrix} 232.9\underline{/28.3} \\ 231.0\underline{/-91.4} \\ 233.1\underline{/148.9} \end{bmatrix}$$

Example 8.8: Figure 8.16 shows three single-phase transformers in a delta – delta connection serving an unbalanced three-phase load connected in delta.

The source voltages at the load are a balanced three phase of 240 volts line to line.

$$[VLL_{abc}] = \begin{bmatrix} 12470\underline{/0} \\ 12470\underline{/-120} \\ 12470\underline{/120} \end{bmatrix} V$$

The loading by phase is

S_{ab} = 100 kVA at 0.9 lagging power factor,

$S_{bc} = S_{ca}$ = 50 kVA at 0.8 lagging power factor.

The ratings of the transformers are as follows:

Phase A-B: 100 kVA, 12,470–240 volts, Z = 0.01 + j0.04 per-unit

Phases B-C and C-A: 50 kVA, 12,470-240 volts, Z = 0.015 + j0.035 per-unit

Determine the following

1. the load line-to-line voltages,
2. the secondary line currents,
3. the primary line currents,
4. the load currents, and
5. the load voltage unbalance.

Before the analysis can start, the transformer impedances must be converted to actual values in ohms and located inside the delta-connected secondary windings.

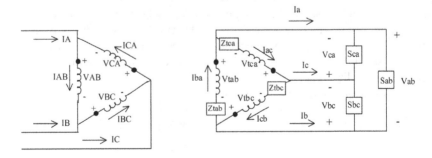

FIGURE 8.16 Delta – delta bank serving an unbalanced delta-connected load.

Phase a-b transformers:

$$Z_{base} = \frac{0.24^2 \cdot 1000}{100} = 0.576\,\Omega,$$

$$Zt_{ab} = (0.01 + j0.04) \cdot 0.576 = 0.0058 + j.023\,\Omega$$

Phase b-c and c-a transformers:

$$Z_{base} = \frac{.24^2 \cdot 1000}{50} = 1.152\,\Omega,$$

$$Zt_{bc} = Zt_{ca} = (.015 + j0.035) \cdot 1.152 = 0.0173 + j0.0403\,\Omega$$

The transformer impedance matrix can now be defined as

$$[Zt_{abc}] = \begin{bmatrix} 0.0058 + j.023 & 0 & 0 \\ 0 & 0.0173 + j0.0403 & 0 \\ 0 & 0 & 0.0173 + j0.0403 \end{bmatrix} \Omega$$

The turns ratio of the transformers is $n_t = \dfrac{12470}{240} = 51.9583$. Define the matrices.

$$[W] = \frac{1}{3} \cdot \begin{bmatrix} 2 & 1 & 0 \\ 0 & 2 & 1 \\ 1 & 0 & 2 \end{bmatrix} \quad [Dv] = \begin{bmatrix} 1 & -1 & 0 \\ 0 & 1 & -1 \\ -1 & 0 & 1 \end{bmatrix} \quad [Di] = \begin{bmatrix} 1 & 0 & -1 \\ -1 & 1 & 0 \\ 0 & -1 & 1 \end{bmatrix}$$

$$[AV] = n_t \cdot \begin{bmatrix} 1 & 0 & 0 \\ 0 & 1 & 0 \\ 0 & 0 & 1 \end{bmatrix} = \begin{bmatrix} 51.9583 & 0 & 0 \\ 0 & 51.9583 & 0 \\ 0 & 0 & 51.9583 \end{bmatrix}$$

$$[AI] = \frac{1}{n_t} \cdot \begin{bmatrix} 1 & 0 & 0 \\ 0 & 1 & 0 \\ 0 & 0 & 1 \end{bmatrix} = \begin{bmatrix} 0.0192 & 0 & 0 \\ 0 & 0.0192 & 0 \\ 0 & 0 & 0.0192 \end{bmatrix}$$

$$[F] = \begin{bmatrix} 1 & 0 & -1 \\ -1 & 1 & 0 \\ 0.0058 + j0.023 & 0.0173 + j0.0403 & 0.0173 + j0.0404 \end{bmatrix}$$

$$[G] = [F]^{-1} = \begin{bmatrix} 0.3941 - j0.0134 & -0.3941 + j0.0134 & 3.2581 - j8.378 \\ 0.3941 - j0.0134 & 0.6059 + j0.0134 & 3.2581 - j8.378 \\ -0.6059 - j0.0134 & -0.3941 + j0.0134 & 3.2581 - j8.378 \end{bmatrix}$$

$$[G1] = \begin{bmatrix} 0.3941 - j0.0134 & -0.3941 + j0.0134 & 0 \\ 0.3941 - j0.0134 & 0.6059 + j0.0134 & 0 \\ -0.6059 - j0.0134 & -0.3941 + j0.0134 & 0 \end{bmatrix}$$

$$[a_t] = [W] \cdot [AV] \cdot [Dv] = \begin{bmatrix} 34.6489 & -17.3194 & -17.3194 \\ -17.3194 & 34.6489 & -17.3194 \\ -17.3194 & -17.3194 & 34.6489 \end{bmatrix}$$

$$[b_t] = [W] \cdot [AV] \cdot [Zt_{abc}] \cdot [G1] = \begin{bmatrix} 0.2166 + j0.583 & 0.0826 + j0.1153 & 0 \\ 0.0826 + j0.1153 & 0.2166 + j0.583 & 0 \\ -0.2993 - j0.6983 & -0.2993 - j0.6983 & 0 \end{bmatrix}$$

$$[d_t] = [AI] = \begin{bmatrix} 0.0192 & 0 & 0 \\ 0 & 0.0192 & 0 \\ 0 & 0 & 0.0192 \end{bmatrix}$$

$$[A_t] = [W] \cdot [AV]^{-1} \cdot [D] = \begin{bmatrix} 0.0128 & -0.0064 & -0.0064 \\ -0.0064 & 0.0128 & -0.0064 \\ -0.0064 & -0.0064 & 0.0128 \end{bmatrix}$$

$$[B_t] = [W] \cdot [Zt_{abc}] \cdot [G1] = \begin{bmatrix} 0.0042 + j0.0112 & 0.0016 + j0.0022 & 0 \\ 0.0016 + j0.0022 & 0.0042 + j0.0112 & 0 \\ -0.0058 - j0.0134 & -0.0058 - j0.0134 & 0 \end{bmatrix}$$

The Mathcad program is modified slightly to account for the delta connections. The modified program is shown in Figure 8.17:

The initial conditions are as follows:

$$[VLL_{ABC}] = \begin{bmatrix} 12470\underline{/30} \\ 12470\underline{/-90} \\ 12470\underline{/150} \end{bmatrix} \quad [Start] = \begin{bmatrix} 0 \\ 0 \\ 0 \end{bmatrix} \quad Tol = 0.000001 \quad VM = 240$$

$$[I_{abc}] = \begin{bmatrix} 540.3\underline{/-19.5} \\ 593.6\underline{/-161.5} \\ 372.8\underline{/81.7} \end{bmatrix}$$

$$[I_{ABC}] = \begin{bmatrix} 10.4\underline{/-19.5} \\ 11.4\underline{/-161.5} \\ 7.2\underline{/81.7} \end{bmatrix}$$

$$[ID_{abc_i}] = \begin{bmatrix} 429.3\underline{/2.4} \\ 216.5\underline{/-128.3} \\ 214.5\underline{/112.0} \end{bmatrix} A$$

Three-Phase Transformer Models

$$X := \begin{array}{|l} I_{abc} \leftarrow \text{Start} \\ V_{old} \leftarrow \text{Start} \\ VLN_{ABC} \leftarrow W \cdot VLL_{ABC} \\ \text{for } n \in 1..200 \\ \quad \begin{array}{|l} VLN_{abc} \leftarrow A_t \cdot VLN_{ABC} - B_t \cdot I_{abc} \\ VLL_{abc} \leftarrow Dv \cdot VLN_{abc} \\ \text{for } j \in 1..3 \\ \quad ID_{abc_j} \leftarrow \dfrac{\overline{SL_j \cdot 1000}}{VLL_{abc_j}} \\ \text{for } k \in 1..3 \\ \quad Error_k \leftarrow \dfrac{\left|VLL_{abc_k} - V_{old_k}\right|}{VM} \\ Error_{max} \leftarrow \max(Error) \\ \text{break if } Error_{max} < Tol \\ V_{old} \leftarrow VLL_{abc} \\ I_{abc} \leftarrow Di \cdot ID_{abc} \\ I_{ABC} \leftarrow d_t \cdot I_{abc} \end{array} \\ Out_1 \leftarrow VLN_{abc} \\ Out_2 \leftarrow VLL_{abc} \\ Out_3 \leftarrow I_{abc} \\ Out_4 \leftarrow I_{ABC} \\ Out_5 \leftarrow n \\ Out \end{array}$$

FIGURE 8.17 Delta – delta Mathcad program.

$$V_{unbalance} = 0.59\%$$

This example demonstrates that a small change in the Mathcad program can be made to represent the delta – delta transformer connection.

8.11 OPEN DELTA – OPEN DELTA

The open delta – open delta transformer connection can be connected in three different ways. Figure 8.18 shows the connection using phases AB and BC.

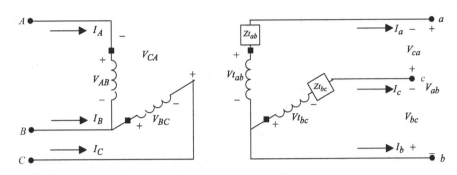

FIGURE 8.18 Open delta – open delta using phases AB and BC.

The relationship between the primary line-to-line voltages and the secondary ideal voltage is given by

$$\begin{bmatrix} V_{AB} \\ V_{BC} \\ V_{CA} \end{bmatrix} = n_t \cdot \begin{bmatrix} 1 & 0 & 0 \\ 0 & 1 & 0 \\ -1 & -1 & 0 \end{bmatrix} \cdot \begin{bmatrix} Vt_{ab} \\ Vt_{bc} \\ Vt_{ca} \end{bmatrix},$$

$$[VLL_{ABC}] = [AV] \cdot [Vt_{abc}],$$

where $[AV] = n_t \cdot \begin{bmatrix} 1 & 0 & 0 \\ 0 & 1 & 0 \\ -1 & -1 & 0 \end{bmatrix}$ \hfill (8.151)

The last row of the matrix [AV] is the result that the sum of the line-to-line voltages must be equal to zero.

The relationship between the secondary and primary line currents is

$$\begin{bmatrix} I_A \\ I_B \\ I_C \end{bmatrix} = \frac{1}{n_t} \cdot \begin{bmatrix} 1 & 0 & 0 \\ -1 & 0 & -1 \\ 0 & 0 & 1 \end{bmatrix} \cdot \begin{bmatrix} I_a \\ I_b \\ I_c \end{bmatrix},$$

$$[I_{ABC}] = [AI] \cdot [I_{abc}],$$

where $[AI] = [d_t] = \frac{1}{n_t} \cdot \begin{bmatrix} 1 & 0 & 0 \\ -1 & 0 & -1 \\ 0 & 0 & 1 \end{bmatrix}$ \hfill (8.152)

The ideal secondary voltages are given by

$$Vt_{ab} = V_{ab} + Zt_{ab} \cdot I_a,$$

Three-Phase Transformer Models

$$Vt_{bc} = V_{bc} + Zt_{bc} \cdot I_c \tag{8.153}$$

The primary line-to-line voltages as a function of the secondary line-to-line voltages are given by

$$V_{AB} = n_t \cdot Vt_{ab} = n_t \cdot V_{ab} + n_t \cdot Zt_{ab} \cdot I_a,$$

$$V_{BC} = n_t \cdot Vt_{bc} = n_t \cdot V_{bc} + n_t \cdot Zt_{bc} \cdot I_c \tag{8.154}$$

The sum of the primary line-to-line voltages must equal zero. Therefore, the voltage V_{CA} is given by

$$V_{CA} = -(V_{AB} + V_{BC}) = -n_t \cdot (V_{ab} + n_t \cdot Zt_{ab} \cdot I_a + V_{bc} + n_t \cdot Zt_{bc} \cdot I_c),$$

$$V_{CA} = -n_t \cdot V_{ab} - n_t \cdot V_{bc} - n_t \cdot Zt_{ab} \cdot I_a - n_t \cdot Zt_{bc} \cdot I_c \tag{8.155}$$

Equations 8.154 and 8.155 can be put into matrix form to create the backward sweep voltage equation.

$$[VLL_{ABC}] = [AV] \cdot [VLL_{abc}] + n_t \cdot [Zt_{abc}] \cdot [I_{abc}],$$

$$\text{where} [Zt_{abc}] = \begin{bmatrix} Zt_{ab} & 0 & 0 \\ 0 & 0 & Zt_{bc} \\ -Zt_{ab} & 0 & -Zt_{bc} \end{bmatrix} \tag{8.156}$$

Equation 8.156 gives the backward sweep equation in terms of line-to-line voltages. To convert the equation to equivalent line-to-neutral voltages, the [W] and [Dv] matrices are applied to Equation 8.156.

$$[VLL_{ABC}] = [AV] \cdot [VLL_{abc}] + n_t \cdot [Zt_{abc}] \cdot [I_{abc}],$$

$$[VLN_{ABC}] = [W] \cdot [VLL_{ABC}] = [W] \cdot [AV] \cdot [Dv] \cdot [VLN_{abc}] + [W] \cdot n_t \cdot [Zt_{abc}] \cdot [I_{abc}],$$

$$[VLN_{ABC}] = [a_t] \cdot [VLN_{abc}] + [b_t] \cdot [I_{abc}],$$

$$\text{where} [a_t] = [W] \cdot [AV] \cdot [Dv],$$

$$[b_t] = [W] \cdot n_t \cdot [Zt_{abc}] \tag{8.157}$$

The forward sweep equation can be derived by defining the ideal voltages as a function of the primary line-to-line voltages.

$$\begin{bmatrix} Vt_{ab} \\ Vt_{bc} \\ Vt_{ca} \end{bmatrix} = \frac{1}{n_t} \cdot \begin{bmatrix} 1 & 0 & 0 \\ 0 & 1 & 0 \\ -1 & -1 & 0 \end{bmatrix} \cdot \begin{bmatrix} V_{AB} \\ V_{BC} \\ V_{CA} \end{bmatrix},$$

$$[Vt_{abc}] = [BV] \cdot [VLL_{ABC}],$$

$$\text{where } [BV] = \frac{1}{n_t} \cdot \begin{bmatrix} 1 & 0 & 0 \\ 0 & 1 & 0 \\ -1 & -1 & 0 \end{bmatrix} \quad (8.158)$$

The ideal secondary voltages as a function of the terminal line-to-line voltages are given by

$$\begin{bmatrix} Vt_{ab} \\ Vt_{bc} \\ Vt_{ca} \end{bmatrix} = \begin{bmatrix} V_{ab} \\ V_{bc} \\ V_{ca} \end{bmatrix} + \begin{bmatrix} Zt_{ab} & 0 & 0 \\ 0 & 0 & Zt_{bc} \\ -Zt_{ab} & 0 & -Zt_{bc} \end{bmatrix} \cdot \begin{bmatrix} I_a \\ I_b \\ I_c \end{bmatrix},$$

$$[Vt_{abc}] = [VLL_{abc}] + [Zt_{abc}] \cdot [I_{abc}],$$

$$\text{where } [Zt_{abc}] = \begin{bmatrix} Zt_{ab} & 0 & 0 \\ 0 & 0 & Zt_{bc} \\ -Zt_{ab} & 0 & -Zt_{bc} \end{bmatrix} \quad (8.159)$$

Equate Equation 8.158 to Equation 8.159.

$$[BV] \cdot [VLL_{ABC}] = [VLL_{abc}] + [Zt_{abc}] \cdot [I_{abc}]$$

$$[VLL_{abc}] = [BV] \cdot [VLL_{ABC}] - [Zt_{abc}] \cdot [I_{abc}] \quad (8.160)$$

Equation 8.160 gives the forward sweep equation in terms of line-to-line voltages. As before, the [W] and [D] matrices are used to convert Equation 8.160 to using line-to-neutral voltages:

$$[VLL_{abc}] = [BV] \cdot [VLL_{ABC}] - [Zt_{abc}] \cdot [I_{abc}],$$

$$[VLN_{abc}] = [W] \cdot [VLL_{abc}] = [W] \cdot [BV] \cdot [Dv] \cdot [VLN_{ABC}] - [W] \cdot [Zt_{abc}] \cdot [I_{abc}],$$

$$[VLN_{abc}] = [A_t] \cdot [VLN_{ABC}] - [B_t] \cdot [I_{abc}],$$

$$\text{where } [A_t] = [W] \cdot [BV] \cdot [Dv],$$

… Three-Phase Transformer Models

$$[B_t] = [W] \cdot [Zt_{abc}] \quad (8.161)$$

Figure 8.19 shows the phasor diagrams (not to scale) for the previously defined voltages. In the phasor diagram, it is clear that there is a voltage drop on phase ab and

Example 8.9: In Example 8.8, remove the transformer connected between phases C and A. This creates an open delta – open delta transformer bank. This transformer bank serves the same loads as in Example 8.8.

Determine the following:

1. the load line-to-line load voltages,
2. the secondary line currents,
3. the primary line currents,
4. the load currents, and
5. load voltage unbalance.

The exact same program from Example 8.8 is used since only the values of the matrices change for this connection. After six iterations, the results are as follows:

$$[VLL_{abc}] = \begin{bmatrix} 229.0/\underline{28.5} \\ 248.2/\underline{-86.9} \\ 255.4/\underline{147.2} \end{bmatrix}$$

$$[I_{abc}] = \begin{bmatrix} 529.9/\underline{-17.9} \\ 579.6/\underline{-161.1} \\ 353.8/\underline{82.8} \end{bmatrix}$$

$$[I_{ABC}] = \begin{bmatrix} 10.2/\underline{-17.9} \\ 11.2/\underline{-161.1} \\ 6.8/\underline{82.8} \end{bmatrix}$$

$$[ID_{abc}] = \begin{bmatrix} 436.7/\underline{2.7} \\ 201.4/\underline{-123.8} \\ 195.8/\underline{110.3} \end{bmatrix}$$

$$V_{unbalance} = 6.2\%$$

An inspection of the line-to-line load voltages should raise a question since two of the three voltages are greater than the no-load voltages of 240 volts. Why is there an apparent voltage rise on two of the phases? This can be explained by computing the voltage drops in the secondary circuit.

$$v_a = Zt_{ab} \cdot I_a = 12.6/\underline{58.0}$$

$$v_b = Zt_{bc} \cdot I_c = 15.5/\underline{149.6}$$

The ideal voltages are

$$Vt_{ab} = 240\underline{/30},$$

$$Vt_{bc} = 240\underline{/-90}$$

The terminal voltages are given by

$$V_{ab} = Vt_{ab} - v_a = 229.0\underline{/28.5},$$

$$V_{bc} = Vt_{bc} - v_b = 248.2\underline{/-86.9},$$

$$V_{ca} = -(V_{ab} + V_{bc}) = 255.4\underline{/147.2}$$

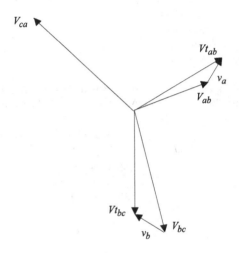

FIGURE 8.19 Voltage phasor diagram

then a voltage rise on phase bc. The voltage on ca also is greater than the rated 240 volts because the sum of the voltages must add to zero.

It is important that when there is a question about the results of a study that basic circuit and transformer theory, along with a phasor diagram, can confirm that the results are correct. This example is a good example of when the results should be confirmed. Notice also that the voltage unbalance is much greater for the open delta – open delta than the closed delta – delta connection.

8.12 THEVENIN EQUIVALENT CIRCUIT

This chapter has developed the general matrices for the forward and backward sweeps for most standard three-phase transformer connections. In Chapter 10, the section for short-circuit analysis will require the Thevenin equivalent circuit referenced for the secondary terminals of the transformer. This equivalent circuit must consider the equivalent impedance between the primary terminals of the transformer

Three-Phase Transformer Models

and the feeder source. Figure 8.20 is a general circuit showing the feeder source down to the secondary bus.

The Thevenin equivalent circuit needs to be determined at the secondary node of the transformer bank. This basically is the same as "referring" the source voltage and the source impedance to the secondary side of the transformer. The desired "Thevenin equivalent circuit" at the transformer secondary node is shown in Figure 8.21.

A general Thevenin equivalent circuit that can be used for all connections defined by the forward and backward sweep matrices.

In Figure 8.20, the primary transformer equivalent line-to-neutral voltages as a function of the source voltages and the equivalent high-side impedance is given by

$$[VLN_{ABC}] = [ELN_{ABC}] - [Zsys_{ABC}] \cdot [I_{ABC}], \quad (8.162)$$

but

$$[I_{ABC}] = [d_t] \cdot [I_{abc}];$$

therefore,

$$[VLN_{ABC}] = [ELN_{ABC}] - [Zsys_{ABC}] \cdot [d_t] \cdot [I_{abc}] \quad (8.163)$$

The forward sweep equation gives the secondary line-to-neutral voltages as a function of the primary line-to-neutral voltages.

$$[VLN_{abc}] = [A_t] \cdot [VLN_{ABC}] - [B_t] \cdot [I_{abc}] \quad (8.164)$$

Substitute Equation 8.163 into Equation 8.164.

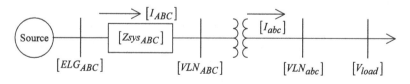

FIGURE 8.20 Equivalent system.

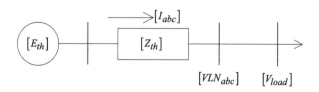

FIGURE 8.21 Thevenin equivalent circuit.

$$[VLN_{abc}] = [A_t] \cdot \{[ELN_{ABC}] - [Zsys_{ABC}] \cdot [d_t] \cdot [I_{abc}]\} - [B_t] \cdot [I_{abc}][VLN_{abc}]$$

$$= [VLN_{abc}] = [A_t] \cdot [ELN_{ABC}]$$

$$-([A_t] \cdot [Zsys_{ABC}] \cdot [d_t] + [B_t]) \cdot [I_{abc}][VLN_{abc}] = [E_{th}] - [Z_{th}] \cdot [I_{abc}],$$

where $[E_{th}] = [A_t] \cdot [ELN_{ABC}]$,

$$[Z_{th}] = ([A_t] \cdot [Zsys_{ABC}] \cdot [d_t] + [B_t]) \tag{8.165}$$

The definitions of the Thevenin equivalent voltages and impedances as given in Equation 8.165 are general and can be used for all transformer connections. Example 8.5 is used to demonstrate the computation and application of the Thevenin equivalent circuit.

Example 8.10: The delta – grounded wye transformer bank of Example 8.2 is connected to a balanced three-phase source of 115 kV through a 5-mile section of a four-wire, three-phase line, as shown in Figure 8.20.

The phase impedance matrix for the 5-mile-long 115 kV line is given by

$$[Zsys_{ABC}] = \begin{bmatrix} 2.2040 + j5.6740 & 0.6415 + j2.2730 & 0.6415 + j1.7850 \\ 0.6415 + j2.2730 & 2.1430 + j5.8920 & 0.6130 + j1.9750 \\ 0.6415 + j1.7850 & 0.6130 + j1.9750 & 2.2430 + j5.8920 \end{bmatrix} \Omega.$$

For the unbalanced load in Example 8.2 using a Mathcad program, the load line-to-neutral voltages and secondary and primary currents are computed as follows:

$$[V_{load}] = \begin{bmatrix} 6477.3/\underline{-66.7} \\ 6764.6/\underline{176.1} \\ 6691./\underline{53.8} \end{bmatrix}$$

$$[I_{abc}] = \begin{bmatrix} 262.5/\underline{-92.6} \\ 177.4/\underline{144.3} \\ 224.2/\underline{35.6} \end{bmatrix}$$

$$[I_{ABC}] = \begin{bmatrix} 24.3/\underline{-70.1} \\ 20.5/\underline{-175.2} \\ 27.4/\underline{63.7} \end{bmatrix}$$

Three-Phase Transformer Models

The Thevenin equivalent voltages and impedances referred to in the secondary terminals of the transformer bank are as follows:

$$[Eth_{abc}] = [A_t] \cdot [ELN_{ABC}] = \begin{bmatrix} 7200/\underline{-60} \\ 7200/\underline{180} \\ 7200/\underline{60} \end{bmatrix} V$$

$$[Zth_{abc}] = [A_t] \cdot [Zsys_{ABC}] \cdot [d_t] + [B_t]$$

$$[Zth_{abc}] = \begin{bmatrix} 0.2424 + j2.6648 & -0.0060 - j0.0140 & -0.0060 - j0.0173 \\ -0.0060 - j0.0140 & 0.2424 + j2.6610 & -0.0060 - j0.0135 \\ -0.0060 - j0.0173 & -0.0060 - j0.0135 & 0.2424 + j2.6643 \end{bmatrix} \Omega$$

The Thevenin equivalent circuit for this case is shown in Figure 8.20.

It is always good to confirm the Thevenin equivalent circuit by using the solved-for load currents and then the Thevenin equivalent circuit to compute the load voltage.

$$[V_{load}] = [Eth_{abc}] - [(Zth_{abc} + Zline_{abc})] \cdot [I_{abc}]$$

$$[V_{load}] = \begin{bmatrix} 6477.3/\underline{-66.7} \\ 6464.6/\underline{176.1} \\ 6691.9/\underline{53.8} \end{bmatrix}$$

The Mathcad program was modified to match the equivalent system in Figure 8.18. The load voltages and load currents were computed. This example is intended to demonstrate that it is possible to compute the Thevenin equivalent circuit at the secondary terminals of the transformer bank. The example shows that using the Thevenin equivalent circuit and the original secondary line currents that the original equivalent line-to-neutral load voltages are computed. The major application of the Thevenin equivalent circuit will be in the short-circuit analysis of a distribution feeder that will be developed in Chapter 10.

8.13 SUMMARY

In this chapter, the forward and backward sweep matrices have been developed for seven common three-phase transformer bank connections. For unbalanced transformer connections, the derivations were limited to just one of at least three ways that the primary phases could be connected to the transformer bank. The methods in the derivation of these transformer banks can be extended to all possible phasing.

One of the major features of the chapter has been to demonstrate how the forward and backward sweep technique (ladder) is used to analyze the operating characteristics of the transformer banks. Several Mathcad programs were used in the examples to demonstrate how the analysis is mostly independent of the transformer connection by using the derived matrices. This approach was first demonstrated with the line models and then continued to the voltage regulators and now the transformer connections. In Chapter 10, the analysis of a total distribution feeder will be developed using the forward and backward sweep matrices for all possible system components.

Many of the examples demonstrated the use of a Mathcad program for the analysis. An extension of this is the use of the student version of the WindMil distribution analysis program that can be downloaded as explained in the preface of this text. When the program is downloaded, a "user's manual" will be included. The user's manual serves two purposes:

- a tutorial on how to get started using WindMil for the first time user, and
- access to WindMil systems for many of the examples in this and other chapters.

It is highly encouraged that the program and manual be downloaded.

PROBLEMS

8.1 A three-phase substation transformer is connected delta – grounded wye and rated:

5,000 kVA, 115 kV delta – 12.47 kV grounded wye, $Z = 1.0 + j7.5\%$.

The transformer serves an unbalanced load of the following:

Phase a: 1384.5 kVA, 89.2% lagging power factor at 6922.5/-33.1 V
Phase b: 1691.2 kVA, 80.2% lagging power factor at 6776.8/-153.4 V
Phase c: 1563.0 kVA, unity power factor at 7104.7/85.9 V

(a) Determine the forward and backward sweep matrices for the transformer.
(b) Compute the primary equivalent line-to-neutral voltages.
(c) Compute the primary line-to-line voltages.
(d) Compute the primary line currents.
(e) Compute the currents flowing in the high-side delta windings.
(f) Compute the real power loss in the transformer for this load condition.

8.2 Write a simple MATLAB program using the ladder technique to solve for the load line-to-ground voltages and line currents in the bank of 8.1 when the source voltages are balanced three-phase of 115 kV line-to-line.

8.3 Create the system in WindMil for Problem 8.2.

8.4 Three single-phase transformers are connected in delta – grounded wye and serve an unbalanced load. The ratings of three transformers are as follows:

Phase A-B: 100 kVA, 12,470 –120 volts, $Z = 1.3 + j1.7\%$
Phase B-C: 50 kVA, 12,470 – 120 volts, $Z = 1.1 + j1.4\%$
Phase C-A: same as Phase B-C transformer

The unbalanced loads are as follows:

Phase a: 40 kVA, 0.8 lagging power factor at $V = 117.5\underline{/-32.5}$ V
Phase b: 85 kVA, 0.95 lagging power factor at $V = 115.7\underline{/-147.3}$ V
Phase c: 50 kVA, 0.8 lagging power factor at $V = 117.0\underline{/95.3}$ V

Modify the MATLAB script M0802.m to
(a) determine the forward and backward sweep matrices for this connection,
(b) compute the load currents,
(c) compute the primary line-to-neutral voltages,
(d) compute the primary line-to-line voltages,
(e) compute the primary currents,
(f) compute the currents in the delta primary windings, and
(g) compute the transformer bank real power loss.

8.5 For the same load and transformers in Problem 8.4, assume that the primary voltages on the transformer bank are a balanced three phase of 12,470 V line to line. Write a MATLAB program to compute the load line-to-ground voltages and the secondary line currents.

8.6 For the transformer connection and loads of Problem 8.4, build the system in WindMil.

8.7 The three single-phase transformers of Problem 8.4 are serving an unbalanced constant impedance load of the following:

Phase a: $0.32 + j0.14$ Ω
Phase b: $0.21 + j0.08$ Ω
Phase c: $0.28 + j0.12$

The transformers are connected to a balanced 12.47 kV source.
(a) Determine the load currents.
(b) Determine the load voltages.
(c) Compute the complex power of each load.
(d) Compute the primary currents.
(e) Compute the operating kVA of each transformer.

8.8 Solve Problem 8.7 using WindMil.

8.9 A three-phase transformer connected wye – delta is rated

$$500 \text{ kVA}, 4160 - 240 \text{ volts}, Z = 1.1 + j3.0\%.$$

The primary neutral is ungrounded. The transformer is serving a balanced load of 480 kW with balanced voltages of 235 volts line to line and a lagging power factor of 0.9. Modify the MATLAB script M0804.m to
(a) compute the secondary line currents,
(b) compute the primary line currents,
(c) compute the currents flowing in the secondary delta windings, and
(d) compute the real power loss in the transformer for this load.

8.10 The transformer of Problem 8.9 is serving an unbalanced delta load of the following:

S_{ab} = 150 kVA, 0.95 lagging power factor
S_{bc} = 125 kVA, 0.90 lagging power factor
S_{ca} = 160 kVA, 0.8 lagging power factor

The transformer bank is connected to a balanced three-phase source of 4,160 volts line to line.
(a) Compute the forward and backward sweep matrices for the transformer bank.
(b) Compute the load equivalent line-to-neutral and line-to-line voltages.
(c) Compute the secondary line currents.
(d) Compute the load currents.
(e) Compute the primary line currents.
(f) Compute the operating kVA of each transformer winding.
(g) Compute the load voltage unbalance.

8.11 Three single-phase transformers are connected in an ungrounded wye – delta connection and serve an unbalanced delta-connected load. The transformers are rated as follows:

Phase A: 15 kVA, 2400 – 240 volts, $Z = 1.3 + j1.0\%$
Phase B: 25 kVA, 2400 – 240 volts, $Z = 1.1 + j1.1\%$
Phase C: Same as phase A transformer

The transformers are connected to a balanced source of 4.16 kV line to line. The primary currents entering the transformer are as follows:

I_A = 4.60 A, 0.95 lagging power factor
I_B = 6.92 A, 0.88 lagging power factor
I_C = 5.37 A, 0.69 lagging power factor

(a) Determine the primary line-to-neutral voltages. Select V_{AB} as a reference.
(b) Determine the line currents entering the delta-connected load.
(c) Determine the line-to-line voltages at the load.
(d) Determine the operating kVA of each transformer.
(e) Is it possible to determine the load currents in the delta-connected load? If so, do it. If not, why not?

8.12 The three transformers of Problem 8.11 are serving an unbalanced delta-connected load of the following:

S_{ab} = 10 kVA, 0.95 lagging power factor
S_{bc} = 20 kVA, 0.90 lagging power factor
S_{ca} = 15 kVA, 0.8 lagging power factor

The transformers are connected to a balanced 4,160 line-to-line voltage source. Determine the load voltages, the primary and secondary line currents for the following transformer connections:
- Ungrounded wye – delta connection

Three-Phase Transformer Models

- Grounded wye – delta connection
- Open wye – open delta connection where the transformer connected to phase C has been removed

This is easily done by modifying the example MATLAB scripts to do this. It is up to the student to decide which programs to modify.

8.13 Three single-phase transformers are connected in grounded wye – grounded wye and serve an unbalance constant impedance load. The transformer bank is connected to a balanced three-phase, 12.47, line-to-line voltage source. The transformers are rated as follows:

Phase A: 100 kVA, 7200 – 120 volts, $Z = 0.9 + j1.8$ %
Phase B and Phase C: 37.5 kVA, 7200 – 120 volts, $Z = 1.1 + j1.4$ %

The constant impedance loads are as follows:

Phase a: $0.14 + j0.08 \, \Omega$
Phase b: $0.40 + j0.14 \, \Omega$
Phase c: $0.50 + j0.20 \, \Omega$

Write a MATLAB program to
(a) compute the generalized matrices for this transformer bank,
(b) determine the load currents,
(c) determine the load voltages,
(d) determine the kVA and power factor of each load,
(e) determine the primary line currents, and
(f) determine the operating kVA of each transformer.

8.14 Three single-phase transformers are connected in delta – delta and are serving a balanced three-phase motor rated 150 kVA, 0.8 lagging power factor, and a single-phase lighting load of 25 kVA, 0.95 lagging power factor connected across phases a-c. The transformers are rated as follows:

Phase A-B: 75 kVA, 4800 – 240 volts, $Z = 1.0 + j1.5\%$
Phase B-C: 50 kVA, 4800 – 240 volts, $Z = 1.1 + j1.4$ %
Phase C-A: same as Phase B-C

The load voltages are a balanced three phase of 240 volts line to line. Modify the MATLAB script M0808.m to
(a) determine the forward and backward sweep matrices,
(b) compute the motor input currents,
(c) compute the single-phase lighting load current,
(d) compute the primary line currents,
(e) compute the primary line-to-line voltages, and
(f) compute the currents flowing in the primary and secondary delta windings.

8.15 In Problem 8.14 the transformers on phase A-B and B-C are connected in an open delta – open delta connection and serve an unbalanced three-phase load of the following:

Phase a-b: 50 kVA at 0.9 lagging power factor

Phase b-c: 15 kVA at 0.85 lagging power factor
Phase c-a: 25 kVA at 0.95 lagging power factor

The source line-to-line voltages are balanced at 4,800 volts line to line. Modify the MATLAB script M0809.m to complete the following:

Determine

(g) the load line-to-line voltages,
(h) the load currents,
(i) the secondary line currents, and
(j) the primary line currents.

WINDMIL ASSIGNMENT

Use System 4 to build new System 5. A 5,000 kVA delta – grounded wye substation transformer is to be connected between the source and Node 1. The voltages for the transformer are 115 kV delta to 12.47 kV grounded wye. The impedance of this transformer is 8.06 %, with an X/R ratio of 8. When installing this substation transformer, be sure to modify the source so that it is 115,000 V rather than the 12.47 V. Follow the steps in the user's manual on how to install the substation transformer.

1. When the transformer has been connected, run voltage drop.
2. What are the node voltages at Node 2?
3. What taps has the regulator gone to?
4. Why did the taps increase when the transformer was added to the system?

9 Load Models

The loads on a distribution system are typically specified by the complex power consumed. With reference to Chapter 2, the specified load will be the "maximum diversified demand". This demand can be specified as kVA and power factor, kW and power factor, or kW and kvar. The voltage specified will always be the voltage at the low voltage bus of the distribution substation. This creates a problem since the current requirement of the loads cannot be determined without knowing the voltage. The iterative power-flow program "ladder" technique, or the "backward/forward sweep" technique, was introduced in Chapter 3.

Loads on a distribution feeder can be modeled as wye connected or delta connected. The loads can be three phase, two phase, or single phase with any degree of unbalance and can be modeled as

- constant real and reactive power (constant PQ),
- constant current,
- constant impedance, or
- any combination of the above.

The load models developed are to be used in the "ladder" technique where the load voltages are initially assumed. One of the results of the "ladder" technique is to replace the assumed voltages with the actual operating load voltages. All models are initially defined by a complex power per phase and an assumed line-to-neutral voltage (wye load) or an assumed line-to-line voltage (delta load). The units of the complex power can be in volt-amperes and volts or per-unit volt-amperes and per-unit voltages. For all loads the line currents entering the load are required to perform the power-flow analysis.

9.1 WYE-CONNECTED LOADS

Figure 9.1 is the model of a wye-connected load.

The notation for the specified complex powers and voltages are as follows:

$$\text{Phase a}: |S_a|/\underline{\theta_a} = P_a + jQ_a \text{ and } |V_{an}|/\underline{\delta_a} \qquad (9.1)$$

$$\text{Phase b}: |S_b|/\underline{\theta_b} = P_b + jQ_b \text{ and } |V_{bn}|/\underline{\delta_b} \qquad (9.2)$$

$$\text{Phase c}: |S_c|/\underline{\theta_c} = P_c + jQ_c \text{ and } |V_{cn}|/\underline{\delta_c} \qquad (9.3)$$

DOI: 10.1201/9781003261094-9

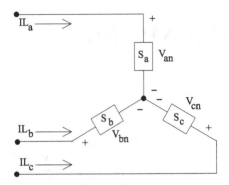

FIGURE 9.1 Wye-connected load.

9.1.1 Constant Real and Reactive Power Loads

The line currents for constant real and reactive power loads (PQ loads) are given by the following:

$$IL_a = \left(\frac{S_a}{V_{an}}\right)^* = \frac{|S_a|}{|V_{an}|} /\delta_a - \theta_a = |IL_a| /\alpha_a$$

$$IL_b = \left(\frac{S_b}{V_{bn}}\right)^* = \frac{|S_b|}{|V_{bn}|} /\delta_b - \theta_b = |IL_b| /\alpha_b$$

$$IL_c = \left(\frac{S_c}{V_{cn}}\right)^* = \frac{|S_c|}{|V_{cn}|} /\delta_c - \theta_c = |IL_c| /\alpha_c \qquad (9.4)$$

In this model, the line-to-neutral voltages will change during each iteration until convergence is achieved.

9.1.2 Constant Impedance Loads

The "constant load impedance" is first determined from the specified complex power and assumed line-to-neutral voltages.

$$Z_a = \frac{|V_{an}|^2}{S_a^*} = \frac{|V_{an}|^2}{|S_a|} /\theta_a = |Z_a| /\theta_a$$

$$Z_b = \frac{|V_{bn}|^2}{S_b^*} = \frac{|V_{bn}|^2}{|S_b|} /\theta_b = |Z_b| /\theta_b$$

$$Z_c = \frac{|V_{cn}|^2}{S_c^*} = \frac{|V_{cn}|^2}{|S_c|} /\theta_c = |Z_c| /\theta_c \qquad (9.5)$$

Load Models

The load currents as a function of the constant impedances are given by the following:

$$IL_a = \frac{V_{an}}{Z_a} = \frac{|V_{an}|}{|Z_a|} /\delta_a - \theta_a = |IL_a| /\alpha_a$$

$$IL_b = \frac{V_{bn}}{Z_b} = \frac{|V_{bn}|}{|Z_b|} /\delta_b - \theta_b = |IL_b| /\alpha_b$$

$$IL_c = \frac{V_{cn}}{Z_c} = \frac{|V_{cn}|}{|Z_c|} /\delta_c - \theta_c = |IL_c| /\alpha_c \qquad (9.6)$$

In this model, the line-to-neutral voltages will change during each iteration, but the impedance computed in Equation 9.5 will remain constant.

9.1.3 Constant Current Loads

In this model, the magnitudes of the currents are computed according to Equations 9.4 and then held constant while the angle of the voltage (δ) changes resulting in a changing angle on the current so that the power factor of the load remains constant.

$$IL_a = |IL_a| /\delta_a - \theta_a,$$

$$IL_b = |IL_b| /\delta_b - \theta_b,$$

$$IL_c = |IL_c| /\delta_c - \theta_c, \qquad (9.7)$$

where
δ_{abc} = line-to-neutral voltage angles,
θ_{abc} = power factor angles.

9.1.4 Combination Loads

Combination loads can be modeled by assigning a percentage of the total load to each of the three aforementioned load models. The current for the constant impedance load is computed assuming nominal load voltage. In a similar manner, the current for the constant current load is computed assuming nominal load voltage. All load currents will change as the load voltage changes in the iterative process. The total line current entering the load is the sum of the three components.

Example 9.1: A combination load is served at the end of a 10,000-foot-long, three-phase distribution line. The impedance of the three-phase line is

$$[Z_{abc}] = \begin{bmatrix} 0.8667 + j2.0417 & 0.2955 + j0.9502 & 0.2907 + j0.7290 \\ 0.2955 + j0.9502 & 0.8837 + j1.9852 & 0.2992 + j0.8023 \\ 0.2907 + j0.7290 & 0.2992 + j0.8023 & 0.8741 + j2.0172 \end{bmatrix}$$

The complex powers of a combination wye-connected loads at nominal voltages are as follows:

$S_{an} = 2240$ at 0.85 power factor

$S_{bn} = 2500$ at 0.95 power factor

$S_{cn} = 2000$ at 0.90 power factor

$$[S_{abc}] = \begin{bmatrix} S_{an} \\ S_{bn} \\ S_{cn} \end{bmatrix} = \begin{bmatrix} 1904.0 + j1180.0 \\ 2375.0 + j780.6 \\ 1800.0 + j871.8 \end{bmatrix} \text{kVA}$$

The load is specified to be 50% constant complex power, 20% constant impedance, and 30% constant current. The nominal line-to-line voltage of the feeder is 12.47 kV.

(a) Assume the nominal voltage and compute the component of load current attributed to each of the loads and the total load current.
The assumed line-to-neutral source voltages at the start of the iterative routine are

$$[ELN_{abc}] = \begin{bmatrix} 7200\underline{/0} \\ 7200\underline{/-120} \\ 7200\underline{/120} \end{bmatrix} \text{V}$$

The complex powers for each of the loads are as follows:
Complex power load:

$$[SP] = [S_{abc}] \cdot 0.5 = \begin{bmatrix} 952.0 + j590.0 \\ 1187.5 + j390.3 \\ 900.0 + j435.9 \end{bmatrix}$$

Constant impedance load:

$$[SZ] = [S_{abc}] \cdot 0.2 = \begin{bmatrix} 380.8 + j236.0 \\ 475.0 + j156.1 \\ 360.0 + j174.4 \end{bmatrix}$$

Constant current load:

$$[SI] = [S_{abc}] \cdot 0.3 = \begin{bmatrix} 571.2 + j354.0 \\ 712.5 + j234.2 \\ 540.0 + j261.5 \end{bmatrix}$$

The currents due to the constant complex power computed at nominal voltages are:

$$Ipq_i = \left(\frac{SP_i \cdot 1000}{VLN_i}\right)^* = \begin{bmatrix} 155.6/\underline{-31.8} \\ 173.6/\underline{-138.2} \\ 138.9/\underline{94.2} \end{bmatrix} A$$

The constant impedances for that part of the load are computed as

$$Z_i = \frac{|VLN_i|^2}{SZ_i^* \cdot 1000} = \begin{bmatrix} 98.4 + j61.0 \\ 98.5 + j33.4 \\ 116.6 + j56.5 \end{bmatrix} \Omega$$

For the first iteration, the currents due to the constant impedance portion of the loads are

$$Iz_i = \left(\frac{VLN_i}{Z_i}\right) = \begin{bmatrix} 62.2/\underline{-31.8} \\ 69.4/\underline{-138.2} \\ 55.6/\underline{94.2} \end{bmatrix} A$$

The magnitudes of the constant current portion of the loads are

$$IM_i = \left|\left(\frac{SL_i \cdot 1000}{VLN_i}\right)^*\right| = \begin{bmatrix} 93.3 \\ 104.2 \\ 83.3 \end{bmatrix} A$$

The contribution of the load currents due to the constant current portion of the loads are

$$II_i = IM_i/\underline{\delta_i - \theta_i} = \begin{bmatrix} 93.3/\underline{-31.8} \\ 104.2/\underline{-138.2} \\ 83.3/\underline{94.2} \end{bmatrix} A$$

The total load currents are the sum of the three components:

$$[I_{abc}] = [Ipq] + [IZ] + [II] = \begin{bmatrix} 311.1/\underline{-31.8} \\ 347.2/\underline{-138.2} \\ 277.8/\underline{94.2} \end{bmatrix} A$$

To check that the computed currents, give the initial complex power:

$$S_{abc_i} = \frac{VLN_i \cdot I_{abc_i}{}^*}{1000} = \begin{bmatrix} 1904.0 + j1180.0 \\ 2375.0 + j780.6 \\ 1800.0 + j871.8 \end{bmatrix}$$

This gives the same complex powers that were given for nominal load voltages.

(b) Determine the currents at the start of the second iteration. The voltages at the load after the first iteration are

$$[VLN] = [ELN] - [Z_{abc}] \cdot [I_{abc}],$$

$$[VLN] = \begin{bmatrix} 6702.2 /\underline{-1.2} \\ 6942.8 /\underline{-123.5} \\ 7006.5 /\underline{118.4} \end{bmatrix} V$$

The steps are repeated with the exception that the impedances of the constant impedance portion of the load will not be changed and the magnitude of the currents for the constant current portion of the load change will not change.

The constant complex power portion of the load currents are

$$Ipq_i = \left(\frac{SP_i \cdot 1000}{VLN_i} \right)^* = \begin{bmatrix} 167.1 /\underline{-33.1} \\ 180.0 /\underline{-141.7} \\ 142.7 /\underline{92.5} \end{bmatrix} A$$

The currents due to the constant impedance portion of the load are

$$Iz_i = \left(\frac{VLN_i}{Z_i} \right) = \begin{bmatrix} 57.9 /\underline{-33.1} \\ 67.0 /\underline{-141.7} \\ 53.1 /\underline{92.5} \end{bmatrix} A$$

The currents due to the constant current portion of the load are

$$II_i = IM_i /\underline{\delta_i - \theta_i} = \begin{bmatrix} 93.3 /\underline{-33.1} \\ 104.2 /\underline{-141.7} \\ 83.3 /\underline{92.5} \end{bmatrix} A$$

The total load currents at the start of the second iteration will be

$$[I_{abc}] = [Ipq] + [IZ] + [II] = \begin{bmatrix} 318.4 /\underline{-33.1} \\ 351.2 /\underline{-141.7} \\ 280.1 /\underline{92.5} \end{bmatrix} A$$

Load Models

The new complex powers of the combination loads:

$$S_{abc_i} = \frac{VLN_i \cdot I_{abc_i}^*}{1000} = \begin{bmatrix} 1813.7 + j1124.0 \\ 2316.2 + j761.3 \\ 1766.4 + j855.5 \end{bmatrix}$$

Because the load voltages have changed, the total complex power has also changed.

Observe how the currents have changed from the original currents. The currents for the constant complex power loads have increased because the voltages are reduced from the original assumption. The currents for the constant impedance portion of the load have decreased because the impedance stayed constant, but the voltages are reduced. Finally, the magnitude of the constant current portion of the load has remained constant. Again, all three components of the loads have the same phase angles since the power factor of the load has not changed.

9.2 DELTA-CONNECTED LOADS

The model for a delta-connected load is shown in Figure 9.2.

The notation for the specified complex powers and voltages in Figure 9.2 is as follows:

$$\text{Phase ab}: |S_{ab}|\underline{/\theta_{ab}} = P_{ab} + jQ_{ab} \text{ and } |V_{ab}|\underline{/\delta_{ab}} \quad (9.8)$$

$$\text{Phase bc}: |S_{bc}|\underline{/\theta_{bc}} = P_{bc} + jQ_{bc} \text{ and } |V_{bc}|\underline{/\delta_{bc}} \quad (9.9)$$

$$\text{Phase ca}: |S_{ca}|\underline{/\theta_{ca}} = P_{ca} + jQ_{ca} \text{ and } |V_{ca}|\underline{/\delta_{ca}} \quad (9.10)$$

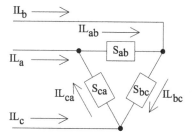

FIGURE 9.2 Delta-connected load.

9.2.1 Constant Real and Reactive Power Loads

The currents in the delta-connected loads are as follows:

$$IL_{ab} = \left(\frac{S_{ab}}{V_{ab}}\right)^* = \frac{|S_{ab}|}{|V_{ab}|} /\underline{\delta_{ab} - \theta_{ab}} = |IL_{ab}| /\underline{\alpha_{ab}}$$

$$IL_{bc} = \left(\frac{S_{bc}}{V_{bc}}\right)^* = \frac{|S_{bc}|}{|V_{bc}|} /\underline{\delta_{bc} - \theta_{bc}} = |IL_{bc}| /\underline{\alpha_{bc}}$$

$$IL_{ca} = \left(\frac{S_{ca}}{V_{ca}}\right)^* = \frac{|S_{ca}|}{|V_{ca}|} /\underline{\delta_{ca} - \theta_{ca}} = |IL_{ca}| /\underline{\alpha_{ca}} \qquad (9.11)$$

In this model, the line-to-line voltages will change during each iteration, resulting in new current magnitudes and angles at the start of each iteration.

9.2.2 Constant Impedance Loads

The "constant load impedance" is first determined from the specified complex power and line-to-line voltages:

$$Z_{ab} = \frac{|V_{ab}|^2}{S_{ab}^*} = \frac{|V_{ab}|^2}{|S_{ab}|} /\underline{\theta_{ab}} = |Z_{ab}| /\underline{\theta_{ab}}$$

$$Z_{bc} = \frac{|V_{bc}|^2}{S_{bc}^*} = \frac{|V_{bc}|^2}{|S_{bc}|} /\underline{\theta_{bc}} = |Z_{bc}| /\underline{\theta_{bc}}$$

$$Z_{ca} = \frac{|V_{ca}|^2}{S_{ca}^*} = \frac{|V_{ca}|^2}{|S_{ca}|} /\underline{\theta_{ca}} = |Z_{ca}| /\underline{\theta_{ca}} \qquad (9.12)$$

The delta load currents as a function of the "constant load impedances" are as follows:

$$IL_{ab} = \frac{V_{ab}}{Z_{ab}} = \frac{|V_{ab}|}{|Z_{ab}|} /\underline{\delta_{ab} - \theta_{ab}} = |IL_{ab}| /\underline{\alpha_{ab}}$$

$$IL_{bc} = \frac{V_{bc}}{Z_{bc}} = \frac{|V_{bc}|}{|Z_{bc}|} /\underline{\delta_{bc} - \theta_{bc}} = |IL_{bc}| /\underline{\alpha_{bc}}$$

$$IL_{ca} = \frac{V_{ca}}{Z_{ca}} = \frac{|V_{ca}|}{|Z_{ca}|} /\underline{\delta_{ca} - \theta_{ca}} = |IL_{ca}| /\underline{\alpha_{ca}} \qquad (9.13)$$

Load Models 295

In this model, the line-to-line voltages will change during each iteration until convergence is achieved.

9.2.3 Constant Current Loads

In this model, the magnitudes of the currents are computed according to Equations 9.11 and then held constant while the angle of the voltage (δ) changes during each iteration. This keeps the power factor of the load constant.

$$IL_{ab} = |IL_{ab}| \underline{/\delta_{ab} - \theta_{ab}}$$

$$IL_{bc} = |IL_{bc}| \underline{/\delta_{bc} - \theta_{bc}}$$

$$IL_{ca} = |IL_{bca}| \underline{/\delta_{ca} - \theta_{ca}} \qquad (9.14)$$

9.2.4 Combination Loads

Combination loads can be modeled by assigning a percentage of the total load to each of the three aforementioned load models. The total delta current for each load is the sum of the three components.

9.2.5 Line Currents Serving a Delta-Connected Load

The line currents entering the delta-connected load are determined by applying KCL at each of the nodes of the delta. In matrix form, the equations are

$$\begin{bmatrix} I_a \\ I_b \\ I_c \end{bmatrix} = \begin{bmatrix} 1 & 0 & -1 \\ -1 & 1 & 0 \\ 0 & -1 & 1 \end{bmatrix} \cdot \begin{bmatrix} IL_{ab} \\ IL_{bc} \\ IL_{ca} \end{bmatrix},$$

$$[I_{abc}] = [Di] \cdot [IL_{abc}]. \qquad (9.15)$$

9.3 TWO-PHASE AND SINGLE-PHASE LOADS

In both the wye- and delta-connected loads, single-phase and two-phase loads are modeled by setting the currents of the missing phases to zero. The currents in the phases present are computed using the same appropriate equations for constant complex power, constant impedance, and constant current.

9.4 SHUNT CAPACITORS

Shunt capacitor banks are commonly used in distribution systems to help in voltage regulation and to provide reactive power support. The capacitor banks are modeled as constant susceptances connected in either wye or delta. Similar to the load model,

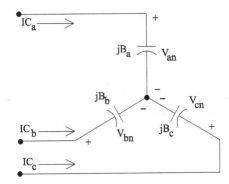

FIGURE 9.3 Wye-connected capacitor bank.

all capacitor banks are modeled as three-phase banks with the currents of the missing phases set to zero for single-phase and two-phase banks.

9.4.1 Wye-Connected Capacitor Bank

The model of a three-phase, wye-connected shunt capacitor bank is shown in Figure 9.3.

The individual phase capacitor units are specified in *kvar* and kV. The constant susceptance for each unit can be computed in Siemens. The susceptance of a capacitor unit is computed by

$$B_c = \frac{kVAr \cdot 1000}{kV_{LN}^2} S \qquad (9.16)$$

With the susceptance computed, the line currents serving the capacitor bank are given by the following:

$$IC_a = jB_a \cdot V_{an}$$

$$IC_b = jB_b \cdot V_{bn}$$

$$IC_c = jB_c \cdot V_{cn} \qquad (9.17)$$

9.4.2 Delta-Connected Capacitor Bank

The model for a delta-connected shunt capacitor bank is shown in Figure 9.4.

The individual phase capacitor units are specified in kVAr and kV. For the delta-connected capacitors, the kV must be the line-to-line voltage. The constant

Load Models

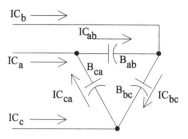

FIGURE 9.4 Delta-connected capacitor bank.

susceptance for each unit can be computed in siemens. The susceptance of a capacitor unit is computed by

$$B_c = \frac{kVAr \cdot 1000}{kV_{LL}^2} \text{ S} \qquad (9.18)$$

With the susceptance computed, the delta currents serving the capacitor bank are given by

$$IC_{ab} = jB_{ab} \cdot V_{ab}$$

$$IC_{bc} = jB_{bc} \cdot V_{bc}$$

$$IC_{ca} = jB_{ca} \cdot V_{ca} \qquad (9.19)$$

The line currents flowing into the delta-connected capacitors are computed by applying KCL at each node. In matrix form, the equations are

$$\begin{bmatrix} IC_a \\ IC_b \\ IC_c \end{bmatrix} = \begin{bmatrix} 1 & 0 & -1 \\ -1 & 1 & 0 \\ 0 & -1 & 1 \end{bmatrix} \cdot \begin{bmatrix} IC_{ab} \\ IC_{bc} \\ IC_{ca} \end{bmatrix} \qquad (9.20)$$

9.5 THREE-PHASE INDUCTION MACHINE

The analysis of an induction machine (motor or generator) when operating with unbalanced voltage conditions has traditionally been performed using the method of symmetrical components. In this section, the symmetrical component analysis method will be used to establish a baseline for the machine operation. Once the sequence currents and voltages in the machine have been determined, they are converted to the phase domain. Direct analysis of the machine in the phase domain is introduced and is employed for the analysis of both motors and generators.

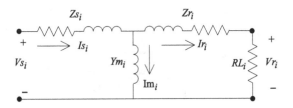

FIGURE 9.5 Sequence network.

9.5.1 Induction Machine Model

The equivalent positive and negative sequence networks for an induction machine can be represented by the circuit in Figure 9.5. Since all induction machines are connected in either an ungrounded wye or delta, there will not be any zero-sequence currents and voltages so only the positive and negative sequence networks are analyzed. In the circuit of Figure 9.5, the power consumed by the resistors (RL_i) represents the electrical power being converted to shaft power.

In Figure 9.5,

$i = 1$ for the positive sequence circuit,

$i = 2$ for the negative sequence circuit.

The given parameters for the induction machine are assumed to be the following:

$$kVA_3 = hp = \text{three}-\text{phase rating},$$

$$kVA_1 = \frac{kVA_3}{3} = \text{single}-\text{phase rating}$$

V_{LL}: rated line – to – line voltage

$$V_{LN} = \frac{V_{LL}}{\sqrt{3}}: \text{rated line}-\text{to}-\text{neutral voltage}$$

$Zs = Rs + jXs$: stator sequence impedance in per – unit

$Zr = Rr + jXr$: rotor sequence impedance in per – unit

$Zm = jXm$: magnetizing impedance in per – unit

Load Models

The impedances must be converted to actual impedances in ohms. Two sets of base values are needed. The impedances in ohms are computed for the wye and delta connections by the following:

Wye

$$IY_{base} = \frac{kVA_1 \cdot 1000}{V_{LN}}$$

$$ZY_{base} = \frac{V_{LN}}{IY_{base}}$$

$$ZY_{ohms} = Z_{pu} \cdot ZY_{base}$$

Delta

$$ID_{base} = \frac{kVA_1 \cdot 1000}{V_{LL}}$$

$$ZD_{base} = \frac{V_{LL}}{ID_{base}}$$

$$ZD_{ohms} = Z_{pu} \cdot ZD_{base} \qquad (9.21)$$

Typically, the sequence networks are assumed to be a wye connection. When the motor is delta connected, the computation for the impedances inside the delta is shown in Equation 9.21. These delta impedances are equal and can be converted to an equivalent wye by dividing by three. As it turns out, these values of the machine impedances in ohms will be the same, as though they were computed using the wye-connected base values.

$$ZY_{ohms} = \frac{ZD_{ohms}}{3} \qquad (9.22)$$

Example 9.2: A three-phase induction machine is rated:

150 kVA, 480 line-to-line volts, P_{FW} = 3.25 kW
$Zs_{pu} = 0.0651 + j0.1627$, $Zr_{pu} = 0.0553 + j0.1136$, $Zm_{pu} = j4.0690$

Determine the wye and delta impedances in ohms.
Set base values:

$$kVA_{base} = 150$$

$$kVA1_{base} = \frac{150}{3} = 50$$

$$kVLL_{base} = \frac{480}{1000} = 0.48$$

$$kVLN_{base} = \frac{kVLL_{base}}{\sqrt{3}} = 0.2771$$

Base values for wye connection:

$$IY_{base} = \frac{kVA1_{base}}{kVLN_{base}} = 180.42$$

$$ZY_{base} = \frac{kVLN_{base} \cdot 1000}{IY_{base}} = 1.536$$

$$ZY_{base} = \frac{kVLN_{base}^2 \cdot 1000}{kVA1_{base}} = \frac{\left(\frac{kVLL_{base}}{\sqrt{3}}\right)^2 \cdot 1000}{\frac{kVA_{base}}{3}}$$

$$ZY_{base} = \frac{kVLL_{base}^2 \cdot 1000}{kVA_{base}} = 1.536$$

Wye-connected impedances in ohms:

$$Zs = Zs_{pu} \cdot ZY_{base} = 0.1 + j0.2499 \text{ ohms}$$

$$Zr = Zr_{pu} \cdot ZY_{base} = 0.0849 + j0.175 \text{ ohms}$$

$$Zm = Zm_{pu} \cdot ZY_{base} = j6.2501 \text{ ohms}$$

Base values for delta connection:

$$ID_{base} = \frac{kVA1_{base}}{kVLL_{base}} = 104.1667$$

$$ZD_{base} = \frac{kVLL_{base} \cdot 1000}{ID_{base}} = 4.608$$

$$ZD_{base} = \frac{kVLL_{base}^2 \cdot 1000}{kVA1_{base}} = \frac{kVLL_{base}^2 \cdot 1000}{\frac{kVA_{base}}{3}}$$

$$ZD_{base} = 3 \cdot \frac{kVLL_{base}^2 \cdot 1000}{kVA_{base}} = 3 \cdot ZY_{base}$$

Delta-connected impedances in ohms:

$$ZD_s = Zs_{pu} \cdot ZD_{base} = 0.3 + j.7497 \text{ ohms}$$

$$ZD_r = Zr_{pu} \cdot ZD_{base} = 0.2548 + j0.5249 \text{ ohms}$$

$$ZD_m = Zm_{pu} \cdot ZD_{base} = 18.7504 \text{ ohms}$$

Note that converting the delta impedances to wye impedances in ohms results in the same values by using the wye base values.

$$Zs_i = \frac{ZD_s}{3} = R_s + jX_s = 0.1 + j0.2499 \text{ ohms,}$$

$$Zr_i = \frac{ZD_r}{3} = R_r + X_r = 0.0849 + j0.175 \, \text{ohms},$$

$$Zm_i = \frac{ZD_m}{3} = jX_m = j6.2501 \, \text{ohms},$$

$$Ym_i = \frac{1}{Zm_i} = -j0.16 \, \text{S},$$

where $i = 1$ positive sequence,

$i = 2$ negative sequence.

9.5.2 Symmetrical Component Analysis of a Motor

In Figure 9.5, the motor sequence resistances are given by

$$RL_i = \frac{1 - s_i}{s_i} \cdot Rr_i,$$

where

positive sequence slip $= s_1 = \dfrac{n_s - n_r}{n_s}$,

negative sequence slip $= s_2 = 2 - s_1$,

$n_s =$ synchronous speed in rpm $= \dfrac{120 \cdot f}{p}$,

$f =$ synchronouse speed,
$p =$ number of poles,

$$n_r = \text{rotor speed in rpm.} \tag{9.23}$$

Note that the negative sequence load resistance will be a negative value that will lead to a negative component of shaft power.

The positive and negative sequence networks can be analyzed individually and then sequence currents and voltages converted to phase components. At this point, it is assumed that the stator line-to-line voltages are known. When only the magnitudes of the line-to-line voltages are known, the Law of Cosines is used to establish the angles on the voltages. The equivalent line-to-neutral voltages are needed for the analysis of the sequence networks. The line-to-neutral voltages are computed by

$$[VLL_{abc}] = \begin{bmatrix} V_{ab} \\ V_{bc} \\ V_{ca} \end{bmatrix},$$

$$[VLN_{abc}] = [W] \cdot [VLL_{abc}] = \begin{bmatrix} V_{an} \\ V_{bn} \\ V_{cn} \end{bmatrix},$$

$$\text{where } [W] = \frac{1}{3} \cdot \begin{bmatrix} 2 & 1 & 0 \\ 0 & 2 & 1 \\ 1 & 0 & 2 \end{bmatrix} \quad (9.24)$$

The computed line-to-neutral voltages are converted to sequence voltages by

$$[Vs_{012}] = [A]^{-1} \cdot [VLN_{abc}] = \begin{bmatrix} 0 \\ Vs_1 \\ Vs_2 \end{bmatrix},$$

$$\text{where } [A] = \begin{bmatrix} 1 & 1 & 1 \\ 1 & a^2 & a \\ 1 & a & a^2 \end{bmatrix},$$

$$a = 1\underline{/120}. \quad (9.25)$$

With the stator sequence voltages computed, the circuits of Figure 9.5 are analyzed to compute the sequence stator and rotor currents. The sequence input impedances for $i = 1$ and 2 are

$$Zin_i = Zs_i + \frac{Zm_i \cdot (Zr_i + RL_i)}{Zm_i + Zr_i + RL_i} \quad (9.26)$$

The stator input sequence currents are

$$Is_i = \frac{Vs_i}{Zin_i}.$$

$$[Is_{012}] = \begin{bmatrix} 0 \\ Is_1 \\ Is_2 \end{bmatrix} \quad (9.27)$$

The rotor currents rotor and voltages are computed by the following:

$$Vm_i = Vs_i - Zs_i \cdot Is_i$$

Load Models

$$Im_i = \frac{Vm_i}{Zm_i}$$

$$Ir_i = Is_i - Im_i$$

$$Vr_i = Vm_i - Zr_i \cdot Ir_i$$

$$\text{Or } Vr_i = RL_i \cdot Ir_i$$

$$[Ir_{012}] = \begin{bmatrix} 0 \\ Ir_1 \\ Ir_2 \end{bmatrix}$$

$$[Vr_{012}] = \begin{bmatrix} 0 \\ Vr_1 \\ Vr_2 \end{bmatrix} \quad (9.28)$$

After the sequence voltages and currents have been computed, they are converted to phase components by

$$[Is_{abc}] = [A] \cdot [Is_{012}],$$

$$[Ir_{abc}] = [A] \cdot [Ir_{012}],$$

$$[Vr_{abc}] = [A] \cdot [Vr_{012}] \quad (9.29)$$

The various complex powers are computed by the following:

$$Ss_{abc_i} = \frac{Vs_{abc_i} \cdot (Is_{abc_i})^*}{1000}$$

$$Ss_{total} = \sum_{k=1}^{3} Ss_{abc_k} = Ps_{total} + jQs_{total} \text{ kVA}$$

$$Sr_{abc_i} = \frac{Vr_{abc_i} \cdot (Ir_{abc_i})^*}{1000}$$

$$Sr_{total} = \sum_{k=1}^{3} Sr_{total_k} = P_{converted} \text{ kW} \quad (9.30)$$

Example 9.3: The motor of Example 9.2 is operating with a positive sequence slip of 0.035 and line-to-line input voltages of

$$[VLL_{abc}] = \begin{bmatrix} 480/\underline{0} \\ 490/\underline{-121.37} \\ 475/\underline{118.26} \end{bmatrix}$$

Compute

- stator and rotor currents,
- load output voltages, and
- input and output complex powers.

The line-to-neutral input phase and sequence voltages are

$$[Vs_{abc}] = [W] \cdot [VLL_{abc}] = \begin{bmatrix} 273.2/\underline{-30.7} \\ 281.9/\underline{-150.4} \\ 279.1/\underline{87.9} \end{bmatrix},$$

$$[Vs_{012}] = [A]^{-1} \cdot [Vs_{abc}] = \begin{bmatrix} 0 \\ 278.1/\underline{-31.0} \\ 5.1/\underline{130.1} \end{bmatrix}$$

The positive and negative sequence stator voltages are

$$[Vs] = \begin{bmatrix} 278.1/\underline{-31.0} \\ 5.1/\underline{130.1} \end{bmatrix}$$

The input stator sequence impedances are

For $k = 1$ and 2,

$$Zin_k = Zs + \frac{Zm \cdot (Zr + RL_k)}{Zm + Zr + RL_k} = \begin{bmatrix} 2.1098 + j1.1786 \\ 0.1409 + j0.4204 \end{bmatrix}$$

The positive and negative sequence stator, magnetizing, and rotor currents are as follows:

$$Is_k = \frac{Vs_k}{Zin_k} = \begin{bmatrix} 115.1/\underline{-60.2} \\ 11.5/\underline{58.6} \end{bmatrix}$$

$$Vm_k = Vs_k - Zs \cdot Is_k = \begin{bmatrix} 254.7/\underline{-35.4} \\ 2.0/\underline{135.1} \end{bmatrix}$$

Load Models

$$Im_k = \frac{Vm_k}{Zm} = \begin{bmatrix} 40.8\underline{/-125.4} \\ 0.3\underline{/45.1} \end{bmatrix}$$

$$Ir_k = Is_k - Im_k = \begin{bmatrix} 104.7\underline{/-39.6} \\ 11.2\underline{/59.0} \end{bmatrix}$$

The sequence current arrays are

$$[Is_{012}] = \begin{bmatrix} 0 \\ Is_1 \\ Is_2 \end{bmatrix} = \begin{bmatrix} 0 \\ 115.1\underline{/-60.2} \\ 11.5\underline{/58.6} \end{bmatrix},$$

$$[Ir_{012}] = \begin{bmatrix} 0 \\ Ir_1 \\ Ir_2 \end{bmatrix} = \begin{bmatrix} 0 \\ 104.7\underline{/-39.6} \\ 11.2\underline{/59.0} \end{bmatrix}$$

The stator and rotor phase currents are

$$[Is_{abc}] = [A] \cdot [Is_{012}] = \begin{bmatrix} 110.0\underline{/-55.0} \\ 126.6\underline{/179.7} \\ 109.6\underline{/54.6} \end{bmatrix},$$

$$[Ir_{abc}] = [A] \cdot [Ir_{012}] = \begin{bmatrix} 103.7\underline{/-33.4} \\ 115.2\underline{/-161.6} \\ 96.2\underline{/76.3} \end{bmatrix}$$

The rotor sequence and phase voltages are as follows:

$$Vr_k = Ir_k \cdot RL_k = \begin{bmatrix} 245.2\underline{/-39.6} \\ 0.5\underline{/-121.0} \end{bmatrix}$$

$$[Vr_{012}] = \begin{bmatrix} 0 \\ 245.2\underline{/-39.6} \\ 0.5\underline{/-121.0} \end{bmatrix}$$

$$[Vr_{abc}] = [A] \cdot [Vr_{012}] = \begin{bmatrix} 245.3\underline{-39.7} \\ 244.7\underline{/-159.5} \\ 245.5\underline{/80.5} \end{bmatrix}$$

The input and output complex powers are the following:

For $i = 1, 2, 3$

$$Ss_{abc_i} = \frac{Vs_{abc_i} \cdot (Is_{abc_k})^*}{1000} = \begin{bmatrix} 27.4 + j12.4 \\ 30.9 + j17.8 \\ 25.6 + j16.8 \end{bmatrix}$$

$$Ss_{total} = \sum_{k=1}^{3} Ss_{abc_k} = 83.87 + j47.00 \, \text{kW} + j\text{kvar}$$

$$Sr_{abc_i} = \frac{Vr_{abc_i} \cdot (I4_{abc_k})^*}{1000} = \begin{bmatrix} 25.28 - j2.76 \\ 28.19 + j1.02 \\ 23.57 + j1.74 \end{bmatrix}$$

$$Sr_{total} = \sum_{k=1}^{3} Sr_{abc_k} = 77.03 \, \text{kW}$$

$$P_{loss} = \text{Re}(Ss_{total}) - \text{Re}(Sr_{total}) = 6.84 \, \text{kW}$$

9.5.3 Phase Analysis of an Induction Motor

In the previous section, the analysis starts by converting known phase voltages to sequence voltages. These sequence voltages are then used to compute the stator and rotor sequence currents, along with the rotor output sequence voltages. The sequence currents and voltages are then converted to phase components. In the following sections, methods will be developed where the total analysis is performed only using the phase domain.

When the positive sequence slip (s_1) is known, the input sequence impedances for the positive and negative sequence networks can be determined as

$$ZM_i = Rs_i + jXs_i + \frac{(jXm_i) \cdot (Rr_i + RL_i + jXr_i)}{Rr_i + RL_i + j(Xm_i + Xr_i)} \quad (9.31)$$

Once the input sequence impedances have been determined, the analysis of an induction machine operating with unbalance voltages requires the following steps:

Step 1: Transform the known line-to-line voltages to sequence line-to-line voltages.

$$\begin{bmatrix} Vab_0 \\ Vab_1 \\ Vab_2 \end{bmatrix} = \frac{1}{3} \begin{bmatrix} 1 & 1 & 1 \\ 1 & a & a^2 \\ 1 & a^2 & a \end{bmatrix} \cdot \begin{bmatrix} V_{ab} \\ V_{bc} \\ V_{ca} \end{bmatrix} \quad (9.32)$$

Load Models

In Equation 9.32, $Vab_0 = 0$ because of KVL.
Equation 9.32 can be written as

$$[VLL_{012}] = [A]^{-1} \cdot [VLL_{abc}] \qquad (9.33)$$

Step 2: Compute the sequence line-to-neutral voltages from the line-to-line voltages:
When the machine is connected either in delta or ungrounded wye, the zero-sequence, line-to-neutral voltage can be assumed to be zero. The sequence line-to-neutral voltages as a function of the sequence line-to-line voltages are given by

$$Van_0 = Vab_0 = 0,$$

$$Van_1 = t^* \cdot Vab_1,$$

$$Van_2 = t \cdot Vab_2,$$

$$\text{where } t = \frac{1}{\sqrt{3}} \underline{/30} \qquad (9.34)$$

Equations 9.34 can be put into matrix form:

$$\begin{bmatrix} Van_0 \\ Van_1 \\ Van_2 \end{bmatrix} = \begin{bmatrix} 1 & 0 & 0 \\ 0 & t^* & 0 \\ 0 & 0 & t \end{bmatrix} \cdot \begin{bmatrix} Vab_0 \\ Vab_1 \\ Vab_2 \end{bmatrix},$$

$$[VLN_{012}] = [T] \cdot [VLL_{012}],$$

$$\text{where } [T] = \begin{bmatrix} 1 & 0 & 0 \\ 0 & t^* & 0 \\ 0 & 0 & t \end{bmatrix} \qquad (9.35)$$

Step 3: Compute the sequence line currents flowing into the machine:

$$Ia_0 = 0$$

$$Ia_1 = \frac{Van_1}{ZM_1}$$

$$Ia_2 = \frac{Van_2}{ZM_2} \qquad (9.36)$$

Step 4: Transform the sequence currents to phase currents:

$$[I_{abc}] = [A] \cdot [I_{012}],$$

$$\text{where } [A] = \begin{bmatrix} 1 & 1 & 1 \\ 1 & a^2 & a \\ 1 & a & a^2 \end{bmatrix},$$

$$a = 1/\underline{120} \qquad (9.37)$$

The four steps outlined above can be performed without computing the sequence voltages and currents. The procedure basically reverses the steps.

$$\text{Define } YM_i = \frac{1}{ZM_i} \qquad (9.38)$$

The sequence currents are

$$I_0 = 0,$$

$$I_1 = YM_1 \cdot Van_1,$$

$$I_2 = YM_2 \cdot Van_2 \qquad (9.39)$$

Since I_0 and Vab_0 are both zero, the following relationship is true:

$$I_0 = Vab_0 \qquad (9.40)$$

Equations 9.39 and 9.40 can be put into matrix form:

$$\begin{bmatrix} I_0 \\ I_1 \\ I_2 \end{bmatrix} = \begin{bmatrix} 1 & 0 & 0 \\ 0 & YM_1 & 0 \\ 0 & 0 & YM_2 \end{bmatrix} \cdot \begin{bmatrix} Van_0 \\ Van_1 \\ Van_2 \end{bmatrix},$$

$$[I_{012}] = [YM_{012}] \cdot [VLN_{012}],$$

Load Models 309

$$\text{where} \begin{bmatrix} YM_{012} \end{bmatrix} = \begin{bmatrix} 1 & 0 & 0 \\ 0 & YM_1 & 0 \\ 0 & 0 & YM_2 \end{bmatrix} \quad (9.41)$$

Substitute Equation 9.35 into Equation 9.41.

$$\begin{bmatrix} I_{012} \end{bmatrix} = \begin{bmatrix} YM_{012} \end{bmatrix} \cdot \begin{bmatrix} T \end{bmatrix} \cdot \begin{bmatrix} VLL_{012} \end{bmatrix} \quad (9.42)$$

From symmetrical component theory,

$$\begin{bmatrix} VLL_{012} \end{bmatrix} = \begin{bmatrix} A \end{bmatrix}^{-1} \cdot \begin{bmatrix} VLL_{abc} \end{bmatrix}, \quad (9.43)$$

$$\begin{bmatrix} I_{abc} \end{bmatrix} = \begin{bmatrix} A \end{bmatrix} \cdot \begin{bmatrix} I_{012} \end{bmatrix} \quad (9.44)$$

Substitute Equation 9.43 into Equation 9.42 and the resultant equation substitute into Equation 9.44 to get

$$\begin{bmatrix} I_{abc} \end{bmatrix} = \begin{bmatrix} A \end{bmatrix} \cdot \begin{bmatrix} T \end{bmatrix} \cdot \begin{bmatrix} YM_{012} \end{bmatrix} \cdot \begin{bmatrix} A \end{bmatrix}^{-1} \cdot \begin{bmatrix} VLL_{abc} \end{bmatrix} \quad (9.45)$$

$$\text{Define} \begin{bmatrix} YM_{abc} \end{bmatrix} = \begin{bmatrix} A \end{bmatrix} \cdot \begin{bmatrix} T \end{bmatrix} \cdot \begin{bmatrix} YM_{012} \end{bmatrix} \cdot \begin{bmatrix} A \end{bmatrix}^{-1} \quad (9.46)$$

$$\text{Therefore,} \begin{bmatrix} I_{abc} \end{bmatrix} = \begin{bmatrix} YM_{abc} \end{bmatrix} \cdot \begin{bmatrix} VLL_{abc} \end{bmatrix} \quad (9.47)$$

The induction machine "phase frame admittance matrix" $[YM_{abc}]$ is defined in Equation 9.46. Equation 9.47 is used to compute the input phase currents of the machine as a function of the phase line-to-line terminal voltages. This is the desired result. Recall that $[YM_{abc}]$ is a function of the slip of the machine so that a new matrix must be computed every time the slip changes.

Equation 9.47 can be used to solve for the line-to-line voltages as a function of the line currents by

$$\begin{bmatrix} VLL_{abc} \end{bmatrix} = \begin{bmatrix} ZM_{abc} \end{bmatrix} \cdot \begin{bmatrix} I_{abc} \end{bmatrix},$$

$$\text{where} \begin{bmatrix} ZM_{abc} \end{bmatrix} = \begin{bmatrix} YM_{abc} \end{bmatrix}^{-1} \quad (9.48)$$

As was done in Chapter 8, it is possible to replace the line-to-line voltages in Equation 9.48 with the "equivalent" line-to-neutral voltages:

$$\begin{bmatrix} VLN_{abc} \end{bmatrix} = \begin{bmatrix} W \end{bmatrix} \cdot \begin{bmatrix} VLL_{abc} \end{bmatrix},$$

where $[W] = [A] \cdot [T] \cdot [A]^{-1} = \dfrac{1}{3} \cdot \begin{bmatrix} 2 & 1 & 0 \\ 0 & 2 & 1 \\ 1 & 0 & 2 \end{bmatrix}$ (9.49)

The matrix [W] is a very useful matrix that allows the determination of the "equivalent" line-to-neutral voltages from the line-to-line voltages. It is important to know that if the feeder serving the motor is grounded wye, then there will be line-to-ground voltages at the motor terminals. Because the motor is either ungrounded wye or delta it will be necessary to convert the feeder line-to-ground voltages to line-to-line voltages and then apply Equation 9.48 to compute the equivalent line-to-neutral voltages of the motor. Equation 9.48 can be substituted into Equation 9.49 to define the "line-to-neutral" impedance equation.

$[VLN_{abc}] = [W] \cdot [ZM_{abc}] \cdot [I_{abc}],$

$[VLN_{abc}] = [ZLN_{abc}] \cdot [I_{abc}],$

where $[ZLN_{abc}] = [W] \cdot [ZM_{abc}]$ (9.50)

The inverse of Equation 9.50 can be taken to determine the line currents as a function of the equivalent line-to-neutral voltages.

$[I_{abc}] = [YLN_{abc}] \cdot [VLN_{abc}],$

where $[YLN_{abc}] = [ZLN_{abc}]^{-1}$ (9.51)

Care must be taken in applying Equation 9.51 to ensure that the voltages used are the equivalent line-to-neutral, not the line-to-ground, voltages. As was pointed out earlier, when the line-to-ground voltages are known, they must first be converted to the line-to-line values and then use Equation 9.49 to compute the line-to-neutral voltages.

Example 9.4: The induction machine of Example 9.2 is operating such that

$[VLL_{abc}] = \begin{bmatrix} Vs_{ab} \\ Vs_{bc} \\ Vs_{ca} \end{bmatrix} = \begin{bmatrix} 480.0/0 \\ 490/-121.4 \\ 475/118.3 \end{bmatrix}$ volts

Positive sequence slip : $s_1 = 0.035$.

Determine the input line currents and complex power input to the machine (motor).

Compute the negative sequence slip.

$$s_2 = 2 - s_1 = 1.965$$

Compute the sequence load resistance values.

$$RL_i = \frac{1 - s_i}{s_i} \cdot Rr_i = \begin{bmatrix} 2.3408 \\ -0.0417 \end{bmatrix}$$

Calculate the input sequence impedances and admittances.

$$ZM_i = Zs_i + \frac{Zm_i \cdot (Zr_i + RL_i)}{Zm_i + Zr_i + RL_i} = \begin{bmatrix} 2.109 + j1.1786 \\ 0.1409 + j0.4204 \end{bmatrix}$$

$$YM_i = \frac{1}{ZM_i} = \begin{bmatrix} 0.3613 - j0.2019 \\ 0.7166 - j2.1385 \end{bmatrix}$$

Define the *T*, *A*, and *W* matrices.

$$t = \frac{1}{\sqrt{3}} \cdot \underline{/30} \quad [T] = \begin{bmatrix} 1 & 0 & 0 \\ 0 & t^* & 0 \\ 0 & 0 & t \end{bmatrix}$$

$$a = 1\underline{/120} \quad [A] = \begin{bmatrix} 1 & 1 & 1 \\ 1 & a^2 & a \\ 1 & a & a^2 \end{bmatrix}$$

$$[W] = \frac{1}{3} \begin{bmatrix} 2 & 1 & 0 \\ 0 & 2 & 1 \\ 1 & 0 & 2 \end{bmatrix}$$

Define the sequence input admittance matrix.

$$[YM_{012}] = \begin{bmatrix} 1 & 0 & 0 \\ 0 & YM_1 & 0 \\ 0 & 0 & YM_2 \end{bmatrix} = \begin{bmatrix} 1 & 0 & 0 \\ 0 & 0.3613 - j0.2019 & 0 \\ 0 & 0 & 0.7166 - j2.1385 \end{bmatrix}$$

Compute the phase input admittance matrix.

$$[YM_{abc}] = [A] \cdot [T] \cdot [YM_{012}] \cdot [A]^{-1}$$

$$[YM_{abc}] = \begin{bmatrix} 0.6993 - j0.3559 & -0.0394 - j0.0684 & 0.34 + j0.4243 \\ 0.34 + j0.4243 & 0.6993 - j0.3559 & -0.0394 - j0.0684 \\ -0.0394 - j0.0684 & 0.34 + j0.4243 & 0.6993 - j0.3559 \end{bmatrix}$$

Compute input line currents.

$$[Is_{abc}] = [YM_{abc}] \cdot [VLL_{abc}] = \begin{bmatrix} 110.0/\underline{-55.0} \\ 126.6/\underline{179.7} \\ 109.6/\underline{54.6} \end{bmatrix}$$

Compute line-to-neutral voltages.

$$[VLN_{abc}] = [W] \cdot [VLL_{abc}] = \begin{bmatrix} 273.2/\underline{-30.7} \\ 281.9/\underline{-150.4} \\ 279.1/\underline{87.9} \end{bmatrix}$$

Compute the stator complex input power.

For $j = 1, 2, 3$

$$Ss_j = \frac{VLN_{abc_j} \cdot (Is_{abc_j})^*}{1000} = \begin{bmatrix} 27.4 + j12.4 \\ 30.9 + j17.8 \\ 25.6 + j16.8 \end{bmatrix}$$

$$Ss_{total} = 83.9 + j47.0 \, kVA$$

Once the machine terminal currents and line-to-neutral voltages are known, the input phase complex powers and total three-phase input complex power can be computed.

$$S_a = V_{an} \cdot (I_a)^*$$

$$S_b = V_{bn} \cdot (I_b)^*$$

$$S_c = V_{cn} \cdot (I_c)^*$$

$$S_{Total} = S_a + S_b + S_c \tag{9.52}$$

Load Models

Many times, the only voltages known will be the magnitudes of the three line-to-line voltages at the machine terminals. When this is the case, the Law of Cosines must be used to compute the angles associated with the measured magnitudes.

Note that these are the same results as in Example 9.3, only fewer steps are required.

Example 9.5: Determine the voltage and current unbalances for the motor in Examples 9.2 and 9.3.

The terminal line-to-line voltages were

$$[VLL_{abc}] = \begin{bmatrix} 480/0 \\ 490/-121.37 \\ 475/118.26 \end{bmatrix} V$$

Step 1: $V_{average} = \dfrac{1}{3} \cdot \sum_{k=1}^{3} |VLL_{abc_k}| = 481.7$

Step 2: $dev_i = \left\| VLL_i \right| - V_{average} \right| = \begin{bmatrix} 1.67 \\ 8.33 \\ 6.67 \end{bmatrix}$

Step 3: $V_{unbalance} = \dfrac{max(dev)}{V_{average}} = \dfrac{8.33}{481.7} \cdot 100 = 1.73\%$

The line currents were

$$[Is_{abc}] = \begin{bmatrix} 110.0/-55.0 \\ 126.6/179.7 \\ 109.8/54.6 \end{bmatrix}$$

Step 1: $I_{average} = \dfrac{1}{3} \cdot \sum_{k=1}^{3} |Is_{abc_k}| = 115.4$

Step 2: $dev_i = \left\| Is_{abc_i} \right| - I_{average} \right| = \begin{bmatrix} 5.39 \\ 11.20 \\ 5.82 \end{bmatrix}$

Step 3: $I_{unbalance} = \dfrac{max(dev)}{I_{average}} = \dfrac{11.20}{115.4} \cdot 100 = 9.71\%$

9.5.4 Voltage and Current Unbalance

Three-phase distribution feeders are unbalanced due to conductor spacings and the unbalanced loads served. Because of this, the line-to-line voltages serving an induction motor will be unbalanced. When a motor operates with unbalanced voltages, it will overheat and draw unbalanced currents that may exceed the rated current of the motor. It has become a rule of thumb to not let the voltage unbalance exceed 3%. A common way of determining voltage and current unbalance is based upon the magnitudes of the line-to-line voltages and line currents. The computation of unbalance involves three steps:

Step 1: Compute the average of the line-to-line voltages.
Step 2: Compute the magnitudes of the deviation (dev) between the phase magnitudes and the average.
Step 3: Compute unbalance: $Unbalance = \dfrac{\max(dev)}{average} \cdot 100\%$

The same three steps are used to compute the current unbalance.

9.5.5 Motor Starting Current

An induction motor under line starting will cause a current to flow that is much greater than rated. Typically, the motor is not line started, but the input voltages will be reduced under starting conditions. The starting current can be computed by setting the positive sequence slip to one. For the motor of Example 9.2 with the same line-to-line voltages applied the starting currents are

$$[Is_{abc}] = \begin{bmatrix} 596.4\underline{/-97.5} \\ 615.3\underline{/142.8} \\ 609.2\underline{/21.1} \end{bmatrix}$$

When the starting voltage is reduced to one-half, the starting currents are

$$[Is_{abc}] = \begin{bmatrix} 298.2\underline{/-97.5} \\ 307.7\underline{/142.8} \\ 304.6\underline{/21.1} \end{bmatrix}$$

Note that the rated current for the motor is 180 amps.

9.5.6 The Equivalent T Circuit

Once the terminal line-to-neutral voltages and currents are known, it is desired to analyze what is happening inside the machine. In particular, the stator and rotor losses are needed in addition to the "converted" shaft power. A method of performing the

Load Models

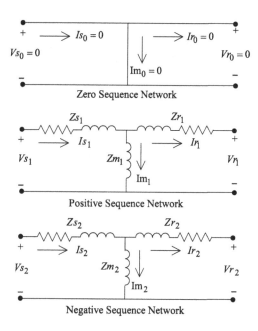

FIGURE 9.6 Induction machine equivalent T sequence networks.

internal analysis can be developed in the phase frame by starting with the equivalent T sequence networks as shown in Figure 9.6.

The three sequence networks of Figure 9.6 can be reduced to the equivalent T sequence circuit shown in Figure 9.7.

Since the zero-sequence voltages and currents are always zero in Figure 9.7, the sequence matrices are defined as follows:

$$\text{Voltages:} \begin{bmatrix} Vs_{012} \end{bmatrix} = \begin{bmatrix} 0 \\ Vs_1 \\ Vs_2 \end{bmatrix} \begin{bmatrix} Vr_{012} \end{bmatrix} = \begin{bmatrix} 0 \\ Vr_1 \\ Vr_2 \end{bmatrix}$$

$$\text{Currents:} \begin{bmatrix} Is_{012} \end{bmatrix} = \begin{bmatrix} 0 \\ Is_1 \\ Is_2 \end{bmatrix} \begin{bmatrix} Im_{012} \end{bmatrix} = \begin{bmatrix} 0 \\ Im_1 \\ Im_2 \end{bmatrix} \begin{bmatrix} Ir_{012} \end{bmatrix} = \begin{bmatrix} 0 \\ Ir_1 \\ Ir_2 \end{bmatrix}$$

$$\text{Impedances:} \begin{bmatrix} Zs_{012} \end{bmatrix} = \begin{bmatrix} 0 & 0 & 0 \\ 0 & Zs_1 & 0 \\ 0 & 0 & Zs_2 \end{bmatrix} \begin{bmatrix} Zr_{012} \end{bmatrix} = \begin{bmatrix} 0 & 0 & 0 \\ 0 & Zr_1 & 0 \\ 0 & 0 & Zr_2 \end{bmatrix} \quad (9.53)$$

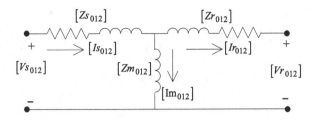

FIGURE 9.7 Sequence equivalent T circuit.

$$[Zm_{012}] = \begin{bmatrix} 0 & 0 & 0 \\ 0 & Zm_1 & 0 \\ 0 & 0 & Zm_2 \end{bmatrix}$$

The sequence voltage drops in the rotor circuit of Figure 9.7 are

$$[vr_{012}] = [Zr_{012}] \cdot [Ir_{012}] \tag{9.54}$$

As an example, the rotor phase voltage drops are given by

$$[vr_{012}] = [Zr_{012}] \cdot [Ir_{012}],$$

$$[vr_{abc}] = [A] \cdot [Vr_{012}],$$

$$[Ir_{012}] = [A^{-1}] \cdot [Ir_{abc}],$$

$$[vr_{abc}] = [A] \cdot [Zr_{012}] \cdot [A]^{-1} \cdot [Ir_{abc}],$$

$$[vr_{abc}] = [Zr_{abc}] \cdot [Ir_{abc}],$$

$$\text{where } [Zr_{abc}] = [A] \cdot [Zr_{012}] \cdot [A]^{-1} \tag{9.55}$$

The same process is used on the other voltage drops so that the circuit in Figure 9.7 can be converted to an equivalent T circuit in terms of the phase components.

The stator voltages and currents of the phase equivalent T circuit as a function of the rotor voltages and currents are defined by

$$\begin{bmatrix} [Vs_{abc}] \\ [Is_{abc}] \end{bmatrix} = \begin{bmatrix} [Am_{abc}] & [Bm_{abc}] \\ [Cm_{abc}] & [Dm_{abc}] \end{bmatrix} \cdot \begin{bmatrix} [Vr_{abc}] \\ [Ir_{abc}] \end{bmatrix},$$

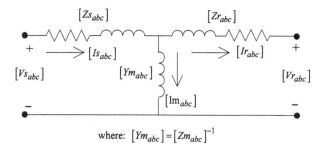

FIGURE 9.8 Phase equivalent T circuit.

where $[Am_{abc}] = [U] + [Zs_{abc}] \cdot [Ym_{abc}]$,

$[Bm_{abc}] = [Zs_{abc}] + [Zr_{abc}] + [Zs_{abc}] \cdot [Ym_{abc}] \cdot [Zr_{abc}]$,

$[Cm_{abc}] = [Ym_{abc}]$,

$[Dm_{abc}] = [U] + [Ym_{abc}] \cdot [Zr_{abc}]$,

$$[U] = \begin{bmatrix} 1 & 0 & 0 \\ 0 & 1 & 0 \\ 0 & 0 & 1 \end{bmatrix} \quad (9.56)$$

The inverse of the ABCD matrices of Equation 9.56 is used to define the rotor voltages and currents as a function of the stator voltages and currents.

$$\begin{bmatrix} [Vr_{abc}] \\ [Ir_{abc}] \end{bmatrix} = \begin{bmatrix} [Am_{abc}] & [Bm_{abc}] \\ [Cm_{abc}] & [Dm_{abc}] \end{bmatrix}^{-1} \cdot \begin{bmatrix} [Vs_{abc}] \\ [Is_{abc}] \end{bmatrix},$$

$$\begin{bmatrix} Vr_{abc} \\ Ir_{abc} \end{bmatrix} = \begin{bmatrix} [Dm_{abc}] & -[Bm_{abc}] \\ -[Cm_{abc}] & [Am_{abc}] \end{bmatrix} \cdot \begin{bmatrix} [Vs_{abc}] \\ [Is_{abc}] \end{bmatrix},$$

because $[Am_{abc}] \cdot [Dm_{abc}] - [Bm_{abc}] \cdot [Cm_{abc}] = [U]$ \quad (9.57)

The power converted to the shaft is given by

$$P_{conv} = Vr_a \cdot (Ir_a)^* + Vr_b \cdot (Ir_b)^* + Vr_c \cdot (Ir_c)^*,$$

$$\text{or } P_{conv} = \sum_{k=1}^{3} Vr_{abc_k} \cdot \left(Ir_{abc_k}\right)^* \tag{9.58}$$

The useful shaft power can be determined from a function of the rotational (FW) losses:

$$P_{shaft} = P_{conv} - P_{FW} \tag{9.59}$$

Example 9.6: For the motor in Example 9.2, determine

1. ABCD matrices for the phase equivalent T circuit,
2. Rotor output voltages and currents,
3. rotor converted and shaft powers,
4. rotor and stator "copper" losses, and
5. total complex power input to the stator.

Define the sequence impedance matrices.

For $Ym_i = \dfrac{1}{Zm_i} = -j0.16$

$$[Zs_{012}] = \begin{bmatrix} 0 & 0 & 0 \\ 0 & Zs_1 & 0 \\ 0 & 0 & Zs_2 \end{bmatrix} = \begin{bmatrix} 0 & 0 & 0 \\ 0 & 0.1+j0.2499 & 0 \\ 0 & 0 & 0.1+j0.2499 \end{bmatrix}$$

$$[Zr_{012}] = \begin{bmatrix} 0 & 0 & 0 \\ 0 & Zr_1 & 0 \\ 0 & 0 & Zr_2 \end{bmatrix} = \begin{bmatrix} 0 & 0 & 0 \\ 0 & 0.0849+j0.175 & 0 \\ 0 & 0 & 0.0849+j0.175 \end{bmatrix}$$

$$[Ym_{012}] = \begin{bmatrix} 0 & 0 & 0 \\ 0 & Ym_1 & 0 \\ 0 & 0 & Ym_2 \end{bmatrix} = \begin{bmatrix} 0 & 0 & 0 \\ 0 & -j0.16 & 0 \\ 0 & 0 & -j0.16 \end{bmatrix}$$

Compute the phase impedance and admittance matrices.

$$[Zs_{abc}] = [A] \cdot [Zs_{012}] \cdot [A]^{-1} =$$

$$[Z_{Sabc}] = \begin{bmatrix} 0.0667+j0.1666 & -0.0333-j0.0833 & -0.0333-j0.0833 \\ -0.0333-j0.0833 & 0.0667+j0.1666 & -0.0333-j0.0833 \\ -0.0333-j0.0833 & -0.0333-j0.0833 & 0.0667+j0.1666 \end{bmatrix}$$

$$[Zr_{abc}] = [A] \cdot [Zr_{012}] \cdot [A]^{-1}$$

Load Models

$$[Z_{rabc}] = \begin{bmatrix} 0.0566 + j0.1167 & -0.0283 - j0.0583 & -0.0283 - j0.0583 \\ -0.0283 - j0.0583 & 0.0566 + j0.1167 & -0.0283 - j0.0583 \\ -0.0283 - j0.0583 & -0.0283 - j0.0583 & 0.0566 + j0.1167 \end{bmatrix}$$

$$[Ym_{abc}] = [A] \cdot [Ym_{012}] \cdot [A]^{-1} = \begin{bmatrix} -j0.1067 & j0.0533 & j0.0533 \\ j0.0533 & -j0.1067 & j0.0533 \\ j0.0533 & j0.0533 & -j0.1067 \end{bmatrix}$$

Compute the phase ABCD matrices.

$$[Am_{abc}] = [U] + [Zs_{abc}] \cdot [Ym_{abc}] =$$

$$[Am_{abc}] = \begin{bmatrix} 1.0267 - j0.0107 & -0.0133 + j0.0053 & -0.0133 + j0.0053 \\ -0.0133 + j0.0053 & 1.0267 - j0.0107 & -0.0133 + j0.0053 \\ -0.0133 + j0.0053 & -0.0133 + j0.0053 & 1.0267 - j0.0107 \end{bmatrix}$$

$$[Bm_{abc}] = [Zs_{abc}] + [Zr_{abc}] + [Zs_{abc}] \cdot [Ym_{abc}] \cdot [Zr_{abc}]$$

$$[Bm_{abc}] = \begin{bmatrix} 0.1274 + j0.2870 & -0.0637 - j0.1435 & -0.0637 - j0.1435 \\ -0.0637 - j0.1435 & 0.1274 + j0.2870 & -0.0637 - j0.1435 \\ -0.0637 - j0.1435 & -0.0637 - j0.1435 & 0.1274 + j0.2870 \end{bmatrix}$$

$$[Cm_{abc}] = [Ym_{abc}] = \begin{bmatrix} -j0.1067 & j0.0533 & j0.0533 \\ j0.0533 & -j0.1067 & j0.0533 \\ j0.0533 & j0.0533 & -j0.1067 \end{bmatrix}$$

$$[Dm_{ABC}] = [U] + [Ym_{abc}] \cdot [Zr_{abc}]$$

$$[Dm_{abc}] = \begin{bmatrix} 1.0187 - j0.0091 & -0.0093 + j0.0045 & -0.0093 + j0.0045 \\ -0.0093 + j0.0045 & 1.0187 - j0.0091 & -0.0093 + j0.0045 \\ -0.0093 + j0.0045 & -0.0093 + j0.0045 & 1.0187 - j0.0091 \end{bmatrix}$$

Compute the rotor output voltages and currents.

$$[Vr_{abc}] = [Dm_{abc}] \cdot [Vs_{abc}] - [Bm_{abc}] \cdot [Is_{abc}] = \begin{bmatrix} 245.3/\underline{-39.7} \\ 244.7/\underline{-159.5} \\ 245.6/\underline{80.5} \end{bmatrix}$$

$$[Ir_{abc}] = -[Cm_{abc}] \cdot [Vs_{abc}] + [Am_{abc}] \cdot [Is_{abc}] = \begin{bmatrix} 103.7/\underline{-33.4} \\ 115.2/\underline{-161.6} \\ 96.2/\underline{76.3} \end{bmatrix}$$

Compute the rotor converted and shaft powers.

$$P_{conv} = \sum_{k=1}^{3} \frac{Vr_{abc_k} \cdot (Ir_{abc_k})^*}{1000} = 77.03 \, kW$$

$$P_{shaft} = P_{conv} - P_{FW} = 77.03 - 3.25 = 73.78 \, kW$$

$$P_{HP} = \frac{73.78}{0.746} = 98.90 \, hp$$

Compute stator and rotor power losses.

$$[vdrop_s] = [Zs_{abc}] \cdot [Is_{abc}] = \begin{bmatrix} 29.6 \underline{/13.2} \\ 34.1 \underline{/-112.1} \\ 29.5 \underline{/122.8} \end{bmatrix}$$

$$P_s = Real \sum_{k=1}^{3} \frac{vdrop_{s_k} \cdot (Is_{abc_k})^*}{1000} = 4.01 \, kW$$

$$[vdrop_r] = [Zr_{abc}] \cdot [Ir_{abc}] = \begin{bmatrix} 20.2 \underline{/30.7} \\ 22.4 \underline{/-97.5} \\ 18.7 \underline{/140.4} \end{bmatrix}$$

$$P_r = Real \sum_{k=1}^{3} \frac{vdrop_{r_k} \cdot (Ir_{abc_k})^*}{1000} = 2.83 \, kW$$

Compute complex power into stator.

$$S_{in} = \sum_{k=1}^{3} \frac{Vs_{abc_k} \cdot (Is_{abc_k})^*}{1000} = 83.87 + j47.00$$

$$P_{conv} = P_{in} - P_s - P_r = 83.87 - 4.01 - 2.83 = 77.03$$

The stator total power loss is

$$[vdrop_s] = [Zs_{abc}] \cdot [Is_{abc}],$$

$$P_{stator} = Real \sum_{k=1}^{3} \left(\frac{vdrop_{s_k} \cdot (Is_{abc_k})^*}{1000} \right) \tag{9.60}$$

Load Models

The rotor total power loss is

$$[vdrop_r] = [Zr_{abc}] \cdot [Ir_{abc}],$$

$$P_{rotor} = \text{Real} \sum_{k=1}^{3} \left(\frac{vdrop_{r_k} \cdot (Ir_{abc_k})^*}{1000} \right) \quad (9.61)$$

The total input complex power is

$$S_{in} = \sum_{k=1}^{3} \left(\frac{Vs_{abc_k} \cdot (Is_{abc_k})^*}{1000} \right) = P_{in} + jQ_{in} \text{ kW} + \text{jkvar} \quad (9.62)$$

9.5.7 Computation of Slip

When the input power to the motor is specified instead of the slip, an iterative process is required to compute the value of slip that will force the input power to be within some small tolerance of the specified input power.

The iterative process for computing the slip that will produce the specified input power starts with assuming an initial value of the positive sequence slip and a change in slip. For purposes here, the initial values are

$$s_{old} = 0.0,$$

$$ds = 0.01. \quad (9.63)$$

The value of slip used in the first iteration is then

$$s_1 = s_{old} + ds,$$

$$\text{where } s_1 = \text{positive sequence slip.} \quad (9.64)$$

With the new value of slip, the input shunt admittance matrix $[YM_{abc}]$ is computed. The given line-to-line voltages are used to compute the stator currents. The $[W]$

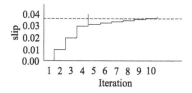

FIGURE 9.9 Slip vs. iteration.

Example 9.7: For the induction motor of Example 9.2, determine the value of the positive sequence slip that will develop 100 kW input power to the motor.

To start the set:

$$s_{old} = 0,$$
$$ds = 0.01,$$
$$tol = 0.01,$$
$$P_{computed} = 100.$$

Figure 9.10 shows a Mathcad program that computes the required slip.

matrix is used to compute the equivalent line-to-neutral voltages. The total three-phase input complex power is then computed. The computed three-phase input power is compared to the specified three-phase input power an error is computed as.

$$error = P_{specified} - P_{computed} \qquad (9.65)$$

If the error is positive, the slip needs to be increased so that the computed power will increase. This is done by

$$s_{old} = \text{value of slip used in previous iteration,}$$

$$s_{new} = s_i = s_{old} + ds \qquad (9.66)$$

The new value of s_1 is used to repeat the calculations for the input power to the motor.

If the error is negative, that means that a bracket has been established. The required value of slip lies between s_{old} and s_{new}. To zero in on the required slip, the old value of slip will be used, and the change in slip will be reduced by a factor of 10.

$$ds = \frac{ds}{10},$$

$$s_i = s_{old} + ds \qquad (9.67)$$

This process is illustrated in Figure 9.9:

When the slip has produced the specified input power within a specified tolerance, the T circuit is used to compute the voltages and currents in the rotor.

$$X := \begin{vmatrix} \text{for } n \in 1..200 \\ \begin{vmatrix} s_1 \leftarrow s_{.old} + ds \\ s_2 \leftarrow 2 - s_1 \\ \text{for } i \in 1..2 \\ \begin{vmatrix} RL_i \leftarrow \dfrac{1-s_i}{s_i} \cdot Rr \\ Zr_i \leftarrow Zxr_i + RL_i \\ ZM_i \leftarrow Zs_i + \dfrac{Zm_i \cdot Zr_i}{Zm_i + Zr_i} \\ YM_i \leftarrow \dfrac{1}{ZM_i} \end{vmatrix} \\ YM_{.012} \leftarrow \begin{pmatrix} 1 & 0 & 0 \\ 0 & YM_1 & 0 \\ 0 & 0 & YM_2 \end{pmatrix} \\ YM_{.abc} \leftarrow A \cdot T \cdot YM_{.012} A^{-1} \\ Is_{.abc} \leftarrow YM_{.abc} \cdot VLL_{.abc} \\ \text{for } i \in 1..3 \\ \quad S_{.abc_i} \leftarrow \dfrac{Vs_{.abc_i} \cdot \overline{Is_{.abc_i}}}{1000} \\ S_{.total} \leftarrow \sum_{k=1}^{3} S_{.abc_k} \\ P_{.computed} \leftarrow Re(S_{.total}) \\ Error \leftarrow |P_{.specified} - P_{.computed}| \\ \text{break if } Error < tol \\ s_1 \leftarrow s_1 - ds \text{ if } P_{.computed} > P_{.specified} \\ ds \leftarrow \dfrac{ds}{10} \text{ if } P_{.computed} > P_{.specified} \\ s_{.old} \leftarrow s_1 \end{vmatrix} \\ Out_1 \leftarrow n \\ Out_2 \leftarrow s_1 \\ Out_3 \leftarrow ds \\ Out_4 \leftarrow P_{.computed} \\ Out_5 \leftarrow Error \\ Out_6 \leftarrow S_{.total} \\ Out \end{vmatrix}$$

FIGURE 9.10 Mathcad program for computing the required slip.

Example 9.8: Using the same induction machine and line-to-line voltages in Examples 9.2 and 9.4, determine the slip of the machine so that it will generate 100 kW. Since the same model is being used with the same assumed direction of currents, the specified power at the terminals of the machine will be

$$P_{gen} = -100$$

As before, the initial "old" value of the slip is set to 0.0. However, since the machine is now a generator, the initial change in the slip will be

$$ds = -0.01$$

As before, the value of slip to be used for the first iteration will be

$$s_1 = s_{old} + ds = -0.01$$

The same Mathcad program as was used in Example 9.7 is used with the exception that the two "if" statements are reversed to determine the new value of slip. The two equations changed are as follows:

$$s_1 \leftarrow s_1 - ds \text{ if } P_{computed} < P_{specified}$$

$$ds \leftarrow \frac{ds}{10} \text{ if } P_{computed} < P_{specified}$$

After 34 iterations, the results are the following:

$$s_1 = -0.03997$$

$$S_{computed} = -99.99 + j59.73 \, kW + jkvar$$

$$Error = 0.0084$$

After 22 iterations, the Mathcad program gives the following results:

$$s_1 = 0.0426$$

$$S_{total} = 100.00 + j52.54$$

$$Error = 0.004$$

Note that the motor is being supplied with reactive power.

Load Models

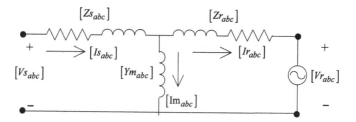

FIGURE 9.11 Induction motor phase equivalent T circuit.

9.5.8 Induction Generator

Three-phase induction generators are becoming common as a source of distributed generation on a distribution system. Windmills generally drive an induction motor. It is, therefore, important that a simple model of an induction generator be developed for power-flow purposes. The same model as was used for the induction motor is used for the induction generator. The only change is that the generator will be driven at a speed more than synchronous speed, which means that the slip will be a negative value. The generator can be modeled with the equivalent admittance matrix from Equation 9.46.

It must be noted that even though the machine is supplying power to the system, it is still consuming reactive power. The point being that even though the induction generator can supply real power to the system, it will still require reactive power from the system. This reactive power is typically supplied by shunt capacitors or a static var supply at the location of the windmill.

9.5.9 Induction Machine Thevenin Equivalent Circuit

When an induction motor is operating under load and a short circuit occurs on the feeder, for a brief period, the motor will supply short-circuit current due to the stored energy in the rotating mass of the motor and load. Figure 9.11 is modified to indicate that there is a voltage at the rotor terminals.

The stator input line-to-neutral voltages and currents are given by

$$[Vs_{abc}] = [Am_{abc}] \cdot [Vr_{abc}] + [Bm_{abc}] \cdot [Ir_{abc}],$$

$$[Is_{abc}] = [Cm_{abc}] \cdot [Vr_{abc}] + [Dm_{abc}] \cdot [Ir_{abc}] \quad (9.68)$$

Solve for the rotor current in Equation 9.66.

$$[Ir_{abc}] = -[Dm_{abc}]^{-1} \cdot [Cm_{abc}] \cdot [Vr_{abc}] + [Dm_{abc}]^{-1} \cdot [Is_{abc}] \quad (9.69)$$

Substitute Equation 9.67 into Equation 9.66.

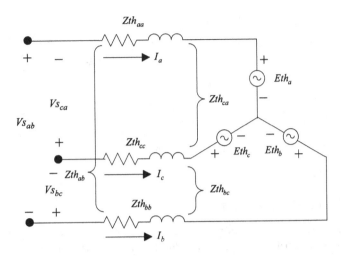

FIGURE 9.12 Motor Thevenin equivalent circuit.

$$[Vs_{abc}] = [Am_{abc}] \cdot [Vr_{abc}] +$$

$$[Bm_{abc}] \cdot \left(-[Dm_{abc}]^{-1} \cdot [Cm_{abc}] \cdot [Vr_{abc}] + [Dm_{abc}]^{-1} \cdot [Is_{abc}] \right)$$

$$[Vs_{abc}] = \left([Am_{abc}] - [Bm_{abc}] \cdot [Dm_{abc}]^{-1} \cdot [Cm_{abc}] \right) \cdot [Vr_{abc}] +$$

$$[Bm_{abc}] \cdot [Dm_{abc}]^{-1} \cdot [Is_{abc}]$$

$$\text{Define}: [Eth_{abc}] = \left([Am_{abc}] - [Bm_{abc}] \cdot [Dm_{abc}]^{-1} \cdot [Cm_{abc}] \right) \cdot [Vr_{abc}]$$

$$[Zth_{abc}] = [Bm_{abc}] \cdot [Dm_{abc}]^{-1}$$

$$\text{Therefore,} [Vs_{abc}] = [Eth_{abc}] + [Zth_{abc}] \cdot [Is_{abc}] \qquad (9.70)$$

The final form of Equation 9.68 reduces the T circuit to the Thevenin equivalent circuit of Figure 9.11 to Figure 9.12.

In Figure 9.12, the Thevenin voltage drops are

$$\begin{bmatrix} v_a \\ v_b \\ v_c \end{bmatrix} = \begin{bmatrix} Zth_{aa} & Zth_{ab} & Zth_{ac} \\ Zth_{ba} & Zth_{bb} & Zth_{bc} \\ Zth_{ca} & Zth_{cb} & Zth_{ca} \end{bmatrix} \cdot \begin{bmatrix} I_a \\ I_b \\ I_c \end{bmatrix},$$

$$[v_{abc}] = [Zth_{abc}] \cdot [I_{abc}] \qquad (9.71)$$

Load Models

The motor terminal Thevenin line-to-line voltages are

$$\begin{bmatrix} Vs_{ab} \\ Vs_{bc} \\ Vs_{ca} \end{bmatrix} = \begin{bmatrix} Eth_a - Eth_b \\ Eth_b - Eth_c \\ Eth_c - Eth_a \end{bmatrix} + \begin{bmatrix} v_a - v_b \\ v_b - v_c \\ v_c - v_a \end{bmatrix}$$

Example 9.9: Use the compute rotor voltages and stator currents from Example 9.3 and the ABCD matrices from Example 9.6 and compute the Thevenin equivalent terminal line-to-line voltages and Thevenin equivalent matrix.

From Example 9.3,

$$[Vr_{abc}] = \begin{bmatrix} 245.3/-39.7 \\ 244.7/-159.5 \\ 245.5/80.5 \end{bmatrix} [Is_{abc}] = \begin{bmatrix} 110.0/-55.0 \\ 126.6/179.7 \\ 109.8/54.6 \end{bmatrix}$$

Compute the Thevenin back emfs and the Thevenin impedance matrix.

$$[Eth_{abc}] = \left([Am_{abc}] - [Bm_{abc}] \cdot [Dm_{abc}]^{-1} \cdot [Cm_{abc}] \right) \cdot [Vr_{abc}] = \begin{bmatrix} 238.6/-38.9 \\ 238.1/-158.8 \\ 238.8/81.3 \end{bmatrix}$$

$$[Zth_{abc}] = [Bm_{abc}] \cdot [Dm_{abc}]^{-1}$$

$$[Zth_{abc}] = \begin{bmatrix} 0.1202 + j0.2808 & -0.0601 - j0.1404 & -0.0601 - j0.1404 \\ -0.0601 - j0.1404 & 0.1202 + j0.2808 & -0.0601 - j0.1404 \\ -0.0601 - j0.1404 & -0.0601 - j0.1404 & 0.1202 + j0.2808 \end{bmatrix}$$

Define the matrix [Dv].

$$[Dv] = \begin{bmatrix} 1 & -1 & 0 \\ 0 & 1 & -1 \\ -1 & 0 & 1 \end{bmatrix}$$

Compute the Thevenin line-to-line voltages and the Thevenin line-to-line impedance matrix.

$$[EthLL_{abc}] = [Dv] \cdot [Eth_{abc}] = \begin{bmatrix} 412.6/-8.9 \\ 412.9/-128.7 \\ 413.8/118.3 \end{bmatrix}$$

$$[Zth_{LL}] = [Dv] \cdot [Zth_{abc}] =$$

$$[Zth_{LL}] = \begin{bmatrix} 0.1803 + j0.4212 & -0.1803 - j0.4212 & 0 \\ 0 & 0.1803 + j0.4212 & -0.1803 - j0.4212 \\ -0.1803 - j0.4212 & 0 & 0.1803 + j0.4212 \end{bmatrix}$$

Compute the terminal Thevenin line-to-line voltages.

$$[VmLL_{abc}] = [EthLL_{abc}] + [Zth_{LL}] \cdot [Is_{abc}] = \begin{bmatrix} 480\underline{/0} \\ 490\underline{/-121.4} \\ 475\underline{/118.3} \end{bmatrix}$$

Note that $\begin{bmatrix} 1 & -1 & 0 \\ 0 & 1 & -1 \\ -1 & 0 & 1 \end{bmatrix} \cdot \begin{bmatrix} v_a \\ v_b \\ v_c \end{bmatrix} = \begin{bmatrix} v_a - v_b \\ v_b - v_c \\ v_c - v_a \end{bmatrix}$

Apply the matrix $[Dv]$.

$$\begin{bmatrix} Vs_{ab} \\ Vs_{bc} \\ Vs_{ca} \end{bmatrix} = \begin{bmatrix} 1 & -1 & 0 \\ 0 & 1 & -1 \\ -1 & 0 & 1 \end{bmatrix} \cdot \begin{bmatrix} Eth_a \\ Eth_b \\ Eth_c \end{bmatrix} + \begin{bmatrix} 1 & -1 & 0 \\ 0 & 1 & -1 \\ -1 & 0 & 1 \end{bmatrix} \cdot \begin{bmatrix} v_a \\ v_b \\ v_c \end{bmatrix}$$

$$[VsLL_{abc}] = [Dv] \cdot [Eth_{abc}] + [Dv] \cdot [v_{abc}] \quad (9.72)$$

Substitute Equation 6.69 into Equation 9.70.

$$\begin{bmatrix} Vs_{ab} \\ Vs_{bc} \\ Vs_{ca} \end{bmatrix} = \begin{bmatrix} 1 & -1 & 0 \\ 0 & 1 & -1 \\ -1 & 0 & 1 \end{bmatrix} \cdot \begin{bmatrix} Eth_a \\ Eth_b \\ Eth_c \end{bmatrix} + \begin{bmatrix} 1 & -1 & 0 \\ 0 & 1 & -1 \\ -1 & 0 & 1 \end{bmatrix} \cdot \begin{bmatrix} Zth_{aa} & Zth_{ab} & Zth_{ac} \\ Zth_{ba} & Zth_{bb} & Zth_{bc} \\ Zth_{ca} & Zth_{cb} & Zth_{ca} \end{bmatrix} \cdot \begin{bmatrix} I_a \\ I_b \\ I_c \end{bmatrix}$$

$$[VsLL_{abc}] = [Dv] \cdot [Eth_{abc}] + [Dv] \cdot [Zth_{abc}] \cdot [I_{abc}]$$

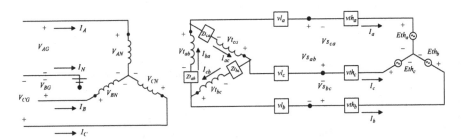

FIGURE 9.13 Ungrounded wye – delta with secondary connected to an induction motor.

Load Models

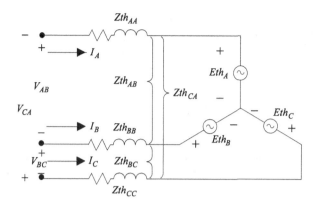

FIGURE 9.14 Primary Thevenin circuit.

$$[VsLL_{abc}] = [EthLL_{abc}] + [Zth_{LL}] \cdot [I_{abc}]$$

where : $[EthLL_{abc}] = [Dv] \cdot [Eth_{abc}]$

$$[Zth_{LL}] = [Dv] \cdot [Zth_{abc}] \quad (9.73)$$

Equation 9.71 gives the terminal Thevenin equivalent line-to-line voltages.

It should come as no surprise that the terminal Thevenin line-to-line voltages are equal to the initial stator line-to-line voltages as specified in Example 9.3. This is a method to prove that the development of the Thevenin equivalent circuit is correct.

9.5.10 The Ungrounded Wye – Delta Transformer Bank with an Induction Motor

In Section 9.5.9, the Thevenin equivalent circuit of a three-phase induction motor was developed and is shown in Figure 9.12.

The Thevenin voltages and impedance matrix are given in Equation 9.68, and the line-to-line terminal voltages of the Thevenin circuit are given in Equation 9.71.

The preferred and most common method of connecting the motor to the distribution feeder is through a three-wire secondary and an ungrounded wye – delta transformer bank. This connection is shown in Figure 9.13.

In Figure 9.13, the voltage drops in the induction motor and secondary are shown rather than the impedances. For short-circuit studies, it is desired to develop a Thevenin equivalent circuit at the primary terminals of the ungrounded wye – delta transformer bank. The resulting primary Thevenin circuit is shown in Figure 9.14.

The equivalent impedance matrix between the motor and the secondary terminals of the transformer are given by

$$[Zeq_{abc}] = [Zth_{abc}] + [Zl_{abc}] \quad (9.74)$$

Carson's equations and the length of the secondary are used to define the 3 × 3 secondary phase impedance matrix as

$$[Zl_{abc}] = \begin{bmatrix} Zl_{aa} & Zl_{ab} & Zl_{ac} \\ Zl_{ba} & Zl_{bb} & Zl_{bc} \\ Zl_{ca} & Zl_{cb} & Zl_{cc} \end{bmatrix} \quad (9.75)$$

The ungrounded wye – delta transformer per-unit impedance, converted to actual impedance is ohms, and referenced to the delta-connected secondary terminals is

$$[Zt_{abc}] = \begin{bmatrix} Zt_{ab} & 0 & 0 \\ 0 & Zt_{bc} & 0 \\ 0 & 0 & Zt_{ca} \end{bmatrix} \quad (9.76)$$

The voltage drops, including the secondary lines and the motor are,

$$[v_{abc}] = [Zeq_{abc}] \cdot [I_{abc}] \quad (9.77)$$

The line-to-line voltages at the secondary terminals of the transformer bank are as follows:

$$\begin{bmatrix} V_{ab} \\ V_{bc} \\ V_{ca} \end{bmatrix} = \begin{bmatrix} Eth_a - Eth_b \\ Eth_b - Eth_c \\ Eth_c - Eth_a \end{bmatrix} + \begin{bmatrix} v_a - v_b \\ v_b - v_c \\ v_c - v_a \end{bmatrix},$$

$$[VLL_{abc}] = [Dv] \cdot [Eth_{abc}] + [Dv] \cdot [Zea_{abc}] \cdot [I_{abc}],$$

$$[VLL_{abc}] = [EthLL_{abc}] + [Zeq_{LL}] \cdot [I_{abc}],$$

$$\text{where}\,[EthLL_{abc}] = [Dv] \cdot [Eth_{abc}],$$

$$[Zeq_{LL}] = [Dv] \cdot [Zea_{abc}] \quad (9.78)$$

The primary currents are

$$[I_{ABC}] = \begin{bmatrix} I_A \\ I_B \\ I_C \end{bmatrix} \quad (9.79)$$

Load Models

In Chapter 8, the currents flowing inside the delta secondary windings were defined as

$$[ID_{abc}] = [AI]^{-1} \cdot [I_{ABC}],$$

$$\text{where } [ID_{abc}] = \begin{bmatrix} I_{ba} \\ I_{cb} \\ I_{ca} \end{bmatrix},$$

$$[AI] = \frac{1}{n_t} \cdot \begin{bmatrix} 1 & 0 & 0 \\ 0 & 1 & 0 \\ 0 & 0 & 1 \end{bmatrix},$$

$$n_t = \frac{kVLN_{high}}{kVLL_{low}} \qquad (9.80)$$

The secondary line currents as a function of the primary line currents are

$$[I_{abc}] = [Di] \cdot [ID_{abc}],$$

$$[I_{abc}] = [Di] \cdot [AI]^{-1} \cdot [I_{ABC}],$$

$$\text{where } [Di] = \begin{bmatrix} 1 & 0 & -1 \\ -1 & 1 & 0 \\ 0 & -1 & 1 \end{bmatrix}. \qquad (9.81)$$

Substitute Equation 9.79 into Equation 9.76.

$$[VLL_{abc}] = [EthLL_{abc}] + [Zeq_{LL}] \cdot [I_{abc}]$$

$$[I_{abc}] = [Di] \cdot [AI]^{-1} \cdot [I_{ABC}]$$

$$[VLL_{abc}] = [EthLL_{abc}] + [Zeq_{LL}] \cdot [Di] \cdot [AI]^{-1} \cdot [I_{ABC}] \qquad (9.82)$$

The voltage drops across the transformer secondary windings are as follows:

$$\begin{bmatrix} Vt_{ab} \\ Vt_{bc} \\ Vt_{ca} \end{bmatrix} = \begin{bmatrix} V_{ab} \\ V_{bc} \\ V_{ca} \end{bmatrix} + \begin{bmatrix} Zt_{ab} & 0 & 0 \\ 0 & Zt_{bc} & 0 \\ 0 & 0 & Zt_{ca} \end{bmatrix} \cdot \begin{bmatrix} I_{ab} \\ I_{bc} \\ I_{ca} \end{bmatrix}$$

Example 9.10: For the system of Figure 9.13, the ungrounded wye – delta transformer bank consists of three single-phase transformers, each rated

$$50\,kVA, 7200/480\,\text{volts}, Zt = 0.011 + j0.018 \text{ per unit}$$

The impedance matrix for the transformer bank relative to the 480-volt side is

$$[Zt_{abc}] = \begin{bmatrix} 0.0507 + j0.0829 & 0 & 0 \\ 0 & 0.0507 + j0.0829 & 0 \\ 0 & 0 & 0.0507 + j0.0829 \end{bmatrix} \text{ohms}$$

The secondary in the system is a quadruplex cable that is 500-feet long. The impedance matrix for the cable is

$$[Zl_{abc}] = \begin{bmatrix} 0.1140 + j0.4015 & 0.0271 + j0.2974 & 0.0271 + j0.2735 \\ 0.0271 + j0.2974 & 0.1140 + j0.4015 & 0.0271 + j0.2974 \\ 0.0271 + j0.2735 & 0.0271 + j0.2974 & 0.1140 + j0.4015 \end{bmatrix} \text{ohms}$$

The induction motor is the motor from Example 9.9 operating at a slip of 0.035. The line-to-line voltages at the primary terminals of the transformer bank are

$$[VLL_{ABC}] = \begin{bmatrix} 12470\underline{/0} \\ 11850\underline{/-118} \\ 12537\underline{/123.4} \end{bmatrix} \text{volts}$$

Determine the Thevenin equivalent circuit of Figure 9.13 relative to the primary side of the transformer bank.

Step 1: Since the system conditions have changed, it is necessary to run a power-flow program to determine the motor stator voltages and currents. For the ungrounded wye – delta transformer bank, the equivalent line-to-neutral voltages must first be computed.

$$[VLN_{ABC}] = [W] \cdot [VLL_{ABC}] = \begin{bmatrix} 7340.4\underline{/-28.4} \\ 6949.6\underline{/-149.9} \\ 6989.6\underline{/93.7} \end{bmatrix} \text{volts}$$

A simple Mathcad routine was run to compute the stator and rotor voltages and currents with the following results:

$$[Vs_{abc}] = \begin{bmatrix} 258.1\underline{/-58.3} \\ 259.9\underline{/-179.0} \\ 250.8\underline{/59.9} \end{bmatrix} \quad [Vr_{abc}] = \begin{bmatrix} 224.9\underline{/-68.3} \\ 225.7\underline{/171.6} \\ 227.1\underline{/51.6} \end{bmatrix} \text{volts}$$

Load Models

$$[Is_{abc}] = \begin{bmatrix} 118.9 \underline{/-85.7} \\ 106.6\underline{/143.2} \\ 94.1\underline{/\,35.6} \end{bmatrix} \quad [Ir_{abc}] = \begin{bmatrix} 110.4\underline{/-67.3} \\ 92.2\underline{/163.6} \\ 88.6\underline{/58.8} \end{bmatrix} \text{ amps}$$

The total three-phase converted power is

$$Sr_{abc} = \sum_{k=1}^{3} \frac{Vr_{abc_k} \cdot Ir_{abc_k}^*}{1000} = 65.4 \text{ kW}$$

The ABCD matrices for the motor are those computed in Example 9.6. The Thevenin equivalent voltages and currents are as follows:

$$[Eth_{abc}] = \left([Am_{abc}] - [Bm_{abc}][Dm_{abc}]^{-1} \cdot [Cm_{abc}]\right) \cdot Vr_{abc} = \begin{bmatrix} 218.7\underline{/-67.4} \\ 219.5\underline{/172.4} \\ 220.9\underline{/52.3} \end{bmatrix}$$

$$[Zth_{abc}] = [Bm_{abc}] \cdot [Dm_{abc}]^{-1} = \begin{bmatrix} 0.1202 + j0.2808 & -0.0601 - j0.1404 & -0.0601 - j0.1404 \\ -0.0601 - j0.1404 & 0.1202 + j0.2808 & -0.0601 - j0.1404 \\ -0.0601 - j0.1404 & -0.0601 - j0.1404 & 0.1202 + j0.2808 \end{bmatrix}$$

It is always a good practice to confirm that the Thevenin equivalent voltages and currents will give the same stator line-to-neutral voltages that were computed at the end of the power-flow program.

$$[Vs_{abc}] = [Eth_{abc}] + [Zth_{abc}] \cdot [Is_{abc}] = \begin{bmatrix} 258.1\underline{/-58.3} \\ 259.9\underline{/179.0} \\ 250.8\underline{/59.9} \end{bmatrix}$$

These voltages match those from the power-flow program, which confirms the accuracy of the Thevenin equivalent circuit for the motor.

From Equations 9.76 and 9.83, the Thevenin equivalent voltages and currents on the primary side of the transformer bank are

$$[Eth_{ABC}] = [AV] \cdot [Dv] \cdot [Eth_{abc}] = \begin{bmatrix} 5699.3\underline{/-37.4} \\ 5722.9\underline{/-157.5} \\ 5702.8\underline{/82.3} \end{bmatrix},$$

$$[Zth_{ABC}] = \begin{bmatrix} 131.7 + j255.0 & -60.1 - j112.8 & -60.1 - j123.6 \\ -60.1 - j112.8 & 131.7 + j255.0 & -60.1 - j123.6 \\ -60.1 - j123.6 & -60.1 - j123.6 & 131.7 + j265.8 \end{bmatrix}$$

Check to confirm that the Thevenin voltages and currents give the initial values of the primary line-to-neutral voltages at the start of the power-flow program.

$$[VLN_{ABC}] = [Eth_{ABC}] + [Zth_{ABC}] \cdot [I_{ABC}]$$

$$[VLN_{ABC}] = \begin{bmatrix} 7340.4/\underline{-28.4} \\ 6949.6/\underline{-149.9} \\ 6989.6/\underline{93.7} \end{bmatrix}$$

These exactly match the initial LN transformer voltages.

$$[Vt_{abc}] = [VLL_{abc}] + [Zt_{abc}] \cdot [ID_{abc}]$$

$$[Vt_{abc}] = [VLL_{abc}] + [Zt_{abc}] \cdot [AI]^{-1} \cdot [I_{ABC}] \quad (9.83)$$

Substitute Equation 9.80 into Equation 9.81.

$$[Vt_{abc}] = [VLL_{abc}] + [Zt_{abc}] \cdot [AI]^{-1} \cdot [I_{ABC}]$$

$$[Vt_{abc}] = \left([EthLL_{abc}] + [Zeq_{LL}] \cdot [Di] \cdot [AI]^{-1} \cdot [I_{ABC}]\right) + [Zt_{abc}] \cdot [AI] \cdot [I_{ABC}]$$

$$[Vt_{abc}] = [EthLL_{abc}] + \left([Zeq_{LL}] \cdot [Di] + [Zt_{abc}]\right) \cdot [AI^{-1}] \cdot [I_{ABC}] \quad (9.84)$$

The primary line-to-neutral voltages are as follows:

$$[VLN_{ABC}] = [AV] \cdot [Vt_{abc}]$$

$$[Vt_{abc}] = [EthLL_{abc}] + \left([Zeq_{LL}] \cdot [Di] + [Zt_{abc}]\right) \cdot [AI]^{-1} \cdot [I_{ABC}]$$

$$[VLN_{ABC}] = [AV] \cdot \left([EthLL_{abc}] + \left([Zeq_{LL}] \cdot [Di] + [Zt_{abc}]\right) \cdot [AI]^{-1} \cdot [I_{ABC}]\right)$$

$$[VLN_{ABC}] = [AV] \cdot [EthLL_{abc}] + [AV] \cdot \left([Zeq_{LL}] \cdot [Di] + [Zt_{abc}]\right) \cdot [AI]^{-1} \cdot [I_{ABC}]$$

But $[EthLL_{abc}] = [Dv] \cdot [Eth_{abc}]$

$$[Zeq_{LL}] = [Dv] \cdot [Zeq_{abc}]$$

$$[VLN_{ABC}] = [AV] \cdot [Dv] \cdot [Eth_{abc}] + [AV] \cdot \left([Dv] \cdot [Zeq_{abc}] \cdot [Di] + [Zt_{abc}]\right) \cdot [AI]^{-1} \cdot [I_{ABC}]$$

$$[VLN_{ABC}] = [Eth_{ABC}] + [Zt_{ABC}] \cdot [I_{ABC}]$$

Load Models

where $\left[Eth_{ABC} \right] = \left[AV \right] \cdot \left[Dv \right] \cdot \left[Eth_{abc} \right]$

$$\left[Zth_{ABC} \right] = \left[AV \right] \cdot \left(\left[Dv \right] \cdot \left[Zeq_{abc} \right] \cdot \left[Di \right] + \left[Zth_{abc} \right] \right) \cdot \left[AI \right]^{-1} \quad (9.85)$$

9.6 ELECTRIC VEHICLE (EV) CHARGERS

In recent years, there have been an increasing number of EVs on the road. These EVs become a significant load on the grid because energy that was previously resourced by petroleum is now transferred over to the electrical sector. It is important for the distribution engineer to understand how EVs impact the distribution system.

There are three levels of EV charging. Level 1 charging is at 120 volts and can be done on any common household 120-volt outlet. Level 1 charging is common for residential EV charging. Level 1 charging can take upwards of 20–40 hours for a full charge depending on the battery and does so at a low current, which does not have a tremendous impact on the distribution system. Level 2 charging is at either 240-volt single phase or 208-volt three phase. Level 2 charging is common in public, workplace, and residential applications. Level 2 charging can take upwards of five to ten hours for a full charge depending on the battery. Charging currents can range from 15 amps to 50 amps, depending on the charger. Level 3 EV chargers are known as DC Fast Chargers. Level 3 charging is at 480+ Vdc and 100+ Adc. Level 3 charging is currently only done in specialized EV charging sites that have the required infrastructure to support it.

The level 2 charger will be focused on due to the impact that it has on common distribution networks and its popularity. It is a common assumption that EV chargers can be modeled as constant power loads with a near-perfect power factor. However, it has been shown in [3] that a combination load model can provide more accurate results and give the distribution engineer a better feel of the system's performance. The combination load for an EV charger should be broken down into active and reactive components for its constant power, constant impedance, and constant current components. The total rated charging load for the EV charger is given by

$$S_{chg} = |S_{chg}| \angle \theta_{chg}$$

$$= S_{Pchg} + S_{Zchg} + S_{Ichg}, \quad (9.86)$$

where θ_{chg} is the power factor angle for the charger. The power factor for EV chargers can be assumed to be 0.95 lagging. Equation 9.84 can be broken up into its real and reactive components.

$$S_{chg} = P_{chg} + jQ_{chg}$$

$$= \left(P_{Pchg} + P_{Zchg} + P_{Ichg} \right) + j \left(Q_{Pchg} + Q_{Zchg} + Q_{Ichg} \right) \quad (9.87)$$

Example 9.11: A residential customer has a constant power demand of 10 kW of connected load. The customer adds a 9.6 kW Level 2 EV charger. They are being fed from a 25 kVA, 2,400/240 distribution transformer with Z_t = 3.8/55%. The transformer feeds the customer through 100 feet of 4 AWG copper conductor. Assuming line voltage on the high side of the transformer is operating at rated voltage, calculate the voltage drop percent from the line to the customer with and without the EV charger.

The transformer turns ratio is

$$n_t = \frac{V_{1-rated}}{V_{2-rated}} = \frac{2400}{240} = 10$$

The low side base impedance for the transformer is

$$Z_{base} = \frac{kV_2^2 \cdot 1000}{kVA} = \frac{.240^2 \cdot 1000}{25} = 2.304\,\Omega$$

Impedance of the transformer in ohms is

$$Z_t = Z_{pu} \cdot Z_{base} = 0.038\,\underline{/55} = 0.0876\,\underline{/55}\,\Omega$$

Impedance of the line is

$$Z_{line} = \frac{100}{5280} \cdot z_{line}$$

$$= \frac{100}{5280} \cdot \left[1.503 + 0.0953 + j0.12134 \cdot \left(\ln\frac{1}{0.00663} + 7.93402\right)\right]$$

$$= 0.0303 + j0.0298\,\Omega$$

It was shown in [3] through computer simulation that the real and reactive components of the charging load can be modeled as

$$S_{chg} = (k_{PP} + k_{PZ} + k_{PI})P_{chg} + j(k_{QP} + k_{QZ} + k_{QI})Q_{chg}, \quad (9.88)$$

where the k_{Pn} constants are the real power combination load coefficients and the k_{Qn} constants are the reactive power combination load coefficients. At rated voltage, the coefficients were determined to be

$$k_{PP} = -0.1773\, k_{QP} = 4.993,$$

$$k_{PZ} = 0.1824\, k_{QZ} = 8.917,$$

$$k_{PI} = 0.9949\, k_{QP} = -12.91$$

Load Models

For the constant power component of the load, we have

$$S_P = |S_P| \underline{/\theta_P} = -0.1773 P_{chg} + j4.993 Q_{chg} \qquad (9.89)$$

For the constant impedance component of the load, we have

$$Z_z = \frac{|V|^2}{S_Z{}^*}, \qquad (9.90)$$

where $S_Z = |S_Z| \underline{/\theta_Z} = 0.1824 P_{chg} + j8.917 Q_{chg}$

For the constant current component of the load, we have

$$I_I = \left(\frac{S_I}{V}\right)^* = \frac{|S_I|}{|V|} \underline{/\delta - \theta_I} = |I_I| \underline{/\delta - \theta_I}, \qquad (9.91)$$

where $S_I = |S_I| \underline{/\theta_I} = 0.9949 P_{chg} - j12.91 Q_{chg}$

As was described in Section 9.1, the constant power component has a constant real and reactive power for each iteration, while the voltage and current can vary. The constant impedance component has a constant impedance for each iteration, while the voltage, current, real power, and reactive power can vary. The constant current component has a constant current magnitude for each iteration, while the voltage, current angle, real power, and reactive power may vary.

It is first necessary to calculate power drawn by the customer, not including the EV charger. This can be done assuming a 0.9 pf lagging.

$$S_{load} = \frac{P_{load}}{pf} \underline{/\cos^{-1} pf} = \frac{10}{0.9} \underline{/\cos^{-1} 0.9} = 10 + j4.8 \, \text{kVA}$$

Next, the EV charging power must be calculated. This is done assuming a power factor of 0.95 lagging.

$$S_{chg} = \frac{P_{chg}}{pf_{chg}} \underline{/\cos^{-1} pf_{chg}} = \frac{9.6}{0.95} \underline{/\cos^{-1} 0.95} = 9.6 + j3.1 \, \text{kVA}$$

The EV charger is best represented by a combination load. It was shown in [3] that the constant power, constant impedance, and constant current loads all have different power factors so their real and reactive power components must be independently solved. Using 9.85 and 9.86, these powers can be calculated.

$$P_{Pchg} = -0.1773 \cdot P_{chg} = -0.1773 \cdot 9.6 = -1.6978 \, \text{kW}$$

$$P_{Zchg} = 0.1824 \cdot P_{chg} = 0.1824 \cdot 9.6 = 1.7467 \, \text{kW}$$

$$P_{Ichg} = 0.9949 \cdot P_{chg} = 0.9949 \cdot 9.6 = 9.5272 \, \text{kW}$$

$$Q_{Pchg} = 4.993 \cdot Q_{chg} = 4.993 \cdot 3.1 = 15.7154 \, \text{kvar}$$

$$Q_{Zchg} = 8.917 \cdot Q_{chg} = 8.917 \cdot 3.1 = 28.0661 \, \text{kvar}$$

$$Q_{Ichg} = -12.91 \cdot Q_{chg} = -12.91 \cdot 3.1 = -40.6340 \, \text{kvar}$$

Checking by applying equation

$$S_{chg} = \left(-1.6978 + 1.7467 + 9.5272\right) + j\left(15.7154 + 28.0661 - 40.6340\right)$$

$$= 9.6 + j3.1 \, \text{kVA},$$

and the remaining connected load is a constant power load. The total connected combination load is then

$$SP = S_{Pchg} + S_{Pload} = 8.3 + j20.6 \, \text{kVA},$$

$$SZ = S_{Zchg} = 1.7 + j28.1 \, \text{kVA},$$

$$SI = S_{Ichg} = 9.5 - j40.6 \, \text{kVA}$$

The constant impedance for the EV charger is computed as

$$Z_{chg} = \frac{|V|^2}{SZ^* \cdot 1000} = \frac{240^2}{(1.7 - j28.1) \cdot 1000} = 0.1272 + j2.0444 \, \Omega$$

The EV charger's current magnitude due to the constant current portion of the EV charge is

$$IM_{chg} = \left|\left(\frac{SI \cdot 1000}{V}\right)^*\right| = \left|\left[\frac{(9.5 - j40.6) \cdot 1000}{240}\right]^*\right| = 173.9 \, \text{A}$$

The current due to the constant complex power computed at nominal voltage is

$$Ipq = \left(\frac{SP \cdot 1000}{V}\right)^* = 92.4 \underline{/68.01} \, \text{A}$$

The constant impedance for that part of the load is computed as

$$Iz = \left(\frac{V}{Z}\right)^* = 117.2 \underline{/-86.44} \, \text{A}$$

Load Models

The contribution of the load current due to the constant current portion of the loads is

$$II = IM \underline{/\delta_i - \theta_i} = 173.9 \underline{/76.80} \text{ A}$$

The total load current is the sum of the three components:

$$I = Ipq + Iz + II = 88.1 \underline{/-22.2} \text{ A}$$

This current is then used to calculate the voltage for the second iteration. This voltage is

$$V = V_t - (Z_t + Z_{line}) \cdot I$$

$$= 240 - (0.0805 + j0.1015) \cdot (88.1 \underline{/-22.2})$$

$$= 230.1 \underline{/-1.39} \text{ V}$$

This process repeats until the voltage for the current iteration is close enough to the voltage for the previous iteration so that the error is within the tolerance as was done in previous power-flow examples. Using a tolerance of 0.001, the program converges with the final voltage and total current of

$$V = 230.0 \underline{/-1.39} \text{ V},$$

$$I = 88.8 \underline{/-22.61} \text{ A}$$

This corresponds to a voltage drop percent of

$$Vdrop = \frac{|Vt| - |V|}{|Vt|} \cdot 100 = \frac{240 - 230}{240} \cdot 100 = 4.16\%$$

If the same power-flow program is run without the EV charging, the final voltage and current are

$$V = 234.5 \underline{/-0.64} \text{ V},$$

$$I = 47.4 \underline{/-26.48} \text{ A}$$

This corresponds to a voltage drop percent of

$$Vdrop = \frac{|Vt| - |V|}{|Vt|} \cdot 100 = \frac{240 - 234.5}{240} \cdot 100 = 2.31\%$$

In addition, the kVA flowing through the transformer with the EV charger was 21.31 kVA and 11.37 kVA without the EV charger. This results in a much greater voltage drop percent from the transformer to the load and twice as much demand on the transformer due to the EV charger. This demand pushes the transformer close to its limit. EVs are becoming more and more prevalent, and distribution engineers should take great consideration in designing the infrastructure to support them.

9.7 SUMMARY

This chapter has developed load models for typical loads on a distribution feeder. It is important to recognize that a combination of constant PW, constant Z, and constant current loads can be modeled using a percentage of each model. An extended model for a three-phase induction machine has been developed with examples of the machine operating as a motor and as a generator. An iterative procedure for the computation of slip to force the input power to the machine to be a specified value was developed and used in examples for both a motor and a generator. Thevenin equivalent circuits have been developed for an induction motor. The Thevenin circuit is used to develop a Thevenin equivalent circuit at the primary terminals of the step-down transformer feeding the secondary and the induction motor. This final Thevenin circuit is used in the next chapter on short-circuit analysis. A model for EV chargers has been developed, and this model shows that these charges can put great stress on the grid, causing transformer overloading and excessive voltage drops.

PROBLEMS

9.1 A 12.47 kV feeder provides service to an unbalanced wye-connected load specified to be one of the following:

Phase a: 1,000 kVA, 0.9 lagging power factor
Phase b: 800 kVA, 0.95 lagging power factor
Phase c: 1,100 kVA, 0.85 lagging power factor

Modify the MATLAB script M0901.m to
(a) compute the initial load currents assuming the loads are modeled as constant complex power.
(b) compute the magnitude of the load currents that will be held constant assuming the loads are modeled as constant current.
(c) compute the impedance of the load to be held constant assuming the loads are modeled as constant impedance.
(d) compute the initial load currents assuming that 60% of the load is complex power, 25% constant current, and 15% constant impedance.

9.2 Using the results of Problem 9.1, rework the problem at the start of the second iteration if the load voltages after the first iteration have been computed to be

$$[VLN_{abc}] = \begin{bmatrix} 6851/\underline{-1.9} \\ 6973/\underline{-122.1} \\ 6886/\underline{117.5} \end{bmatrix} V$$

9.3 A 12.47 kV feeder provides service to an unbalanced delta-connected load specified to be one of the following:

Phase a: 1500 kVA, 0.95 lagging power factor
Phase b: 1000 kVA, 0.85 lagging power factor
Phase c: 950 kVA, 0.9 lagging power factor

Modify the MATLAB script M0901.m to
(a) compute the load and line currents if the load is modeled as constant complex power,
(b) compute the magnitude of the load current to be held constant if the load is modeled as constant current,
(c) compute the impedance to be held constant if the load is modeled as constant impedance, and
(d) compute the line currents if the load is modeled as 25% constant complex power, 20% constant current, and 55% constant impedance.

9.4 After the first iteration of the system of Problem 9.5, the load voltages are

$$[VLL_{abc}] = \begin{bmatrix} 11,981/\underline{28.3} \\ 12,032/\underline{-92.5} \\ 11,857/\underline{147.7} \end{bmatrix} V$$

(a) Compute the load and line currents if the load is modeled as constant complex power.
(b) Compute the load and line currents if the load is modeled as constant current.
(c) Compute the load and line current if the load is modeled as constant impedance.
(d) Compute the line currents if the load mix is 25% constant complex power, 20% constant current, and 55% constant impedance.

9.5 A three-phase induction motor has the following data:

25 hp, 240 volt
$Z_s = 0.0336 + j0.08$ pu
$Z_r = 0.0395 + j0.08$ pu
$Z_m = j3.12$ pu

(a) The motor is operating with a slip of 0.03 with balanced three-phase voltages of 240 volts line-to-line. Modify the MATLAB script M0906.m to determine the following:

(b) the input line currents and complex three-phase input complex power,
(c) the currents in the rotor circuit, and
(d) the developed shaft power in hp.

9.6 The motor of Problem 9.5 is operating with a slip of 0.03, and the line-to-line voltage magnitudes are

$$V_{ab} = 240, V_{bc} = 230, V_{ca} = 250 \text{ volts}.$$

(a) Compute the angles for the line-to-line voltages, assuming the voltage a-b is the reference.
(b) For the given voltages and slip, determine the input line currents and complex input complex power.
(c) Compute the rotor currents.
(d) Compute the developed shaft power in hp.

9.7 The motor of Problem 9.5 is operating with line-to-line voltages of

$$V_{ab} = 240/0, V_{bc} = 233.4/-118.1, V_{ca} = 240/122.1 \text{ volts}.$$

The motor input kW is to be 20 kW.
Determine the following:
(a) required slip,
(b) the input kW and kvar, and
(c) the converted shaft power.

9.8 A three-phase 100-hp, 480-volt, wye-connected induction motor has the following per-unit impedances:

$$Zpu_s = 0.043 + j0.089, Zpu_r = 0.034 + j0.081, Zpu_m = j3.11$$

The rotating loss is $P_{FW} = 2.75$ kW.
(a) Determine the impedances in ohms.
(b) The motor is operating with a slip of 0.035. Determine the input shunt admittance matrix $[YM_{abc}]$.

9.9 The motor of Problem 9.8 is operating at a slip of 0.035 with line-to-line input voltages of

$$[VLL_{abc}] = \begin{bmatrix} V_{ab} \\ V_{bc} \\ V_{ca} \end{bmatrix} = \begin{bmatrix} 480.0/0 \\ 475.0/-121.5 \\ 466.7/119.8 \end{bmatrix}$$

Determine the following:
(a) the input stator currents,
(b) the per-phase complex input power,
(c) the total three-phase complex input power, and

(d) the stator voltage and current unbalances.
9.10 For the motor in Problem 9.8, modify the MATLAB script M0901.m as follows:
   (a) For the T equivalent circuit the matrices $[Am_{abc}]$, $[Bm_{abc}]$, $[Cm_{abc}]$, $[Dm_{abc}]$
   (b) For the results of Problem 9.9, determine
      1. the rotor currents and output voltages,
      2. the rotor converted and shaft powers, and
      3. rotor and stator "copper" losses.
9.11 For the induction motor of Problem 9.8, determine the value of the positive sequence slip that will deliver 75 kW of input power to the motor.
9.12 The induction motor of Problem 9.9 is operating as a generator with a positive sequence slip of −0.04. Determine the stator output complex power.
9.13 Using the results of Problem 9.10, determine the line-to-line motor Thevenin voltages and Thevenin equivalent matrix.
9.14 The motor of Problem 9.8 is connected through a three-phase distribution line to three single-phase transformers, as shown in Figure 9.15.

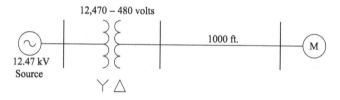

FIGURE 9.15 Simple system.

The three-phase induction motor is that of Problem 9.8. The secondary impedance matrix is

$$[zl_{abc}] = \begin{bmatrix} 0.4013 + j1.4133 & 0.0953 + j1.0468 & 0.0953 + j0.9627 \\ 0.0953 + j1.0468 & 0.4013 + j1.4133 & 0.0953 + j1.0468 \\ 0.0953 + j0.9627 & 0.0953 + j1.0468 & 0.4013 + j1.4133 \end{bmatrix} \text{ohms/mile}$$

The three-phase transformer bank consists of three single-phase transformers, each rated

$kVA = 100, kVLN_{hi} = 7.2, kVLL_{low} = .48, Zpu = 0.0133 + j0.019$.

The motor is operating at a slip of 0.035. Modify the MATLAB script M0910.m to determine the Thevenin equivalent line-to-neutral voltages referenced to the high-voltage side of the transformers $[Eth_{ABC}]$ and Thevenin equivalent line impedance matrix $[Zth_{ABC}]$.

9.15 It has been shown that the combination load coefficients used for EV charger modeling can vary from simulation to laboratory test results. An experimental model for the charging coefficients is determined to be the following:

$$k_{PP} = 0.7934 \; k_{QP} = -0.8973$$

$$k_{PZ} = -0.0886 \; k_{QZ} = -0.3994$$

$$k_{PI} = 0.2952 \; k_{QP} = 2.2967$$

Modify the MATLAB script M0911.m to determine the difference in complex power for each of the combination loads using the model in Example 9.11 and again using the previous model.

REFERENCES

1. *American National Standard for Electric Power Systems and Equipment – Voltage Ratings (60 Hertz)*, ANSI C84.1-1995, National Electrical Manufacturers Association, Rosslyn, Virginia, 1996.
2. Kersting, W. H., Phillips, W. H., Phase Frame Analysis of the Effects of Voltage Unbalance on Induction Machines, *IEEE Transactions on Industry Applications*, March/April 1997.
3. Haidar, A., Muttaqi, K., Behavioral Characterization of Electric Vehicle Charging Loads in a Distribution Power Grid Through Modeling of Battery Chargers, *IEEE Transactions on Industry Applications*, January/February 2016

10 Distribution Feeder Analysis

The analysis of a distribution feeder will typically consist of a study of the feeder under normal steady-state operating conditions (power-flow analysis) and a study of the feeder under short-circuit conditions (short-circuit analysis). Models of all components of a distribution feeder have been developed in previous chapters. These models will be applied for the analysis under steady-state and short-circuit conditions.

10.1 POWER-FLOW ANALYSIS

The power-flow analysis of a distribution feeder is like that of an interconnected transmission system. Typically, what will be known prior to the analysis will be the three-phase voltages at the substation, the impedance matrix of all lines, and the complex power of all loads (constant complex power, constant impedance, constant current, or a combination). Sometimes, the input complex power supplied to the feeder from the substation is also known.

The LIT was introduced in Chapter 3. This technique considers the various line configurations and load types. The ladder technique will be used throughout this chapter.

In Chapters 6–8, phase frame models were developed for the series components of a distribution feeder. In Chapter 9, models were developed for the shunt components (static loads, induction machines, and capacitor banks). These models are used in the LIT analysis of a distribution feeder.

The ladder computer program analysis of a feeder can determine the following by phase and total three-phase values:

- Voltage magnitudes and angles at all nodes of the feeder
- Line flow in each line section specified in kW and kVAr, amps and degrees, or amps and power factor
- Power loss in each line section
- Total feeder input kW and kVAr
- Total feeder power losses
- Load kW and kVAr based upon the specified model for the **load**

10.1.1 General Feeder

A typical distribution feeder will consist of the "primary main" with laterals tapped off the primary main and sublaterals tapped off the laterals, etc. Figure 10.1 shows an example of a typical feeder.

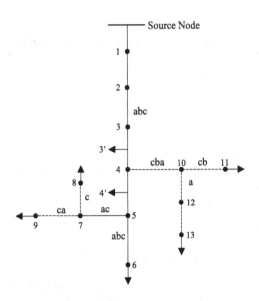

FIGURE 10.1 Typical distribution feeder.

In Figure 10.1, no distinction is made as to what type of element is connected between nodes. However, the phasing is shown, and this is a must. Note in Figure 10.1 that there are loads in the center of lines 3–4 and 4–5, named 3′ and 4′, respectively.

10.1.2 Uniformly Distributed Loads

Many times, it can be assumed that loads are uniformly distributed along a line where the line can be a three-phase, two-phase, or single-phase feeder or lateral. This is certainly the case on single-phase laterals where the same rating transformers are spaced uniformly over the length of the lateral. When the loads are uniformly distributed, it is not necessary to model each load to determine the total voltage drop from the source end to the last load.

Figure 10.2 shows a single-phase line with n uniformly spaced loads dx miles apart. The loads are all equal and will be treated as constant current loads with a value of di. The total current into the feeder is I_T. It is desired to determine the total voltage drop from the source Node (**S**) to the last Node-**n**.

Let
l = length of the feeder
$z = r + jx$ = impedance of the line in Ω/mile
dx = length of each line section
di = load currents at each node
n = number of nodes and number of line sections
I_T = total current into the feeder

Distribution Feeder Analysis

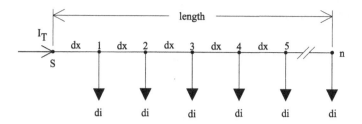

FIGURE 10.2 Uniformly distributed loads.

The load currents are given by

$$di = \frac{I_T}{n} \qquad (10.1)$$

The voltage drop in the first line segment is given by

$$Vdrop_1 = \text{Re}\{z \cdot dx \cdot (n \cdot di)\} \qquad (10.2)$$

The voltage drop in the second line segment is given by

$$Vdrop_2 = \text{Re}\{z \cdot dx \cdot [(n-1) \cdot di]\} \qquad (10.3)$$

The total voltage drop from the source node to the last node is then given by

$$Vdrop_{total} = Vdrop_1 + Vdrop_2 + \ldots\ldots + Vdrop_n,$$

$$Vdrop_{total} = \text{Re}\{z \cdot dx \cdot di \cdot [n + (n-1) + (n-2) + \cdots + (1)]\} \qquad (10.4)$$

Equation 10.4 can be reduced by recognizing the series expansion:

$$1 + 2 + 3 + \cdots + n = \frac{n(n+1)}{2} \qquad (10.5)$$

Using the expansion, Equation 10.4 becomes

$$Vdrop_{total} = Real\left\{z \cdot dx \cdot di \cdot \left[\frac{n \cdot (n+1)}{2}\right]\right\} \qquad (10.6)$$

The incremental distance is

$$dx = \frac{l}{n} \qquad (10.7)$$

The incremental current is

$$di = \frac{I_T}{n} \quad (10.8)$$

Substitute Equations 10.7 and 10.8 into Equation 10.6 results in

$$Vdrop_{total} = \text{Re}\left\{z \cdot \frac{l}{n} \cdot \frac{I_T}{n} \cdot \left[\frac{n \cdot (n+1)}{2}\right]\right\},$$

$$Vdrop_{total} = \text{Re}\left\{z \cdot l \cdot I_T \cdot \frac{1}{2}\left(\frac{n+1}{n}\right)\right\}$$

$$Vdrop_{total} = \text{Re}\left\{\frac{1}{2} \cdot Z \cdot I_T \cdot \left(1 + \frac{1}{n}\right)\right\}, \quad (10.9)$$

where $Z = z \cdot l$.

Equation 10.9 gives the general equation for computing the total voltage drop from the source to the last Node-**n** for a line of length l. In the limiting case where **n** goes to infinity, the final equation becomes

$$Vdrop_{total} = \text{Re}\left\{\frac{1}{2} \cdot Z \cdot I_T\right\} \quad (10.10)$$

In Equation 10.10, Z represents the total impedance from the source to the end of the line. The voltage drop is the total from the source to the end of the line. The equation can be interpreted in two ways. The first is to recognize that the total line distributed load can be lumped at the midpoint of the lateral, as shown in Figure 10.3.

The system in Figure 10.3 will work for calculating the total voltage drop down the line. However, this model does not give the correct power loss down the line. The equation for computing the three-phase power loss down the line is given in Equation 10.11:

$$Ploss_{total} = 3\left(\frac{1}{3} \cdot R \cdot |I_T|^2\right) \text{watts} \quad (10.11)$$

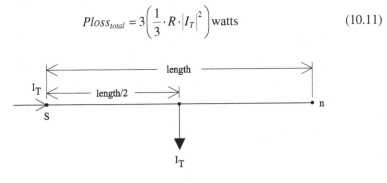

FIGURE 10.3 Load at the midpoint.

Distribution Feeder Analysis

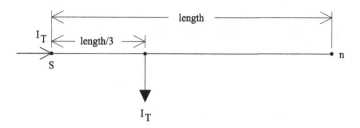

FIGURE 10.4 Power loss model.

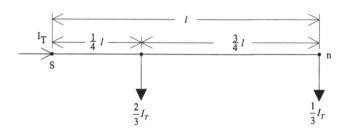

FIGURE 10.5 Exact lumped load model.

The circuit model for Equation 10.11 is given in Figure 10.4.

The power loss on a distribution line is small compared to the total system loss. However, if both the voltage drop and power loss are to be computed correctly on a line, the system in Figure 10.5 is used.

10.1.3 SERIES FEEDER

All series elements of a distribution feeder can be modeled as shown in Figure 10.6.

The LIT uses the forward and backward sweep models for the series elements. With reference to Figure 10.6, the forward and backward sweep equations are as follows:

$$\text{Forward sweep}: \left[VLN_{abc}\right]_n = \left[A\right]\cdot\left[VLN_{abc}\right]_m - \left[B\right]\cdot\left[I_{abc}\right]_n$$

$$\text{Backward sweep}: \left[I_{abc}\right]_n = \left[c\right]\cdot\left[VLN_{abc}\right]_m + \left[d\right]\cdot\left[I_{abc}\right]_n \quad (10.11)$$

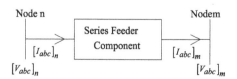

FIGURE 10.6 Standard feeder series component model.

In most cases, the [c] matrix will be zero. Long underground lines will be the exception. It was also shown that for the grounded wye – delta transformer bank, the backward sweep equation is:

$$[I_{abc}]_n = [x_t] \cdot [VLN_{abc}]_n + [d] \cdot [I_{abc}]_m \qquad (10.12)$$

The reason for this is that the currents flowing in the secondary delta windings are a function of the primary line-to-ground voltages. The matrices used in Equations 10.11 and 10.12 are derived in previous chapters.

Referring to Figure 10.1, Nodes 4, 10, 5, and 7 are referred to as "junction nodes". In both the forward and backward sweeps, the junction nodes must be recognized. In the forward sweep, the voltages at all nodes down the lines from the junction nodes must be computed. In the backward sweep, the currents at the junction nodes must be summed before proceeding toward the source. In developing a program to apply the ladder method, it is necessary for the ordering of the lines and nodes to be such that all node voltages in the forward sweep are computed, and all currents in the backward sweep are computed.

10.1.4 The Unbalanced Three-Phase Distribution Feeder

The previous section outlined the general procedure for performing the LIT. This section will address how that procedure can be used for an unbalanced three-phase feeder.

Figure 10.7 is the one-line diagram of an unbalanced three-phase feeder.

The topology of the feeder in Figure 10.7 is the same as the feeder in Figure 10.1. Figure 10.7 shows more detail of the feeder with step regulators at the source,

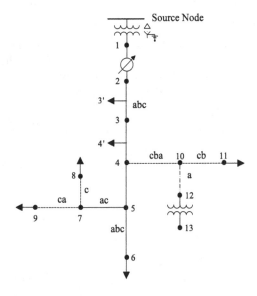

FIGURE 10.7 Unbalanced three-phase distribution feeder.

Distribution Feeder Analysis

distributed loads on lines 3–4 and 4–5, and a transformer bank at Node 12. The feeder in Figure 10.7 can be broken into the "series" components and the "shunt" components. The series components have been shown in Section 10.1.3.

10.1.4.1 Shunt Components

The shunt components of a distribution feeder are:

- spot static loads,
- spot induction machines, and
- capacitor banks.

Spot static loads are located at a node and can be three phase, two phase, or single phase and connected in either a wye or a delta connection. The loads can be modeled as constant complex power, constant current, constant impedance, or a combination of the three.

An induction machine at a node is modeled using the shunt admittance matrix, as defined in Chapter 9. The machine can be modeled as a motor with a positive slip or as an induction generator with a negative slip. The input power (positive for a motor, negative for a generator) can be specified and the required slip computed using the iterative process described in Chapter 9.

Capacitor banks are located at a node and can be three phase, two phase, or single phase and can be connected in a wye or delta. Capacitor banks are modeled as constant admittances.

In Figure 10.7, the solid line segments represent overhead lines, while the dashed lines represent underground lines. Note that the phasing is shown for the line segments. In Chapter 4, the application of Carson's equations for computing the line impedances for overhead and underground lines was presented. In that chapter, it was pointed out that two-phase and single-phase lines are represented by a three-by-three matrix with zeros set in the rows and columns of the missing phases.

In Chapter 5, the method for the computation of the shunt capacitive susceptance for overhead and underground lines was presented. Most of the time, the shunt capacitance of the line segment can be ignored; however, for long underground line segments, the shunt capacitance should be included.

The "node" currents may be three phase, two phase, or single phase and consist of the sum of the spot load currents and one-half of the distributed load currents (if any) at the node plus the capacitor current (if any) at the node. It is possible that for a line, the distributed load can be one-half of the distributed load in the "from" segment plus one-half of the distributed load is connected to the "to" segment. In Figure 10.7, a "dummy" node is created in the center of a line, and the total distributed load is connected to this node.

10.1.5 Applying the Iterative Technique

In the LIT for the forward and backward sweeps, the required matrices have been developed in Chapters 6–8 for the series devices. By applying these matrices, the computation of the voltage drops along a segment will always be the same regardless of whether the segment represents a line, voltage regulator, or transformer.

In the preparation of data for a power-flow study using the ladder technique, it is extremely important that the impedances and admittances of the line segments are computed using the exact spacings and phasing. Because of the unbalanced loading and resulting unbalanced line currents, the voltage drops due to the mutual coupling of the lines become very important. It is not unusual to observe a voltage rise on a lightly loaded phase of a line segment that has an extreme current unbalance.

The real power loss in a device can be computed in two ways. The first method is to compute the power loss in each phase by taking the phase current squared times the total resistance of the phase. Care must be taken to not use the resistance value from the phase impedance matrix. The actual phase resistance that was used in Carson's equations must be used. In developing a computer program calculating power loss, this way requires that the conductor resistance be stored in the active database for each line segment. Unfortunately, this method does not give the total power loss in a line segment since the power loss in the neutral conductor and ground are not included. To determine the losses in the neutral and ground, the method outlined in Chapter 4 must be used to compute the neutral and ground currents and then the power losses.

A second, and preferred, method is to compute power loss as the difference from real power into a line segment minus the real power output of the line segment. Because the effects of the neutral conductor and ground are included in the phase impedance matrix, the total power loss in this method will give the same results as mentioned earlier where the neutral and ground power losses are computed separately. This method can lead to some interesting numbers for very unbalanced line flows in that it is possible to compute what appears to be a negative phase power loss. This is a direct result of the accurate modeling of the mutual coupling between phases. Remember that the effects of the neutral conductor and the ground resistance are included in Carson's equations. In reality, there cannot be a negative-phase power loss. Using this method, the algebraic sum of the line power losses will equal the total three-phase power loss that was computed using the current squared times resistance for the phase and neutral conductors along with the ground current.

10.1.6 Let's Put It All Together

At this point, the models for all components of a distribution feeder have been developed. The LIT has also been developed. It is time to put them all together and demonstrate the power-flow analysis of a very simple system. Example 10.2 will demonstrate how the models of the components work together in applying the ladder technique to achieve a final solution of the operating characteristics of an unbalanced feeder.

Distribution Feeder Analysis

Example 10.1: A very simple distribution feeder is shown in Figure 10.8. This system is the IEEE 4 node test feeder that can be found on the IEEE website [3].

For the system in Figure 10.8, the infinite bus voltages are a balanced three phase of 12.47 kV line to line. The "source" line segment from Node 1 to Node 2 is a three-wire delta 2,000-feet-long line and is constructed on the pole configuration of Figure 4.7 without the neutral. The "load" line segment from Node 3 to Node 4 is 2,500-feet long and also is constructed on the pole configuration of Figure 4.7 but is a four-wire wye so the neutral is included. Both line segments use 336,400 26/7 ACSR phase conductors and the neutral conductor on the four-wire wye line is 4/0 6/1 ACSR. Since the lines are short, the shunt admittance will be neglected. The 50°C resistance is used for the phase and neutral conductors:

336,400 26/7 ACSR: resistance at 50°C = 0.306 Ω/mile

4/0 6/1 ACSR: resistance at 50°C = 0.592 Ω/mile

The phase impedance matrices for the two-line segments are

$$[Zeq_S] = \begin{bmatrix} 0.1520 + j0.5353 & 0.0361 + j0.3225 & 0.0361 + j0.2752 \\ 0.0361 + j0.3225 & 0.1520 + j0.5353 & 0.0361 + j0.2955 \\ 0.0361 + j0.2752 & 0.0361 + j0.2955 & 0.1520 + j0.5353 \end{bmatrix} \Omega,$$

$$[Zeq_L] = \begin{bmatrix} 0.2166 + j0.5140 & 0.0738 + j0.2375 & 0.0727 + j0.1823 \\ 0.0738 + j0.2375 & 0.2209 + j0.4963 & 0.0748 + j0.2006 \\ 0.0727 + j0.1823 & 0.0748 + j0.2006 & 0.2185 + j0.5043 \end{bmatrix} \Omega$$

The transformer bank is connected to delta – grounded wye, and the three-phase ratings are

$$kVA = 6000, kVLL_S = 12.47, kVLL_L = 4.16, Z_{pu} = 0.01 + j0.06$$

The feeder serves an unbalanced three-phase, wye-connected constant PQ load of

S_a = 750 kVA at 0.85 lagging power factor,

S_b = 900 kVA at 0.90 lagging power factor,

S_c = 1,100 kVA at 0.95 lagging power factor.

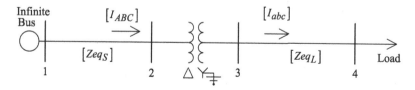

FIGURE 10.8 Example 10.1 IEEE 4 node test feeder.

Before starting the iterative solution, the forward and backward sweep matrices must be computed for each series element. The ladder method is going to be employed, so only the [A], [B], and [d] matrices need to be computed

Source line segment with shunt admittance neglected:

$$[U] = \begin{bmatrix} 1 & 0 & 0 \\ 0 & 1 & 0 \\ 0 & 0 & 1 \end{bmatrix}$$

$$[A_1] = [U] = \begin{bmatrix} 1 & 0 & 0 \\ 0 & 1 & 0 \\ 0 & 0 & 1 \end{bmatrix}$$

$$[B_1] = [Zeq_S] = \begin{bmatrix} 0.1414 + j0.5353 & 0.0361 + j0.3225 & 0.0361 + j0.2752 \\ 0.0361 + j0.3225 & 0.1414 + j0.5353 & 0.0361 + j0.2955 \\ 0.0361 + j0.2752 & 0.0361 + j0.2955 & 0.1414 + j0.5353 \end{bmatrix}$$

$$[d_1] = [U] = \begin{bmatrix} 1 & 0 & 0 \\ 0 & 1 & 0 \\ 0 & 0 & 1 \end{bmatrix}$$

Load line segment:

$$[A_2] = [U] = \begin{bmatrix} 1 & 0 & 0 \\ 0 & 1 & 0 \\ 0 & 0 & 1 \end{bmatrix}$$

$$[B_2] = [Zeq_L] = \begin{bmatrix} 0.1907 + j0.5035 & 0.0607 + j0.2302 & 0.0598 + j0.1751 \\ 0.0607 + j0.2302 & 0.1939 + j0.4885 & 0.0614 + j0.1931 \\ 0.0598 + j0.1751 & 0.0614 + j0.1931 & 0.1921 + j0.4970 \end{bmatrix}$$

$$[d_2] = [U] = \begin{bmatrix} 1 & 0 & 0 \\ 0 & 1 & 0 \\ 0 & 0 & 1 \end{bmatrix}$$

Transformer

The transformer impedance must be converted to the actual value in ohms referenced to the low-voltage windings.

$$Z_{base} = \frac{kVLL_L^2 \cdot 1000}{kVA} = 2.88\,\Omega$$

Distribution Feeder Analysis

$$Zt_{low} = (0.01 + j0.06) \cdot 2.88 = 0.0288 + j0.1728 \, \Omega$$

The transformer phase impedance matrix is

$$[Zt_{abc}] = \begin{bmatrix} 0.0288 + j0.1728 & 0 & 0 \\ 0 & 0.0288 + j0.1728 & 0 \\ 0 & 0 & 0.0288 + j0.1728 \end{bmatrix} \Omega$$

The "turns" ratio: $n_t = \dfrac{kVLL_S}{kVLN_L} = 5.1958$.

The ladder sweep matrices are as follows:

$$[A_t] = \frac{1}{n_t} \cdot \begin{bmatrix} 1 & 0 & -1 \\ -1 & 1 & 0 \\ 0 & -1 & 1 \end{bmatrix} = \begin{bmatrix} 0.1925 & 0 & -0.1925 \\ -0.1925 & 0.1925 & 0 \\ 0 & -0.1925 & 0.1925 \end{bmatrix}$$

$$[B_t] = [Zt_{abc}] = \begin{bmatrix} 0.0288 + j0.1728 & 0 & 0 \\ 0 & 0.0288 + j0.1728 & 0 \\ 0 & 0 & 0.0288 + j0.1728 \end{bmatrix}$$

$$[d_t] = \frac{1}{n_t} \cdot \begin{bmatrix} 1 & -1 & 0 \\ 0 & 1 & -1 \\ -1 & 0 & 1 \end{bmatrix} = \begin{bmatrix} 0.1925 & -0.1925 & 0 \\ 0 & 0.1925 & -0.1925 \\ -0.1925 & 0 & 0.1925 \end{bmatrix}$$

Define the Node 4 loads.

$$[S4] = \begin{bmatrix} 750/\mathrm{acos}(0.85) \\ 900/\mathrm{acos}(0.90) \\ 1100/\mathrm{acos}(0.95) \end{bmatrix} = \begin{bmatrix} 750/\underline{31.79} \\ 900/\underline{25.84} \\ 1100/\underline{18.19} \end{bmatrix} = \begin{bmatrix} 637.5 + j395.1 \\ 810.0 + j392.3 \\ 1045.0 + j343.5 \end{bmatrix} \mathrm{KVA}$$

Define infinite bus line-to-line and line-to-neutral voltages.

$$[ELL_s] = \begin{bmatrix} 12,470/\underline{30} \\ 12,470/\underline{-90} \\ 12,470/\underline{150} \end{bmatrix} \mathrm{V}$$

$$[ELN_s] = \begin{bmatrix} 7199.6/\underline{0} \\ 7199.6/\underline{-120} \\ 7199.6/\underline{120} \end{bmatrix} \mathrm{V}$$

The initial conditions are

$$start = \begin{bmatrix} 0 \\ 0 \\ 0 \end{bmatrix} \quad tol = 0.00001 \quad VM = 7199.6$$

A program is shown in Figure 10.9.

$$X := \begin{vmatrix} I_{.abc} \leftarrow start \\ I_{.ABC} \leftarrow start \\ V_{.old} \leftarrow start \\ \text{for } n \in 1..200 \\ \quad \begin{vmatrix} VLN_{.2} \leftarrow A_{.1} \cdot E_{.s} - B_{.1} \cdot I_{.ABC} \\ VLN_{.3} \leftarrow A_{.t} \cdot VLN_{.2} - B_{.t} \cdot I_{.abc} \\ VLN_{.4} \leftarrow A_{.2} \cdot VLN_{.3} - B_{.2} \cdot I_{.abc} \\ \text{for } i \in 1..3 \\ \quad Error_i \leftarrow \dfrac{\left|\left|VLN_{.4_i}\right| - \left|V_{.old_i}\right|\right|}{VM} \\ \text{break if } \max(Error) < Tol \\ \text{for } i \in 1..3 \\ \quad I_{.abc_i} \leftarrow \dfrac{\overline{SL_i \cdot 1000}}{VLN_{.4_i}} \\ V_{.old} \leftarrow VLN_{.4} \\ I_{.ABC} \leftarrow d_{.t} \cdot I_{.abc} \end{vmatrix} \\ Out_1 \leftarrow VLN_{.2} \\ Out_2 \leftarrow I_{.ABC} \\ Out_3 \leftarrow VLN_{.3} \\ Out_4 \leftarrow I_{.abc} \\ Out_5 \leftarrow VLN_{.4} \\ Out_6 \leftarrow n \\ Out \end{vmatrix}$$

FIGURE 10.9 Mathcad program.

Distribution Feeder Analysis

The Mathcad program is used to analyze the system, and after seven iterations, the load voltages on a 120-volt base are

$$[V4_{120}] = \begin{bmatrix} 113.3 \\ 111.2 \\ 111.5 \end{bmatrix} V$$

The voltages at Node 4 are below the desired 120 volts. These low voltages can be corrected with the installation of three step-voltage regulators connected in wye on the secondary bus (Node 3) of the transformer. The new configuration of the feeder is shown in Figure 10.10.

For the regulator, the potential transformer ratio will be 2,400–120 volts ($N_{pt} = 20$), and the CT ratio is selected to carry the rated current of the transformer bank. The rated current is

$$I_{rated} = \frac{6,000}{\sqrt{3} \cdot 2.4} = 832.7 \text{A}$$

The CT ratio is selected to be 1,000:5 = CT = 200.

The potential transformer ratio is $N_{pt} = \dfrac{VLN_{rated}}{120} = \dfrac{2400}{120} = 20$

The equivalent phase impedance between Node 3 and Node 4 is computed using the converged voltages at the two nodes. This is done so that the R and X settings of the compensator can be determined.

$$[V3] = \begin{bmatrix} 2349.8/\underline{-31.2} \\ 2348.2/\underline{-151.5} \\ 2343.6/\underline{88.0} \end{bmatrix} V \quad [V4] = \begin{bmatrix} 2266.3/\underline{-32.0} \\ 2223.8/\underline{-153.1} \\ 2230.7/\underline{84.1} \end{bmatrix} V \quad [I3] = \begin{bmatrix} 330.9/\underline{-63.7} \\ 404.7/\underline{-179.0} \\ 493.1/\underline{65.9} \end{bmatrix} A$$

$$Zeq_i = \frac{V3_i - V4_i}{I3_i} = \begin{bmatrix} 0.1628 + j0.2147 \\ 0.2023 + j0.2820 \\ 0.1063 + j0.3745 \end{bmatrix} \Omega$$

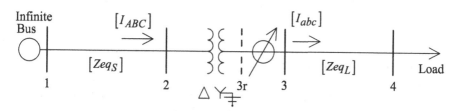

FIGURE 10.10 Voltage regulator added to the system.

The three regulators are to have the same R and X compensator settings. The average value of the computed impedances will be used.

$$Z_{avg} = \frac{1}{3} \cdot \sum_{k=1}^{3} Zeq_k = 0.1571 + j0.2904\,\Omega$$

The value of the compensator impedance in volts is given by Equation 7.78:

$$R' + jX' = (0.1571 + j0.2904) \cdot \frac{1000}{20} = 7.9 + j14.5\,\text{V}$$

The value of the compensator settings in ohms is

$$Zcomp = R_\Omega + jX_\Omega = \frac{7.9 + j14.5}{5} = 1.46 + j2.84\,\Omega$$

With the regulator in the neutral position, the voltages being input to the compensator circuit for the given conditions are

$$Vreg_i = \frac{V3_i}{N_{pt}} = \begin{bmatrix} 117.5/-31.2 \\ 117.4/-151.5 \\ 117.2/88.0 \end{bmatrix}\,\text{V}$$

The compensator currents are

$$Icomp_i = \frac{Iabc_i}{CT} = \begin{bmatrix} 1.6535/-63.8 \\ 2.0174/-179.1 \\ 2.4560/64.9 \end{bmatrix}\,\text{A}$$

With the input voltages and compensator currents, the voltages across the voltage relays in the compensator circuit are computed to be

$$[V_{relay}] = [Vreg] - [Zcomp] \cdot [Icomp] = \begin{bmatrix} 113.0/-32.6 \\ 112.2/-153.5 \\ 111.4/85.3 \end{bmatrix}\,\text{V}$$

Notice how close these compare to the actual voltages on a 120-volt base at Node 4.

Assume that the voltage level has been set at 120 volts with a bandwidth of 2 volts. This means that the relay voltages must lie between 119 and 121 volts. In order to model this system, the flow chart of Figure 10.12 is slightly modified in the

Distribution Feeder Analysis

forward and backward sweeps. The initial matrices for the regulator are computed with the regulator taps set at zero.

Forward Sweep

$$[VLN_2] = [A_1] \cdot [E_s] - [B_1] \cdot [I_{ABC}]$$

$$[VLN_{3r}] = [A_t] \cdot [VLN_2] - [B_t] \cdot [I_{in}]$$

$$[VLN_3] = [A_{reg}][VLN_{3r}] - [B_{reg}] \cdot [I_{abc}]$$

$$[VLN_4] = [A_2] \cdot [VLN_3] - [B_2] \cdot [I_{abc}]$$

Backward Sweep

$$[V_{old}] = [VLN_4]$$

$$[I_{in}] = [d_{reg}] \cdot [I_{abc}]$$

$$[I_{ABC}] = [d_t] \cdot [I_{in}]$$

After the analysis routine has converged, a new routine will compute whether tap changes need to be made. The computer routine for computing the new taps is shown in Figure 10.11. Recall that in Chapter 7 it was shown that each tap changes the relay voltages by 0.75 volts.

The computational sequence for the determination of the final tap settings and convergence of the system is shown in the flow chart of Figure 10.12.

The tap changing routine changes individual regulators so that the relay voltages lie within the voltage bandwidth. For this simple system, the initial change in taps become the final tap settings of

$$[Tap] = \begin{bmatrix} 9 \\ 10 \\ 11 \end{bmatrix}$$

The final relay voltages are

$$[V_{relay}] = \begin{bmatrix} 120.2 \\ 120.4 \\ 120.4 \end{bmatrix} V$$

$$XY := \begin{vmatrix} \text{for } i \in 1..3 \\ \quad Vreg_i \leftarrow \dfrac{VLN_{3_i}}{N_{pt}} \\ \quad Icomp_i \leftarrow \dfrac{I_{abc_i}}{CT} \\ \quad Vrelay_i \leftarrow Vreg_i - Z_{comp_{i,i}} \cdot Icomp_i \\ \quad Tap_{old_i} \leftarrow Tap_i \\ \quad dTap_i \leftarrow \dfrac{120 - |Vrelay_i|}{.75} \\ \quad Tap_i \leftarrow Tap_{old_i} + dTap_i \\ \quad Tap_i \leftarrow \text{Round}(Tap_i, 1) \\ \quad a_{R_i} \leftarrow 1 - 0.00625 Tap_i \\ Out_1 \leftarrow Vrelay \\ Out_2 \leftarrow Tap \\ Out_3 \leftarrow a_{.R} \\ Out \end{vmatrix}$$

FIGURE 10.11 Tap changing routine.

The final voltages on a 120-volt base at the load center (Node 4) are

$$[VLN_{4_{120}}] = \begin{bmatrix} 120.5 \\ 119.5 \\ 121.7 \end{bmatrix} V$$

The compensator relay voltages and the actual load center voltages are very close to each other.

The same system was built in WindMil with the following results and the regulator installed:

10.1.7 Load Allocation

The input complex power (kW and kvar) to a feeder is known because of the metering at the substation. This information can be either a total three phase or for each individual phase. Sometimes the metered data is the current and power factor in each phase.

It is desirable to force the computed input complex power to the feeder, match the metered input. This can be accomplished (following a converged iterative solution) by computing the ratio of the metered input to the computed input. The phase loads

Distribution Feeder Analysis 361

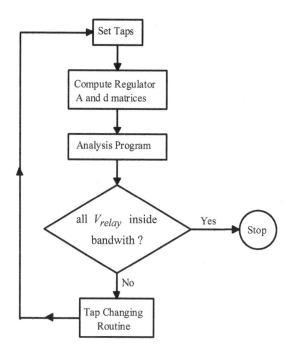

FIGURE 10.12 Computational sequence.

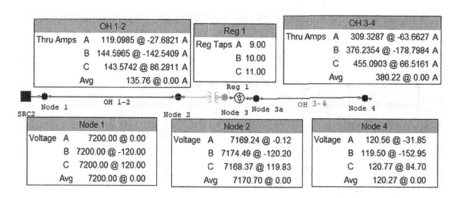

FIGURE 10.13 WindMil results.

can now be modified by multiplying the loads by this ratio. Because the losses of the feeder will change when the loads are changed, it is necessary to go through the ladder iterative process to determine a new input to the feeder. This computed input will be closer to the metered input but, most likely, not within a specified tolerance. Again, a ratio can be determined, and the loads modified followed. This process is repeated until the computed input is within a specified tolerance of the metered input.

Load allocation does not have to be limited to matching metered readings just at the substation. The same process can be performed at any point on the feeder where metered data is available. The only difference is that now only the "downstream" nodes from the metered point will be modified.

10.1.8 Loop Flow

The ladder technique has proven to be a fast and efficient method for performing power-flow studies on radial distribution feeders. The shortcoming for this method is that there are cases where a feeder is not totally radial and therefore a different method must be applied. Many times, the feeder may have just a few loops, in which case the ladder method can be modified to take into account the small, looped feeder [5]. A method called "loop flow" will be developed that will allow for loops in a predominately radial feeder.

10.1.8.1 Single-Phase Feeder

Figure 10.14 shows two small single-phase systems operating independently but with a switch between the two that once closed the two small systems become one looped system. When the switch is closed, the difference voltage (dV_{34}) will be zero. Something has to be done to the system in order to force the difference voltage to be zero.

A way to simulate the closed switch is illustrated in Figure 10.15.

To simulate the closed switch in Figure 10.15, it is necessary to determine the correct value of IT to be injected into Node 3 and the negative of IT to be injected into Node 4 that will force the voltage dV_{34} to be zero. The circuit of Figure 10.14 is modified to include the injected currents at Nodes 3 and 4.

In Figure 10.16, the voltages at Nodes 3 and 4 are given by the following:

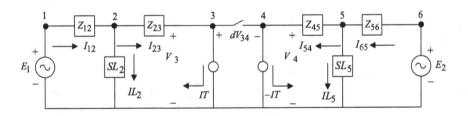

FIGURE 10.14 Single-phase system with a loop.

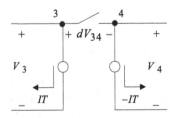

FIGURE 10.15 Simulation of closed switch.

Distribution Feeder Analysis

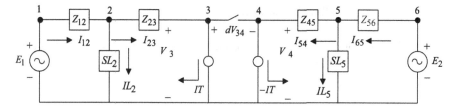

FIGURE 10.16 Modified circuit.

$$V_3 = E_1 - Z_{12} \cdot (IL_2 + IT) - Z_{23} \cdot IT$$

$$V_3 = E_1 - Z_{12} \cdot IL_2 - (Z_{12} + Z_{23}) \cdot IT$$

Define:

$$V_{3V} = E_1 - Z_{12} \cdot IL_2$$

$$V_{3I} = -(Z_{12} + Z_{23}) \cdot IT$$

Therefore : $V_3 = V_{3V} + V_{3I}$ \hfill (10.9)

In a similar manner, the voltage at Node 4 is computed as follows:

$$V_4 = E_2 - Z_{56} \cdot (IL_5 - IT) - Z_{45} \cdot (-IT)$$

$$V_4 = E_2 - Z_{56} \cdot IL_5 + (Z_{56} + Z_{45}) \cdot IT$$

Define:

$$V_{4V} = E_2 - Z_{56} \cdot IL_5$$

$$V_{4I} = +(Z_{56} + Z_{45}) \cdot IT$$

Therefore : $V_4 = V_{4V} + V_{4I}$ \hfill (10.10)

The voltage drop between Nodes 3 and 4 consists of a component due to the source voltages and a component due to the injected currents. Using the final form of the node voltages, the difference voltage between Nodes 3 and 4 is the following:

$$V_3 = V_{3V} + V_{3I}$$

$$V_4 = V_{4V} + V_{4I}$$

$$dV_{34} = V_3 - V_4 = V_{3V} - V_{4V} + (V_{3I} - V_{4I})$$

$$dV_{34} = dV_{34V} + dV_{34I} \qquad (10.11)$$

Applying Equations 10.9 and 10.10, the difference voltage resulting from the application of the injection currents is the following:

$$dV_{34I} = V_{3I} - V_{4I} = -(Z_{12} + Z_{23} + Z_{56} + Z_{45}) \cdot IT$$

$$dV_{34I} = -Thev \cdot IT$$

where : $ThevZ = Z_{12} + Z_{23} + Z_{56} + Z_{45}$ \qquad (10.12)

For this simple system, *ThevZ* is the sum of the line impedances around the closed loop. This impedance is referred to as the Thevenin equivalent impedance. For a general system, the impedance is computed by the principle of superposition where the voltage sources are set to zero, the loads neglected, and only the injected currents are applied to the system. For this simple system, that circuit is shown in Figure 10.17.

Equation 10.12 applies KVL around the looped system in Figure 10.17. With the voltage drop dV_{34I} computed, the Thevenin equivalent impedance is

$$ThevZ = \frac{-dV_{34I}}{IT} \qquad (10.13)$$

With *ThevZ* computed, the final goal is to determine the value of *IT* that will force the difference voltage to be zero, as shown in Equation 10.14.

$$0 = dV_{34V} + dV_{34I}$$

$$dV_{34I} = -dV_{34V}$$

where : $dV_{34I} = -ThevZ \cdot IT$

therefore : $-dV_{34V} = -ThevZ \cdot IT$

$$IT = \frac{dV_{34V}}{ThevZ} \qquad (10.14)$$

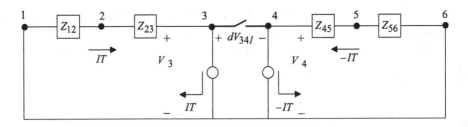

FIGURE 10.17 Thevenin equivalent impedance circuit.

Distribution Feeder Analysis

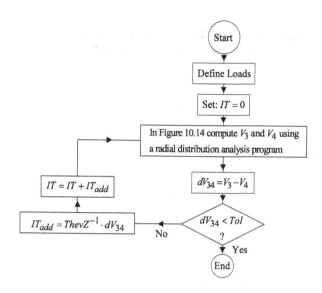

FIGURE 10.18 Flow chart for the solution with injected currents.

Because the loads on a distribution system are non-linear, an iterative routine is used to compute the needed injection currents. Figure 10.18 shows a simple flow chart of an iterative routine used to compute the injected currents.

It is noted in Figure 10.18 that the initial injection currents are set to zero. During the first iteration, the difference voltage will be a function of only the source voltages and the load currents. With this difference voltage computed, the first value of the injection currents is computed according to Equation 10.14 and then added to the initial value of $IT = 0$. With the new value of injection currents, the circuit of Figure 10.15 is evaluated to compute the new difference voltage which now includes the effect of the injection currents. The new difference voltage is checked to see if it is within a specified tolerance of the desired zero. If it is not, an additional injection current is computed and added to the most recent value of the injection.

10.1.8.2 IEEE 13 Bus Test Feeder

The IEEE 13 bus test feeder was developed to allow distribution analysis programs to be tested with the results compared to the published results [3]. This feeder will be used to demonstrate the simulation of a looped system using the method presented in the previous section. A one-line diagram of the IEEE 13 bus test feeder is shown in Figure 10.18. Note that in this figure, the buses have been renumbered from the original document and a new node is added to the other side of the regulator.

The voltage on phase c at Bus 6 has gone below the ANSI minimum of 114 volts. The voltage at Bus 10 phase c is also below the ANSI standard. The real problem with Bus 10 is that the current capacity of 260 amps on the underground concentric neutral cable between Bus 9 and Bus 10 is exceeded. To solve these problems and to demonstrate the looped feeder simulation, two new lines will be added to the system. The one-line diagram of the modified IEEE 13 feeder is shown in Figure 10.20.

Example 10.2: Determine the values of the injected currents for the system of Figure 10.16 to simulate loop flow between the two sources.

For the system, the following data is given:

$$E_1 = 7200\,\text{V}, E_2 = 7560\,\text{V}$$

Impedance of the lines:	$z1 = 0.5152 + j1.1359$ ohms/mile
Length of lines:	$L_{12} = 5,\ L_{23} = 2,\ L_{45} = 3,\ L_{56} = 4$ miles
Line impedances:	$Z_{12} = 0.2576 + j0.5679\ \Omega$
	$Z_{23} = 1.0304 + j2.2718\ \Omega$
	$Z_{45} = 1.5456 + j3.4077\ \Omega$
	$Z_{56} = 2.0608 + j4.5436\ \Omega$
Loads:	$SL_2 = 1500 + j1250$ kVA
	$SL_5 = 1000 + j750$ kVA

Step 1: Set the voltage sources to zero and apply the positive and negative injection currents.
The base values are

$$VA_{base} = 1000,\ kVLN_{base} = 7.2,$$

$$I_{base} = \frac{kVA_{base}}{kVLN_{base}} = 138.9\,\text{A}$$

The injected current magnitude is set to the base current:

$$IT_1 = I_{base} = 138.9\,\text{A}$$

$$IT_2 = -IT_1 = -138.9\,\text{A}$$

The analysis of the system with the two injected currents gives the following:

$$V_{3I} = 433.1\underline{/-114.4}\,\text{V}$$

$$V_{4I} = 1212.6\underline{/65.6}\,\text{V}$$

$$dV_{34I} = V_{3I} - V_{4I} = 1645.7\underline{/-114.4}\,\text{V}$$

The Thevenin equivalent impedance is

$$Z_{th} = \frac{-dV_{34I}}{IT} = \frac{1645.7\underline{/-114.4}}{138.9\underline{/0}} = 4.8944 + j10.7911\,\Omega$$

The new line from Bus 4 to Bus 16 is phase c and consists of 1/0 ACSR 6/1 constructed on a single pole, as shown in Figure 10.21. The length of the line is 600 feet.

Distribution Feeder Analysis

Step 2: Set the injected currents to zero and compute dV_{34}.

With the injected currents equal to zero the voltages are computed as follows:

$$V_{3V} = 7043.97\underline{/-0.6}\,\text{V}$$

$$V_{4V} = 6734.96\underline{/-3.4}\,\text{V}$$

$$dV_{34V} = V_{3V} - V_{4V} = 454.86\underline{/45.2}\,\text{V}$$

With the equivalent Thevenin impedance computed, the first value of the required injected current is the following:

$$IT_{add} = \frac{V_{34V}}{Z_{th}} = 38.39\underline{/-20.4}\,\text{A}$$

$$IT = IT + IT_{add} = 0 + 38.39\underline{/-20.4} = 38.39\underline{/-20.4}\,\text{A}$$

The difference voltage is now computed with the voltage sources and injected currents. After this first iteration, the difference voltage is computed to be the following:

$$V_3 = 6958.3\underline{/-1.3}\,\text{V}$$

$$V_4 = 7007.5\underline{/-1.4}\,\text{V}$$

$$dV_{34} = 50.6\underline{/164.9}\,\text{V}$$

The added injection current and new total injected current are as follows:

$$IT_{add} = \frac{50.6\underline{/164.9}}{11.85\underline{/65.60}} = 4.28\underline{/99.27}\,\text{A}$$

$$IT = IT + IT_{add} = 38.39\underline{/-20.4} + 4.28\underline{/99.27} = 36.46\underline{/-14.52}\,\text{A}$$

The difference voltage is again computed using the new value of injected current. This process continues until after the fifth iteration, the difference voltage is the following:

$$V_3 = 6970.42.\underline{/-1.33}\,\text{V}$$

$$V_4 = 6970.42.\underline{/-1.33}\,\text{V}$$

$$dV_{34} = V_3 - V_4 = 0\,\text{V}$$

The line currents flowing with injected currents are the following:

$I_{23} = 36.8842/\underline{-14.65}$ A

$I_{45} = 36.8842/\underline{-14.65}$ A

Since these two currents are identical, the two systems are now operating as one system, with system 1 providing current to system 2 just as though the switch between Nodes 3 and 4 was closed.

Example 10.3: **The system of Figure 10.19 was created in WindMil. With the original data, partial results are shown in Table 10.1. The currents are in amps and the voltages are line-to-neutral on a 120-volt base.**

As seen in Table 10.1, the voltages are very unbalanced at Bus 10. This unbalance was purposely created so that distribution analysis programs could be tested for convergence in a very unbalanced feeder. No effort will be made to balance the voltages. Even though the voltages are unbalanced, they are still within the ANSI standard [5] of all voltages being between 114 and 126 volts.

Rather than balance the voltages, two new loads are going to be added to the existing feeder, which will lead to the need for the looped feeder. The following loads are added to the system:

Bus 5: Phase c: 200 + j100 kVA
Bus 10: Three-phase load: 750 + j525 kVA

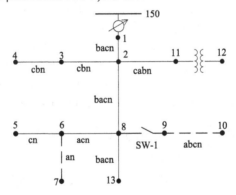

FIGURE 10.19 IEEE 13 bus test feeder.

With the new loads and the voltage regulator operating, the partial results are shown in Table 10.2.

The three-phase line from Bus 14 to Bus 13 consists of 4/0 ACSR 6/1 phase and neutral conductors with a pole configuration, as shown in Figure 10.22. The length of the line is 800 feet.

In order to simulate, the loop flow currents must be injected into Buses 5, 6, 13, and 11, as shown in Figure 10.23.

TABLE 10.1
Original Feeder Results

	Phase a	Phase b	Phase c
Reg. Taps	11	9	12
I 1–2	597	451	648
I 3–4	0	65	65
I 9–10	221	36	156
V Bus 6	118		115
V Bus 10	118	126	115

TABLE 10.2
New Loads Added

	Phase a	Phase b	Phase c
Reg Taps	12	10	14
I 1–2	731	578	903
I 3–4		65	65
I 9–10	350	159	303
V Bus 6		118	109
V Bus 10	117	125	113

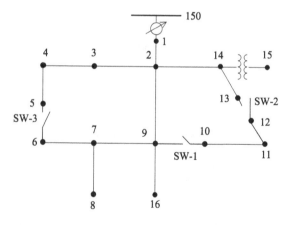

FIGURE 10.20 13 bus test feeder.

The line from Bus 2 to Node 9 has a distributed load, which is modeled as two-thirds of the distributed load at Bus 2a, which is one-third the length of the line, and the remaining distributed load is connected at the end of the line at the new Bus 9a. In Figure 10.23, SW-1 is a three-phase switch that can only be modeled as open or closed. SW-2 is a three-phase switch that can be modeled as open, closed, or looped. SW-3 is a single-phase switch on phase c and can only be modeled as open or looped.

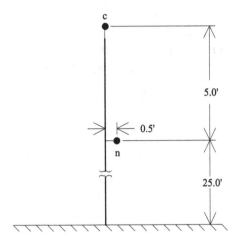

FIGURE 10.21 Single-phase line.

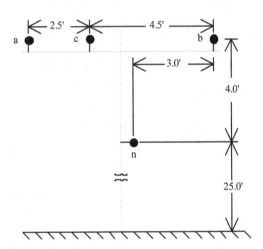

FIGURE 10.22 Three-phase line.

The first step in simulating loop flow is to compute the Thevenin equivalent impedance at each looped switch for each phase. The currents injected into Bus 15 ($IT1$) will be in phases a, b, and c. The injected currents at Bus 16 ($IT3$) will be the negative of ($IT1$). Switch SW-3 is a single-phase switch on phase c. Therefore, the injected current at Bus 5 ($IT4$) will only be a phase c current. The injected current at Bus 5 ($IT4$) will be the negative of ($IT2$). There will be a Thevenin equivalent impedance computed at each switch for each phase injected at each of the injection nodes. This will lead to a 4 × 4 Thevenin equivalent matrix ($ThevZ$). For this process, the value of the injected current is assumed to be the base current of the system 694.4 amps.

Distribution Feeder Analysis

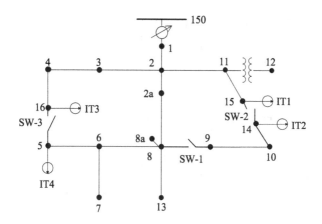

FIGURE 10.23 Injected currents.

For the computation of *ThevZ* the source voltages and all loads are set to zero. Refer to Figure 10.21 with SW-1 closed. The line-to-neutral voltages at Buses 15, 10, 16, and 5 are computed with only the phase a currents injected at Buses 13 and 12. A Mathcad program is used to compute the voltages. With only the phase a currents, the vectors for the injected currents at the switch buses are:

$$[IT1] = \begin{bmatrix} 694.4 \\ 0 \\ 0 \end{bmatrix} A \quad [IT3] = \begin{bmatrix} -694.4 \\ 0 \\ 0 \end{bmatrix} A \quad (10.15)$$

The bus voltages with just the phase a injected current at Bus 13 are computed to be

$$[V_{15}] = \begin{bmatrix} 239.5/-122.5 \\ 77.3/-110.5 \\ 89.8/-107.3 \end{bmatrix} \quad [V_{14}] = \begin{bmatrix} 330.0/64.2 \\ 147.8/65.2 \\ 125.9/61.4 \end{bmatrix} \quad (10.16)$$

$$V[V_{5c}] = 0 \, V, [V_{6c}] = 118.9/-110.5 \, V$$

The difference voltages are

$$[dV_{1514}] = \begin{bmatrix} 568.6/-118.6 \\ 224.9/-113.3 \\ 214.7/-113.9 \end{bmatrix} V,$$

$$dV_{56c} = 118.93/-110.5 \, V \quad (10.17)$$

Equation 10.17 is used to compute the first column elements of the *ThevZ* matrix.

$$\begin{bmatrix} ThevZ_{1,1} \\ ThevZ_{2,1} \\ ThevZ_{3,1} \end{bmatrix} = -\frac{1}{IT} \cdot \begin{bmatrix} dV_{1514a} \\ dV_{1514b} \\ dV_{1514c} \end{bmatrix} = \begin{bmatrix} 0.3921 + j0.7187 \\ 0.1282 + j0.2974 \\ 0.1252 + j0.2826 \end{bmatrix}$$

$$Thevz_{4,1} = \frac{-dV_{56c}}{IT} = 0.0598 + j0.1605 \, \Omega \qquad (10.18)$$

This process is repeated with the injected currents in phase b at Buses 15 and 14, followed by the injected currents in phase c at Buses 15 and 14. The last step is to have the phase c currents injected at Buses 16 and 5. The final $ThevZ$ matrix in ohms is

$$[ThevZ] = \begin{bmatrix} 0.3921 + j0.7178 & 0.1282 + j0.2974 & 0.1252 + j0.2826 & 0.0598 + j0.1605 \\ 0.1282 + j0.2974 & 0.3866 + j0.7302 & 0.1262 + j0.2437 & 0.0581 + j0.1458 \\ 0.1252 + j0.2826 & 0.1262 + j0.2437 & 0.3880 + j0.7324 & 0.1293 + j0.3920 \\ 0.0598 + j0.1605 & 0.0581 + j0.1458 & 0.1293 + j0.3920 & 0.6328 + j0.9023 \end{bmatrix} \quad (10.19)$$

With $ThevZ$ computed, the values of the needed injection currents must be determined with the loads and capacitors included. The method is to apply Equation 10.18 in matrix form for the IEEE 13 bus feeder and solve for the $[IT]$ array. The equation is

$$\begin{bmatrix} IT1_a \\ IT1_b \\ IT1_c \\ IT2_c \end{bmatrix} = [ThevZ]^{-1} \cdot \begin{bmatrix} dV_{1510a} \\ dV_{1510b} \\ dV_{1510c} \\ dV_{165c} \end{bmatrix} \qquad (10.20)$$

To determine the injection currents, the power-flow program must be run where all of the loads and capacitors, along with the injection currents, are modeled. The tap settings for the regulator are set at the same taps as in Table 10.2. As was done with the single-phase system, initially all the injection currents are set to zero, and the difference voltages in Equation 10.20 are computed. Equation 10.20 is used to compute the initial change in injection currents. The computed difference voltages for this first iteration are

$$\begin{bmatrix} dV_{1510a} \\ dV_{1510b} \\ dV_{1510c} \\ dV_{165c} \end{bmatrix} = \begin{bmatrix} 181.2 \underline{/50.8} \\ 39.4 \underline{/-17.4} \\ 206.2 \underline{/151.8} \\ 228.7 \underline{/153.6} \end{bmatrix} V \qquad (10.21)$$

With the difference voltages computed, Equation 10.20 is used to compute the injection currents that will force the difference voltages to zero. During the first iteration, the injection currents were set to zero. Applying Equation 10.14 will give the

Distribution Feeder Analysis

"added" injection currents needed. After the first iteration, the computed injection currents are

$$\begin{bmatrix} IT_{1a} \\ IT_{1b} \\ IT_{1c} \\ IT_{2c} \end{bmatrix} = \begin{bmatrix} 304.6/-22.0 \\ 145.5/-122.4 \\ 261.7/\underline{102.8} \\ 147.4/\underline{99.0} \end{bmatrix} A \qquad (10.22)$$

With the currents in Equation 10.22 injected into the buses, the power-flow program is run again. Because the current flows on the lines will now be different the bus voltages will also change. Since many of the loads are modeled as constant PQ, those load currents are subject to change. The second iteration is necessary to recalculate the bus voltages and determine if additional injection currents are needed. After the second iteration, the difference voltages across the looped switches are

$$\begin{bmatrix} dV_{1510a} \\ dV_{1510b} \\ dV_{1510c} \\ dV_{165c} \end{bmatrix} = \begin{bmatrix} 8.2/-165.1 \\ 4.4/-108.2 \\ 17.1/-76.8 \\ 18.1/-76.3 \end{bmatrix} V \qquad (10.23)$$

Since the difference voltages are not close to zero, additional injection currents are needed. The additional currents are

$$[IT_{add}] = \begin{bmatrix} 15.1/\underline{97.9} \\ 1.2/-8.5 \\ 18.6/-129.3 \\ 11.0/-132.5 \end{bmatrix} A \qquad (10.24)$$

The total injection currents for the next iteration are

$$\begin{bmatrix} IT_{1a} \\ IT_{1b} \\ IT_{1c} \\ IT_{2} \end{bmatrix} = \begin{bmatrix} 304.6/-22.0 \\ 145.5/-122.4 \\ 261.7/\underline{102.8} \\ 147.4/\underline{99.0} \end{bmatrix} + \begin{bmatrix} 15.1/\underline{97.9} \\ 1.2/-8.5 \\ 18.6/-129.3 \\ 11.0/-132.5 \end{bmatrix} = \begin{bmatrix} 297.4/-19.4 \\ 145.0/-122.0 \\ 250.7/\underline{106.2} \\ 140.8/\underline{102.5} \end{bmatrix} A \qquad (10.25)$$

Following this procedure, after the fourth iteration, the difference voltages are very small and the process stops. The injected currents and difference voltages are

$$\begin{bmatrix} IT_{1a} \\ IT_{1b} \\ IT_{1c} \\ IT_{2c} \end{bmatrix} = \begin{bmatrix} 298.0/-19.5 \\ 145.0/-121.9 \\ 251.7/\underline{106.0} \\ 141.4/\underline{102.4} \end{bmatrix} A,$$

TABLE 10.3
Looped Switches

	Phase a	Phase b	Phase c
Reg Taps	11	8	14
I 1–2	675.1	536.2	798.8
I 6–5			17.2
I 9–10	20.6	19.6	41.1
V Bus 5			119.3
V Bus 10	120.5	123.8	119.4

$$\begin{bmatrix} dV_{1510a} \\ dV_{1510b} \\ dV_{1510c} \\ dV_{165c} \end{bmatrix} = \begin{bmatrix} 0.0276 /\underline{-121.2} \\ 0.0343 /\underline{-93.9.6} \\ 0.0931 /\underline{-62.4} \\ 0.0963 /\underline{-62.2} \end{bmatrix} V \quad (10.26)$$

In Table 10.2, it was shown that when the new loads were added at Buses 5 and 10, the bus voltage at Bus 5 was below the ANSI standard. Also, the concentric neutral line was overloaded. After adding the looped switches, Table 10.3 shows the results.

A comparison of Tables 10.2 and 10.3 shows the following:

- The voltage at Bus 5 is now above the minimum ANSI standard of 114 volts.
- The current flowing in the concentric neutral cable from Bus 9 to Bus 10 is much lower than the cable current rating.
- The voltages at Bus 10 are not as unbalanced and somewhat higher.
- The current out of the substation is basically the same since the net injection currents are zero, so the slight change in this current is due to the change in load currents for the constant PQ loads.
- The regulator tap positions have not changed since the current into the line drop compensator is basically the same as before the looped switches were closed.

10.1.8.3 Summary of Loop Flow

A method of simulating a looped distribution system has been presented. The loop is simulated by the installation of a switch between two existing buses in the feeder. To simulate the loop flow, the voltage at buses on the two sides of the switch must be equal. This is accomplished by the injection of a positive current at one of the buses and a negative value on the other bus. A method of calculating the necessary injection currents to force the difference voltage across the loop switch to be zero has been presented. Initially, a simple single-phase system was used to develop the technique.

Distribution Feeder Analysis

Following that the IEEE 13 bus test feeder was used to demonstrate the closing of a three-phase loop switch and a single-phase loop switch.

10.1.9 Summary of Power-Flow Studies

This section has developed a method for performing power-flow studies on a distribution feeder. Models for the various components of the feeder have been developed in previous chapters. The purpose of this section has been to develop and demonstrate the modified LIT using the forward and backward sweep matrices for the series elements. It should be obvious that a study of a large feeder with many laterals and sublaterals can not be performed without the aid of a complex computer program. In addition to the LIT, a method of modeling a feeder with closed loops was presented under the name "loop flow".

The development of the models and examples in this text use actual values of voltage, current, impedance, and complex power. When per-unit values are used, it is imperative that all values be converted to per unit using a common set of base values. In the usual application of per unit, there will be a base line-to-line voltage and a base line-to-neutral voltage, also, there will be a base line current and a base delta current. For both the voltage and current, there is a square root of three relationships between the two base values. In all of the derivations of the models, and in particular those for the three-phase transformers, the square root of three has been used to relate the difference in magnitudes between line-to-line and line-to-neutral voltages and between the line and delta currents. Because of this, when using the per-unit system, there should be only one base voltage and that should be the base line-to-neutral voltage. When this is done, for example, the per-unit positive and negative sequence voltages will be the square root of three times the per-unit positive and negative sequence line-to-neutral voltages. Similarly, the positive and negative sequence per-unit line currents will be the square of three times the positive and negative sequence per-unit delta currents. By using just one base voltage and one base current, the per-unit generalized matrices for all system models can be determined.

10.2 SHORT-CIRCUIT STUDIES

The computation of short-circuit currents for unbalanced faults in a normally balanced three-phase system has traditionally been accomplished by the application of symmetrical components. However, this method is not well suited to a distribution feeder that is inherently unbalanced. The unequal mutual coupling between phases leads to mutual coupling between sequence networks [4]. When this happens, there is no advantage to using symmetrical components. Another reason for not using symmetrical components is that the phases between which faults occur are limited. For example, using symmetrical components, line-to-ground faults are limited to phase a to ground. What happens if a single-phase lateral is connected to phase b or c and the short-circuit current is needed? This section will develop a method for short-circuit analysis of an unbalanced three-phase distribution feeder using the phase frame.

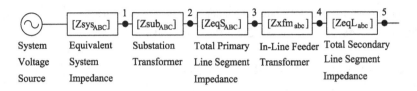

FIGURE 10.24 Unbalanced feeder short-circuit analysis model.

10.2.1 General Short-Circuit Theory

Figure 10.24 shows the unbalanced feeder as modeled for short-circuit calculations.

Short circuits can occur at any one of the five points shown in Figure 10.23. Point 1 is the high-voltage bus of the distribution substation transformer. The values of the short-circuit currents at point 1 are normally determined from a transmission system short-circuit study. The results of these studies are supplied in terms of the three-phase and single-phase, short-circuit MVAs. Using the short-circuit MVAs, the positive and zero-sequence impedances of the equivalent system can be determined. These values are needed for the short-circuit studies at the other four points in Figure 10.24.

Given the three-phase, short-circuit MVA magnitude and angle, the positive sequence equivalent system impedance in ohms is determined by

$$Z_+ = \frac{kVLL^2}{\left(MVA_{3-phase}\right)^*} \Omega \qquad (10.27)$$

Given the single-phase, short-circuit MVA magnitude and angle, the zero-sequence equivalent system impedance in ohms is determined by

$$Z_0 = \frac{3 \cdot kVLL^2}{\left(MVA_{1-phase}\right)^*} - 2 \cdot Z_+ \; \Omega \qquad (10.28)$$

In Equations 10.27 and 10.28, kVLL is the nominal line-to-line voltage in kV of the transmission system.

The computed positive and zero-sequence impedances need to be converted into the phase impedance matrix using the symmetrical component transformation matrix defined in Equation 6.51.

$$[Z_{012}] = \begin{bmatrix} Z_0 & 0 & 0 \\ 0 & Z_1 & 0 \\ 0 & 0 & Z_1 \end{bmatrix}$$

$$[Z_{abc}] = [A_s] \cdot [Z_{012}] \cdot [A_s]^{-1} \qquad (10.29)$$

Distribution Feeder Analysis

For short circuits at points 2, 3, 4, and 5, it is going to be necessary to compute the Thevenin equivalent three-phase circuit at the short circuit point. The Thevenin equivalent voltages will be the nominal line-to-ground voltages with the appropriate angles. For example, assume the equivalent system line-to-ground voltages are balanced three phase of nominal voltage with the phase a voltage at zero degrees. The Thevenin equivalent voltages at points 2 and 3 will be computed by multiplying the system voltages by the generalized transformer matrix $[A_t]$ of the substation transformer. Carrying this further, the Thevenin equivalent voltages at points 4 and 5 will be the voltages at Node 3 multiplied by the generalized matrix $[A_t]$ for the in-line transformer.

The Thevenin equivalent phase impedance matrices will be the sum of the Thevenin phase impedance matrices of each device between the system voltage source and the point of fault. Step-voltage regulators are assumed to be set in the neutral position, so they do not enter into the short-circuit calculations. Anytime a three-phase transformer is encountered, the total phase impedance matrix on the primary side of the transformer must be referred to the secondary side using Equation 8.160.

Figure 10.25 illustrates the Thevenin equivalent circuit at the faulted node.[3]

In Figure 10.25, the voltage sources E_a, E_b, and E_c represent the Thevenin equivalent line-to-ground voltages at the faulted node. The matrix [ZTOT] represents the Thevenin equivalent phase impedance matrix at the faulted node. The fault impedance is represented by Z_f in Figure 10.24.

KVL in matrix form can be applied to the circuit of Figure 10.24.

$$\begin{bmatrix} E_a \\ E_b \\ E_c \end{bmatrix} = \begin{bmatrix} Z_{aa} & Z_{ab} & Z_{ac} \\ Z_{ba} & Z_{bb} & Z_{bc} \\ Z_{ca} & Z_{cb} & Z_{cc} \end{bmatrix} \cdot \begin{bmatrix} If_a \\ If_b \\ If_c \end{bmatrix} + \begin{bmatrix} Z_f & 0 & 0 \\ 0 & Z_f & 0 \\ 0 & 0 & Z_f \end{bmatrix} \cdot \begin{bmatrix} If_a \\ If_b \\ If_c \end{bmatrix} + \begin{bmatrix} V_{ax} \\ V_{bx} \\ V_{cx} \end{bmatrix} + \begin{bmatrix} V_{xg} \\ V_{xg} \\ V_{xg} \end{bmatrix} \quad (10.30)$$

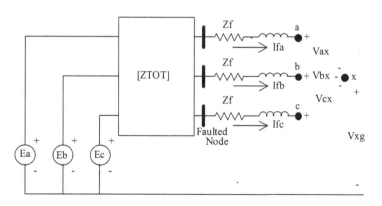

FIGURE 10.25 Thevenin equivalent circuit.

Equation 10.30 can be written in compressed form as

$$[E_{abc}] = [ZTOT] \cdot [If_{abc}] + [ZF] \cdot [If_{abc}] + [V_{abcx}] + [V_{xg}] \quad (10.31)$$

Combine terms in Equation 10.31.

$$[E_{abc}] = [ZEQ] \cdot [If_{abc}] + [V_{abcx}] + [V_{xg}], \quad (10.32)$$

$$\text{where } [ZEQ] = [ZTOT] + [ZF] \quad (10.33)$$

Solve Equation 10.15 for the fault currents.

$$[If_{abc}] = [Y] \cdot [E_{abc}] - [Y] \cdot [V_{abcx}] - [Y] \cdot [V_{xg}], \quad (10.34)$$

$$\text{where } [Y] = [ZEQ]^{-1} \quad (10.35)$$

Since the matrices $[Y]$ and $[E_{abc}]$ are known, define

$$[IP_{abc}] = [Y] \cdot [E_{abc}] \quad (10.36)$$

Substituting Equation 10.19 into Equation 10.17 and rearranging results in

$$[IP_{abc}] = [If_{abc}] + [Y] \cdot [V_{abcx}] + [Y] \cdot [V_{xg}] \quad (10.37)$$

Expanding Equation 10.37,

$$\begin{bmatrix} IP_a \\ IP_b \\ IP_c \end{bmatrix} = \begin{bmatrix} If_a \\ If_b \\ If_c \end{bmatrix} + \begin{bmatrix} Y_{aa} & Y_{ab} & Y_{ac} \\ Y_{ba} & Y_{bb} & Y_{bc} \\ Y_{ca} & Y_{cb} & Y_{cc} \end{bmatrix} \cdot \begin{bmatrix} V_{ax} \\ V_{bx} \\ V_{cx} \end{bmatrix} + \begin{bmatrix} Y_{aa} & Y_{ab} & Y_{ac} \\ Y_{ba} & Y_{bb} & Y_{bc} \\ Y_{ca} & Y_{cb} & Y_{cc} \end{bmatrix} \cdot \begin{bmatrix} V_{xg} \\ V_{xg} \\ V_{xg} \end{bmatrix} \quad (10.38)$$

Perform the matrix operations in Equation 10.38.

$$IP_a = If_a + (Y_{aa} \cdot V_{ax} + Y_{ab} \cdot V_{bx} + Y_{ac} \cdot V_{cx}) + Ys_a \cdot V_{xg}$$

$$IP_b = If_b + (Y_{ba} \cdot V_{ax} + Y_{bb} \cdot V_{bx} + Y_{bc} \cdot V_{cx}) + Ys_b \cdot V_{xg}$$

$$IP_c = If_a + (Y_{ca} \cdot V_{ax} + Y_{cb} \cdot V_{bx} + Y_{cc} \cdot V_{cx}) + Ys_c \cdot V_{xg} \quad (10.39)$$

Distribution Feeder Analysis

where : $Ys_a = Y_{aa} + Y_{ab} + Y_{ac}$

$Ys_b = Y_{ba} + Y_{bb} + Y_{bc}$

$Ys_c = Y_{ca} + Y_{cb} + Y_{cc}$

or : $Ys_i = \sum_{k=1}^{3} Y_{i,k}$ (10.40)

Equation 10.39 becomes the general equation that is used to simulate all types of short circuits. Basically, there are three equations and seven unknowns (If_a, If_b, If_c, V_{ax}, V_{bx}, V_{cx}, and V_{xg}). The other three variables in the equation (IP_a, IP_b, and IP_c) are functions of the total impedance and the Thevenin voltages and are therefore known. To solve Equation 10.22, it will be necessary to specify four additional independent equations. This equation is a function of the type of fault being simulated. The additional required four equations for various types of faults are given next. These values are determined by placing short circuits in Figure 10.13 to simulate the fault. For example, a three-phase fault is simulated by placing a short circuit from Node a to x, Node b to x, and Node c to x. That gives three voltage equations. The fourth equation comes from applying KCL at Node x, which gives the sum of the fault currents to be zero.

10.2.2 Specific Short Circuits

Three-phase faults

$$V_{ax} = V_{bx} = V_{cx} = 0$$

$$I_a + I_b + I_c = 0 \quad (10.41)$$

Three-phase-to-ground faults

$$V_{ax} = V_{bx} = V_{cx} = V_{xg} = 0 \quad (10.42)$$

Line-to-line faults (assume i-j fault with phase k unfaulted)

$$V_{ix} = V_{jx} = 0$$

$$If_k = 0$$

$$If_i + If_j = 0 \quad (10.43)$$

Line-to-line-to-ground faults (assume i-j –g fault with phase k unfaulted)

$$V_{ix} = V_{jx} = 0$$

$$V_{xg} = 0$$

$$I_k = 0 \tag{10.44}$$

Line-to-ground Faults (assume phase k fault with phases i and j unfaulted)

$$V_{kx} = V_{xg} = 0$$

$$If_i = If_j = 0 \tag{10.45}$$

Notice that Equations 10.43, 10.44, and 10.45 will allow the simulation of line-to-line faults, line-to-line-to-ground, and line-to-ground faults for all phases. There is no limitation to b-c faults for line to line and a-g for line to ground, as is the case when the method of symmetrical components is employed.

A good way to solve the seven equations is to set them up in matrix form.

$$\begin{bmatrix} IP_a \\ IP_b \\ IP_c \\ 0 \\ 0 \\ 0 \\ 0 \end{bmatrix} = \begin{bmatrix} 1 & 0 & 0 & Y_{1,1} & Y_{1,2} & Y_{1,3} & Ys_1 \\ 0 & 1 & 0 & Y_{2,1} & Y_{2,2} & Y_{2,3} & Ys_2 \\ 0 & 0 & 1 & Y_{3,1} & Y_{3,2} & Y_{3,3} & Ys_3 \\ - & - & - & - & - & - & - \\ - & - & - & - & - & - & - \\ - & - & - & - & - & - & - \\ - & - & - & - & - & - & - \end{bmatrix} \cdot \begin{bmatrix} If_a \\ If_b \\ If_c \\ V_{ax} \\ V_{bx} \\ V_{cx} \\ V_{xg} \end{bmatrix} \tag{10.46}$$

Equation 10.29 in condensed form:

$$[IP_s] = [C] \cdot [X] \tag{10.47}$$

Equation 10.47 can be solved for the unknowns in matrix [X]:

$$[X] = [C]^{-1} \cdot [IP_s] \tag{10.48}$$

The blanks in the last four rows of the coefficient matrix in Equation 10.46 are filled in with the known variables depending upon what type of fault is to be simulated. For example, the elements in the [C]matrix simulating a three-phase fault would be the following:

$$C_{4,4} = C_{5,5} = C_{6,6} = 1$$

Distribution Feeder Analysis

$$C_{7,1} = C_{7,2} = C_{7,3} = 1$$

All the other elements in the last four rows will be set to zero.

Example 10.4: Use the system of Example 10.2 and compute the short-circuit currents for a bolted ($Z_f = 0$) line-to-line fault between phases *a* and *b* at Node 4.

The infinite bus balanced line-to-line voltages are 12.47 kV, which leads to balanced line-to-neutral voltages at 7.2 kV.

$$[ELL_s] = \begin{bmatrix} 12,470\underline{/30} \\ 12,470\underline{/-90} \\ 12,470\underline{/150} \end{bmatrix} V$$

$$[ELN_s] = [W] \cdot [ELL_s] \begin{bmatrix} 7,199.6\underline{/0} \\ 7,199.6\underline{/-120} \\ 7,199.6\underline{/120} \end{bmatrix} V$$

The line-to-neutral Thevenin circuit voltages at Node 4 are determined using Equation 8.165.

$$[Eth_4] = [A_t] \cdot [ELN_s] = \begin{bmatrix} 2,400\underline{/-30} \\ 2,400\underline{/-150} \\ 2,400\underline{/150} \end{bmatrix} V$$

The Thevenin equivalent impedance at the secondary terminals (Node 3) of the transformer consists of the primary line impedances referred across the transformer plus the transformer impedances. Using Equation 8.165.

$$[Zth_3] = [A_t] \cdot [ZeqS_{ABC}] \cdot [d_t] + [Zt_{abc}]$$

$$[Zth_3] = \begin{bmatrix} 0.0366 + j0.1921 & -.0039 - j0.0086 & -0.0039 - j0.0106 \\ -0.0039 - j0.0086 & 0.0366 + j0.1886 & -0.0039 - j0.0071 \\ -0.0039 - j0.0106 & -0.0039 - j0.0071 & 0.0366 + j0.1906 \end{bmatrix} \Omega$$

Note that the Thevenin impedance matrix is not symmetrical. This is a result, once again, of the unequal mutual coupling between the phases of the primary line segment.

The total Thevenin impedance at Node 4 is

$$[Zth_4] = [ZTOT] = [Zth_3] + [ZeqL_{abc}],$$

$$[ZTOT] = \begin{bmatrix} 0.2273 + j0.6955 & 0.0568 + j0.2216 & 0.0559 + j0.1645 \\ 0.0568 + j0.2216 & 0.2305 + j0.6771 & 0.0575 + j0.1860 \\ 0.0559 + j0.1645 & 0.0575 + j0.1860 & 0.2287 + j0.6876 \end{bmatrix} \Omega$$

The equivalent admittance matrix at Node 4 is

$$[Yeq_4] = [ZTOT]^{-1}$$

$$[Yeq_4] = \begin{bmatrix} 0.5031 - j1.4771 & -0.1763 + j0.3907 & -0.0688 + j0.2510 \\ -0.1763 + j0.3907 & 0.5501 - j1.5280 & -0.1148 + j0.3133 \\ -0.0688 + j0.2510 & -0.1148 + j0.3133 & 0.4843 - j1.4532 \end{bmatrix} S$$

Using Equation 10.36, the equivalent injected currents at the point of fault are

$$[IP] = [Yeq_4] \cdot [Eth_4] = \begin{bmatrix} 4466.8/\underline{-96.3} \\ 4878.9/\underline{138.0} \\ 4440.9/\underline{16.4} \end{bmatrix} A$$

The sums of each row of the equivalent admittance matrix are computed according to Equation 10.40.

$$Ys_i = \sum_{k=1}^{3} Yeq_{i,k} = \begin{bmatrix} 0.2580 - j0.8354 \\ 0.2590 - j0.8240 \\ 0.3008 - j0.8889 \end{bmatrix} S$$

For the a-b fault at Node 4, according to Equation 10.43:

$$V_{ax} = 0$$

$$V_{bx} = 0$$

$$If_c = 0$$

$$If_a + If_b = 0$$

The coefficient matrix [C] using Equation 10.46:

$$[C] = \begin{bmatrix} 1 & 0 & 0 & 0.5031 - j1.4771 & -0.1763 + j0.3907 & -0.0688 + 0.2510 & 0.2580 - j0.8354 \\ 0 & 1 & 0 & -0.1763 + j.3907 & 0.5501 - j1.5280 & -0.1148 + j.3133 & 0.2590 - j0.8240 \\ 0 & 0 & 1 & -0.0688 + j.2510 & -0.1148 + j.3133 & 0.4843 - j1.4532 & 0.3008 - j0.8890 \\ 0 & 0 & 0 & 1 & 0 & 0 & 0 \\ 0 & 0 & 0 & 0 & 1 & 0 & 0 \\ 0 & 0 & 1 & 0 & 0 & 0 & 0 \\ 1 & 1 & 0 & 0 & 0 & 0 & 0 \end{bmatrix}$$

The injected current matrix is

$$[IP_s] = \begin{bmatrix} 4466.8/\underline{-96.3} \\ 4878.9/\underline{138.0} \\ 4440.9/\underline{16.4} \\ 0 \\ 0 \\ 0 \\ 0 \end{bmatrix} V$$

Distribution Feeder Analysis

The unknowns are computed by

$$[X] = [C]^{-1} \cdot [IP_s] = \begin{bmatrix} 4193.7/-69.7 \\ 4193.7/110.3 \\ 0 \\ 0 \\ 0 \\ 3646.7/88.1 \\ 1220.2/-91.6 \end{bmatrix} A$$

The interpretation of the results are as follows:

$$If_a = X_1 = 4193.7/-69.7 \text{ A}$$

$$If_b = X_2 = 4193.7/110.3 \text{ A}$$

$$If_c = X_3 = 0$$

$$V_{ax} = X_4 = 0$$

$$V_{bx} = X_5 = 0$$

$$V_{cx} = X_6 = 3646.7/88.1 \text{ V}$$

$$V_{xg} = X_7 = 1220.2/-91.6 \text{ V}$$

Using the line-to-ground voltages at Node 4 and the short-circuit currents and working back to the source using the generalized matrices will check the validity of these results.

The line-to-ground voltages at Node 4 are

$$[VLG_4] = \begin{bmatrix} V_{ax} + V_{xg} \\ V_{bx} + V_{xg} \\ V_{cx} + V_{xg} \end{bmatrix} = \begin{bmatrix} 1220.2/-91.6 \\ 1220.2/-91.6 \\ 2426.5/88.0 \end{bmatrix} V$$

The short-circuit currents in matrix form:

$$[I_4] = [I_3] = \begin{bmatrix} 4193.7/-69.7 \\ 4193.7/110.3 \\ 0 \end{bmatrix} A$$

The line-to-ground voltages at Node 3 are

$$[VLG_3] = [a_2] \cdot [VLG_4] + [b_1] \cdot [I_4] = \begin{bmatrix} 1814.0/-47.3 \\ 1642.1/-139.2 \\ 2405.1/89.7 \end{bmatrix} V$$

The equivalent line-to-neutral voltages and line currents at the primary terminals (Node 2) of the transformer are

$$[VLN_2] = [a_t] \cdot [VLG_3] + [b_t] \cdot [I_3] = \begin{bmatrix} 6784.3/\underline{0.2} \\ 7138.8/\underline{-118.7} \\ 7080.6/\underline{118.3} \end{bmatrix} V,$$

$$[I_2] = [d_t] \cdot [I_3] = \begin{bmatrix} 1614.3/\underline{-69.7} \\ 807.1/\underline{110.3} \\ 807.1/\underline{110.3} \end{bmatrix} A$$

Finally, the equivalent line-to-neutral voltages at the infinite bus can be computed.

$$[VLN_1] = [a_1] \cdot [VLN_2] + [b_1] \cdot [I_2] = \begin{bmatrix} 7201.2/\underline{0} \\ 7198.2/\underline{-120} \\ 7199.3/\underline{120} \end{bmatrix} V$$

The source line-to-line voltages are

$$[VLL_1] = [Dv] \cdot [VLN_1] = \begin{bmatrix} 12470/\underline{30} \\ 12470/\underline{-90} \\ 12470/\underline{150} \end{bmatrix} V$$

These are the same line-to-line voltages that were used to start the short-circuit analysis.

10.2.3 BACK-FEED, GROUND-FAULT CURRENTS

The wye-delta transformer bank is the most common transformer connection used to serve three-phase loads or a combination of single-phase lighting loads and a three-phase load. With this connection, a decision has to be made as to whether or not to ground the neutral of the primary wye connection. The neutral can be directly connected to ground or grounded through a resistor or left floating (ungrounded wye – delta). When the neutral is grounded, the transformer bank becomes a grounding bank that provides a path for zero-sequence fault currents. In particular, the grounded connection will provide a path for a line-to-ground fault current (back-feed current) for a fault upstream from the transformer bank. A method for the analysis of the upstream fault currents will be presented.

10.2.3.1 One Downstream Transformer Bank

A simple system consisting of one downstream grounded wye – delta transformer bank is shown in Figure 10.26.

In Figure 10.26, the substation transformer is connected to a high-voltage equivalent source consisting of a three-phase voltage source and an equivalent impedance

Distribution Feeder Analysis 385

matrix. The equivalent source and substation transformer bank combination can be represented as shown in Figure 10.27.

In Chapter 8, it was shown that the combination of the equivalent source and substation transformer can be reduced to a Thevenin equivalent circuit, as shown in Figure 10.28.

The Thevenin equivalent voltages and impedance matrix are given by whte following:

$$[E_{th}] = [A_t] \cdot [ELG_{123}]$$

$$[Zth_{ABC}] = [A_t] \cdot [Zsys_{123}] \cdot [d_t] + [B_t],$$

where for the delta – grounded wye transformer

$$n_t = \frac{kVLL_{primary}}{kVLN_{secondary}}$$

$$[A_t] = \frac{1}{n_t} \cdot \begin{bmatrix} 1 & 0 & -1 \\ -1 & 1 & 0 \\ 0 & -1 & 1 \end{bmatrix}$$

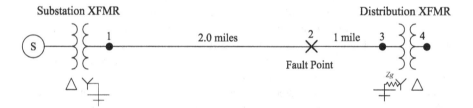

FIGURE 10.26 Simple system.

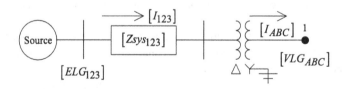

FIGURE 10.27 Equivalent source and substation transformer.

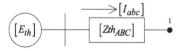

FIGURE 10.28 Thevenin equivalent circuit.

$$[B_t] = \begin{bmatrix} Zt_a & 0 & 0 \\ 0 & Zt_b & 0 \\ 0 & 0 & Zt_c \end{bmatrix}$$

$$[d_t] = \frac{1}{n_t} \cdot \begin{bmatrix} 1 & -1 & 0 \\ 0 & 1 & -1 \\ -1 & 0 & 1 \end{bmatrix}$$

$$[Zsys_{123}] = \begin{bmatrix} Zl_{1,1} & Zl_{1,2} & Zl_{1,3} \\ Zl_{2,1} & Zl_{2,2} & Zl_{2,3} \\ Zl_{3,1} & Zl_{3,2} & Zl_{3,3} \end{bmatrix} \tag{10.49}$$

Applying the Thevenin equivalent circuit, Figure 10.25 is modified to that of Figure 10.29:

In Figure 10.28, the equivalent impedance matrix is

$$[Zeq] = [Zth_{ABC}] + [Z12_{ABC}] \tag{10.50}$$

A question that comes up when a wye-delta transformer ban is to be installed is, Should the neutral be grounded? If the neutral is to be grounded it can either be a direct ground or grounded through a resistance. The other option is to just leave the neutral floating. Figure 10.30 shows the three-phase circuit for the modified simple system of Figure 10.29 when a line-to-ground fault has occurred at Node 2. Note the question mark on grounding of the neutral.

For future reference, the source voltages are $[Es_{ABC}] = [E_{th}]$. When the neutral is left floating in Figure 10.30, there is no path for the currents to flow from the transformer back to the fault. In this case, the only short-circuit current will be from the substation to the faulted node.

Figure 10.31 represents the transformer bank grounded neutral through a resistance.

Before getting into the derivation of the computation of the short-circuit currents of Figure 10.31, it is important to do a visual analysis of the circuit. The most important observation is that there is a path for the current It_A to flow from the phase A transformer through the fault and back to the neutral through the grounding resistance (Z_g). Note that this resistance can be set to zero for the direct grounding of the neutral. Since It_A flows, then there must be a current in the delta transformer secondary. This current is given by

$$I_{ab} = n_t \cdot It_A \tag{10.51}$$

Since the line currents out of the delta are zero, then all of the currents flowing in the delta must be equal.

Distribution Feeder Analysis

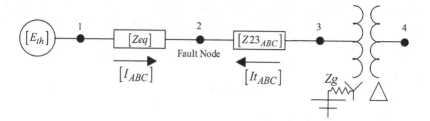

FIGURE 10.29 Modified simple system.

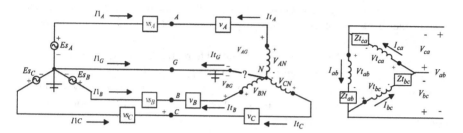

FIGURE 10.30 Three-phase circuit with floating transformer neutral.

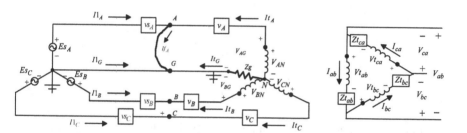

FIGURE 10.31 Three-phase circuit with grounded transformer neutral.

$$I_{ab} = I_{bc} = I_{ca} = n_t \cdot It_A \qquad (10.52)$$

Because the delta currents are equal, define the currents out of the transformer:

$$It_A = It_B = It_C = It \qquad (10.53)$$

The sum of the line-to-line secondary voltages must add to zero:

$$V_{ab} + V_{bc} + V_{ca} = 0 \qquad (10.54)$$

10.2.3.2 Complete Three-Phase Circuit Analysis

A method to calculate the short-circuit currents is to apply basic circuit and transformer analysis to determine all voltages and currents is developed in [6].

A three-phase circuit showing an A-G fault at Node 2 is shown in Figure 10.31. There are 28 unknowns, which will require 28 independent equations. Without going into detail, the 28 equations are

- 13 KVL
- six basic transformer primary/secondary
- five KCL
- four unique to type of fault

The 28 independent equations will be reduced to eight independent equations that will compute the voltages and currents in the fault and the back-feed current from the transformer bank. All other system voltages and currents can be computed knowing these eight variables.

In Figure 10.32, a Node X has been installed to represent the fault at Node 2. With the Node X, there are four voltages defined as Vf_{AX}, Vf_{BX}, Vf_{CX}, Vf_{XG} and three fault currents defined as If_A, If_B, If_C. The different types of faults are modeled by setting the appropriate voltages and currents to zero. For example, an A-G fault the following conditions are set:

$$Vf_{AX} = 0$$

$$Vf_{XG} = 0$$

$$If_B = 0$$

$$If_C = 0 \qquad (10.55)$$

In Figure 10.30, three loop equations can be written between the source and the faulted node.

$$Es_A = Vf_{AX} + Vf_{XG} + Zeq_{1,1} \cdot Il_A + Zeq_{1,2} \cdot Il_B + Zeq_{1,3} \cdot Il_C$$

$$Es_B = Vf_{BX} + Vf_{XG} + Zeq_{2,1} \cdot Il_A + Zeq_{2,2} \cdot Il_B + Zeq_{2,3} \cdot Il_C$$

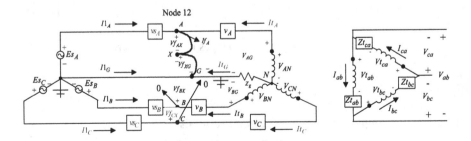

FIGURE 10.32 Three-phase circuit with AG fault.

Distribution Feeder Analysis

$$EsC = Vf_{CX} + Vf_{XG} + Zeq_{3,1} \cdot I1_A + Zeq_{3,2} \cdot I1_B + Zeq_{3,3} \cdot I1_C \tag{10.56}$$

KCL can be applied at the faulted node point:

$$I1_A = If_A - It$$

$$I1_B = If_B - It$$

$$I1_C = If_C - It \tag{10.57}$$

Substitute Equation 10.57 into Equation 10.56.

$$Es_A = Vf_{AX} + Vf_{XG} + Zeq_{1,1} \cdot If_A + Zeq_{1,2} \cdot If_B + Zeq_{1,3} \cdot If_C - Zx_1 \cdot It$$

$$Es_B = Vf_{BX} + Vf_{XG} + Zeq_{2,1} \cdot If_A + Zeq_{2,2} \cdot If_B + Zeq_{2,3} \cdot If_C - Zx_2 \cdot It$$

$$EsC = Vf_{CX} + Vf_{XG} + Zeq_{3,1} \cdot If_A + Zeq_{3,2} \cdot If_B + Zeq_{3,3} \cdot If_C - Zx_3 \cdot It$$

For $i = 1, 2, 3$

$$Zx_i = \sum_{k=1}^{3} Zeq_{i,k} \tag{10.58}$$

The line-to-ground voltages at the transformer primary terminals are as follows:

$$V_{AG} = Vf_{AX} + Vf_{XG} + \left(Z23_{11} + Z23_{12} + Z23_{13}\right) \cdot It$$

$$V_{BG} = Vf_{BX} + Vf_{XG} + \left(Z23_{21} + Z23_{22} + Z23_{23}\right) \cdot It$$

$$V_{CG} = Vf_{CX} + Vf_{XG} + \left(Z23_{31} + Z23_{32} + Z23_{33}\right) \cdot It$$

$$V_{AG} = Vf_{AX} + Vf_{XG} + Zy_1 \cdot It$$

$$V_{BG} = Vf_{BX} + Vf_{XG} + Zy_2 \cdot It$$

$$V_{CG} = Vf_{CX} + Vf_{XG} + Zy_2 \cdot It$$

$$\text{where}: Zy_i = \sum_{k=1}^{3} Z23_{i,k}$$

$$\text{also}: Zy_{sum} = Zy_1 + Zy_2 + Z_3 \tag{10.59}$$

The line-to-neutral ideal transformer voltages are the following:

$$V_{AN} = V_{AG} + 3 \cdot Z_g \cdot It$$

$$V_{BN} = V_{BG} + 3 \cdot Z_g \cdot It$$

$$V_{CN} = V_{CG} + 3 \cdot Z_g \cdot It \tag{10.60}$$

Substitute Equation 10.59 into Equation 10.60.

$$V_{AN} = V_{AG} + 3 \cdot Z_g \cdot It = Vf_{AX} + Vf_{XG} + (Zy_1 + 3 \cdot Z_g) \cdot It$$

$$V_{BN} = V_{BG} + 3 \cdot Z_g \cdot It = Vf_{BX} + Vf_{XG} + (Zy_2 + 3 \cdot Z_g) \cdot It$$

$$V_{CN} = V_{CG} + 3 \cdot Z_g \cdot It = Vf_{CX} + Vf_{XG} + (Zy_3 + 3 \cdot Z_g) \cdot It \tag{10.61}$$

From the currents flowing in the delta secondary are the following:

$$I_{ab} = I_{bc} = I_{ca} = n_t \cdot It_A$$

$$I_{ab} = I_{bc} = I_{ca} = n_t \cdot It$$

$$\text{since}: It_A = It \tag{10.62}$$

The line-to-line secondary voltages are as follows:

$$V_{ab} = Vt_{ab} + Zt_{ab} \cdot I_{ab}$$

$$V_{bc} = Vt_{bc} + Zt_{bc} \cdot I_{bc}$$

$$V_{ca} = Vt_{ca} + Zt_{ca} \cdot I_{ca} \tag{10.63}$$

In Equation 10.63,

$$Vt_{ab} = \frac{1}{n_t} \cdot V_{AN}$$

$$Vt_{bc} = \frac{1}{n_t} \cdot V_{BN}$$

$$Vt_{ca} = \frac{1}{n_t} \cdot V_{CN}$$

$$I_{ab} = I_{bc} = I_{ca} = n_t \cdot It \tag{10.64}$$

Distribution Feeder Analysis

Substitute Equation 10.64 into Equation 10.63.

$$V_{ab} = \frac{1}{n_t} \cdot V_{AN} + Zt_{ab} \cdot n_t \cdot It$$

$$V_{bc} = \frac{1}{n_t} \cdot V_{BN} + Zt_{bc} \cdot n_t \cdot It$$

$$V_{ca} = \frac{1}{n_t} \cdot V_{CN} + Zt_{ca} \cdot n_t \cdot It$$

$$\text{but}: V_{ab} + V_{bc} + V_{ca} = 0$$

$$\text{therefore}: 0 = \frac{1}{n_t} \cdot (V_{AN} + V_{BN} + V_{CN}) + n_t \cdot Zt_{sum} \cdot It$$

$$\text{where}: Zt_{sum} = Zt_{ab} + Zt_{bc} + Zt_{ca} \tag{10.65}$$

Substitute Equation 10.61 into Equation 10.65.

$$0 = \frac{1}{n_t} \cdot (V_{AN} + V_{BN} + V_{CN}) + n_t \cdot Zt_{sum} \cdot It$$

$$V_{AN} = Vf_{AX} + Vf_{XG} + (Zy_1 + 3 \cdot Z_g) \cdot It$$

$$V_{BN} = Vf_{BX} + Vf_{XG} + (Zy_2 + 3 \cdot Z_g) \cdot It$$

$$V_{CN} = Vf_{CX} + Vf_{XG} + (Zy_3 + 3 \cdot Z_g) \cdot It$$

$$0 = \frac{1}{n_t} \cdot \left((Vf_{AX} + Vf_{BX} + Vf_{CX} + 3 \cdot Vf_{XG}) + Zy_{sum} + 9 \cdot Z_g\right) \cdot It + n_t \cdot Zt_{sum} \cdot It$$

$$0 = \frac{1}{n_t} \cdot (Vf_{AX} + Vf_{BX} + Vf_{CX} + 3 \cdot Vf_{XG}) + \frac{1}{n_t} \cdot \left(Zy_{sum} + n_t^2 \cdot Zt_{sum} + 9 \cdot Z_g\right) \cdot It$$

$$0 = \frac{1}{n_t} \cdot (Vf_{AX} + Vf_{BX} + Vf_{CX} + 3 \cdot Vf_{XG}) + Z_{total} \cdot It$$

$$\text{where}: Z_{total} = \frac{1}{n_t} \cdot \left(Zy_{sum} + n_t^2 \cdot Zt_{sum} + 9 \cdot Z_g\right) \tag{10.66}$$

Combining Equations 10.58 and 10.66, give four equations with eight unknowns:

$$Es_A = Vf_{AX} + Vf_{XG} + Zeq_{1,1} \cdot If_A + Zeq_{1,2} \cdot If_B + Zeq_{1,3} \cdot If_C - Zx_1 \cdot It$$

$$Es_B = Vf_{BX} + Vf_{XG} + Zeq_{2,1} \cdot If_A + Zeq_{2,2} \cdot If_B + Zeq_{2,3} \cdot If_C - Zx_2 \cdot It$$

$$Es_C = Vf_{CX} + Vf_{XG} + Zeq_{3,1} \cdot If_A + Zeq_{3,2} \cdot If_B + Zeq_{3,3} \cdot If_C - Zx_3 \cdot It$$

$$0 = \frac{1}{n_t} \cdot \left(Vf_{AX} + Vf_{BX} + Vf_{CX} + 3 \cdot Vf_{XG}\right) + Z_{total} \cdot It$$

where: $Z_{total} = \dfrac{1}{n_t} \cdot \left(Zy_{sum} + Zt_{sum} + 9 \cdot Z_g\right)$

$$Zy_{sum} = Zy_1 + Zy_2 + Zy_3$$

$$Zt_{sum} = Zt_{ab} + Zt_{bc} + Zt_{ca} \tag{10.67}$$

Equation 10.67 gives four independent equations. The final four independent equations are unique for the type of fault to be modeled. The four equations that model each of the various types of faults are defined in Equations 10.41 to 10.45.

A general matrix equation for modeling a B-C-G fault is shown next. The first four rows in the coefficient matrix [C] come from Equation 10.69, and the last four rows are for the B-C-G fault as specified in Section 10.2.2.

$$\begin{bmatrix} Es_A \\ Es_B \\ Es_C \\ 0 \\ 0 \\ 0 \\ 0 \\ 0 \end{bmatrix} = \begin{bmatrix} Zeq_{1,1} & Zeq_{1,2} & Zeq_{1,3} & -Zx_1 & 1 & 0 & 0 & 1 \\ Zeq_{2,1} & Zeq_{2,2} & Zeq_{2,3} & -Zx_2 & 0 & 1 & 0 & 1 \\ Zeq_{3,1} & Zeq_{3,2} & Zeq_{3,3} & -Zx_3 & 0 & 0 & 1 & 1 \\ 0 & 0 & 0 & Z_{total} & \dfrac{1}{n_t} & \dfrac{1}{n_t} & \dfrac{1}{n_t} & \dfrac{3}{n_t} \\ 1 & 0 & 0 & 0 & 0 & 0 & 0 & 0 \\ 0 & 0 & 0 & 0 & 1 & 0 & 0 & 0 \\ 0 & 0 & 0 & 0 & 0 & 1 & 0 & 0 \\ 0 & 0 & 0 & 0 & 0 & 0 & 0 & 1 \end{bmatrix} \cdot \begin{bmatrix} If_A \\ If_B \\ If_C \\ It \\ Vf_{AX} \\ Vf_{BX} \\ Vf_{CX} \\ Vf_{XG} \end{bmatrix}$$

$$[Ex] = [C] \cdot [X] \tag{10.68}$$

10.2.3.3 Back-Feed Currents Summary

When a wye-delta transformer connection is used, the basic question is, Should the neutral be grounded? In this section, a method of analyzing a simple system was developed for analysis and then demonstrated with an example. The conclusion

Distribution Feeder Analysis

Example 10.6: The system of Figure 10.29 is to be analyzed for a B-C-G fault at Node 2. The given information for the system is as follows:

Equivalent System:

$$[ELL_{sys}] = \begin{bmatrix} 230,000\underline{/60} \\ 230,000\underline{/-60} \\ 230,000\underline{/180} \end{bmatrix} [ELN_{sys}] = [W] \cdot [ELL_{sys}] = \begin{bmatrix} 132,790.6\underline{/30} \\ 132,790.6\underline{/-90} \\ 132,790.6\underline{/150} \end{bmatrix} V$$

Line length = 200 miles

$$[Zsys_{123}] = \begin{bmatrix} 88.1693 + j226.9581 & 25.6647 + j90.9260 & 25.6647 + j71.3971 \\ 25.6647 + j90.9260 & 87.7213 + j235.6914 & 24.5213 + j78.2847 \\ 25.6647 + j71.3971 & 24.5213 + j78.2847 & 85.7213 + j235.6814 \end{bmatrix} \Omega$$

Substation transformer:

$$kVA = 2500 \, kVLL_{pri} = 230 \, kVLL_{sec} = 12.47 \, Z_{sub} = 0.005 + j0.06 \, \text{per unit}$$

$$Zbase_{sec} = \frac{kVLL_{zec}^2 \cdot 1000}{kVA} = 62.2004$$

$$Zt = Zt_{sub} \cdot Zbase_{sec} = 0.311 + j3.732$$

$$[Zsub_{ABC}] = \begin{bmatrix} 0.311 + j3.732 & 0 & 0 \\ 0 & 0.311 + j3.732 & 0 \\ 0 & 0 & 0.311 + j3.732 \end{bmatrix} \Omega$$

Compute substation transformer matrices:

$$kVLN = \frac{kVLL}{\sqrt{3}} = 7.1996$$

$$n_t = \frac{kVLL_{high}}{kVLN_{sec}} = 31.9464$$

$$[A_t] = \frac{1}{n_t} \cdot \begin{bmatrix} 1 & -1 & 0 \\ 0 & 1 & -1 \\ -1 & 0 & 1 \end{bmatrix}$$

$$[B_t] = [Zsub_{ABC}]$$

$$[d_t] = \frac{1}{n_t} \cdot \begin{bmatrix} 1 & -1 & 0 \\ 0 & 1 & -1 \\ -1 & 0 & 1 \end{bmatrix}$$

Compute substation transformer Thevenin equivalent circuit relative to secondary (Chapter 8 Section 8.12):

$$[E_{th}] = [A_t] \cdot [ELN_{sys}] = \begin{bmatrix} 7,199.5579/\underline{0} \\ 7,199.5579/\underline{-120} \\ 7,199.5579/\underline{120} \end{bmatrix} V$$

$$[Zth_{ABC}] = [A_t] \cdot [Zsys_{123}] \cdot [d_t] + [B_t]$$

$$[Zth_{ABC}] = \begin{bmatrix} 0.4311 + j4.40454 & -0.0601 - j0.1400 & -0.0600 - j0.1734 \\ -0.0601 - j0.1400 & 0.4311 + j4.0071 & -0.0600 - j0.1351 \\ -0.0600 - j0.1734 & -0.0600 - j0.1351 & 0.4309 + j4.0405 \end{bmatrix} \Omega$$

Given distribution line impedance matrices:

$$[Z12_{ABC}] = \begin{bmatrix} 0.9151 + j2.1561 & 0.3119 + j1.0033 & 0.3070 + j0.7699 \\ 0.3119 + j1.0033 & 0.9333 + j2.0963 & 0.3160 + j0.8473 \\ 0.3070 + j0.7699 & 0.3160 + j0.8473 & 0.9229 + j2.1301 \end{bmatrix} \Omega$$

$$[Z23_{ABC}] = \begin{bmatrix} 0.4576 + j1.0780 & 0.1559 + j0.5017 & 0.1535 + j0.3849 \\ 0.1559 + j0.5017 & 0.4666 + j1.0482 & 0.1580 + j0.4236 \\ 0.1535 + j0.3849 & 0.1580 + j0.4236 & 0.4615 + j1.0651 \end{bmatrix} \Omega$$

Grounded wye – delta transformer data:

$kVA_1 = 50 \; kVA_2 = 100 \; kVA_3 = 50$

$Ztpu_1 = 0.011 + j0.018 \; Ztpu_2 = 0.01 + j0.021 \; Ztpu_3 = 0.011 + j0.018$

$kVLN_{pri} = 7.2 \; kVLL_{sec}$

$$n_t = \frac{kVLN_{pri}}{kVLL_{sec}}$$

$$\text{For } i = 1, 2, 3 \; ZDbase_i = \frac{kVLL_{lo}^2 \cdot 1000}{kVA_i} = \begin{bmatrix} 4.608 \\ 2.304 \\ 4.608 \end{bmatrix} \Omega$$

$$Ztdel_i = Ztpu_i \cdot ZDbase_i = \begin{bmatrix} 0.0507 + j0.0829 \\ 0.0230 + j0.0484 \\ 0.0507 + j0.0829 \end{bmatrix} \Omega$$

$Zt_{ab} = Ztdel_1 \; Zt_{bc} = Ztdel_2 \; Zt_{ca} = Ztdel_3$

Distribution Feeder Analysis

Grounding resistance: $Z_g = 5$

Compute impedance terms for Equation 10.50:

$$[Zeq] = [Zth_{ABC}] + [Z12_{ABC}]$$

$$[Zeq] = \begin{bmatrix} 1.3462 + j6.2015 & 0.2518 + j0.8633 & 0.2470 + j0.5965 \\ 0.2518 + j0.8633 & 1.3643 + j6.1035 & 0.2560 + j0.7122 \\ 0.2470 + j0.5965 & 0.2560 + j0.7122 & 1.3539 + j6.1706 \end{bmatrix} \Omega$$

For $i = 1, 2, 3$

$$[Zx_i] = \sum_{k=1}^{3} Zeq_{i,k} = \begin{bmatrix} 1.8450 + j76613 \\ 1.8722 + j7.6790 \\ 1.8569 + j7.4793 \end{bmatrix} \Omega$$

$$[Zy_i] = \sum_{k=1}^{3} Z23_{ABC_{i,k}} = \begin{bmatrix} 0.7670 + j1.9646 \\ 0.7806 + j1.9735 \\ 0.7730 + j1.8736 \end{bmatrix} \Omega$$

$$Zy_{sum} = \sum_{k=1}^{3} Zy_k = 2.3205 + j5.8118 \, \Omega$$

$$Zt_{sum} = Zt_{ab} + Zt_{bc} + Zt_{ca} = 0.1244 + j0.2143 \, \Omega$$

$$Z_{total} = \frac{1}{n_t} \cdot \left(Zy_{sum} + n_t^2 \cdot Zt_{sum} + 9 \cdot Z_g\right) = 5.0209 + j3.6015 \, \Omega$$

Using the Thevenin source voltages and the numerical values from above, create the Equation 10.68 matrices for the B-C-G fault at Node 2. Remember that the last four rows of the [C] matrix represent the type of fault specified in Section 10.2.2.

$$[Ex] = \begin{bmatrix} 7199.5579 \underline{/0} \\ 7199.5579 \underline{/-120} \\ 7199.5579 \underline{/120} \\ 0 \\ 0 \\ 0 \\ 0 \\ 0 \end{bmatrix} V$$

$$[C] = \begin{bmatrix} 1.3462+j6.2015 & 0.2518+j0.8633 & 0.2470+j0.5965 & -1.8450-7.6613 \\ 0.2518+j0.8633 & 1.3643+j6.1035 & 0.2560+j0.7122 & -1.8722-j7.6790 \\ 0.2470+j0.5965 & 0.2560+j0.7122 & 1.3539+j6.1706 & -1.8569-j7.4793 \\ 0 & 0 & 0 & 5.0209+j3.6015 \\ 1 & 0 & 0 & 0 \\ 0 & 0 & 0 & 0 \\ 0 & 0 & 0 & 0 \\ 0 & 0 & 0 & 0 \end{bmatrix}$$

$$\begin{bmatrix} 1 & & & 1 \\ 0.0667 & 0.0667 & 0.0667 & 0.2 \\ 0 & 1 & 0 & 0 \\ 0 & 0 & 1 & 0 \\ 0 & 0 & 0 & 1 \end{bmatrix}$$

Solve for the unknown matrix $[X]$:

$$[X] = [C]^{-1} \cdot [Ex]$$

The computed short-circuit currents are as follows:

$$\begin{bmatrix} If_A \\ If_B \\ If_C \end{bmatrix} = \begin{bmatrix} X_1 \\ X_2 \\ X_3 \end{bmatrix} = \begin{bmatrix} 0 \\ 1,342.2/\underline{165.84} \\ 1,194.6/\underline{40.15} \end{bmatrix} A$$

$$It = X_4 = 81.5/\underline{143.53} \, A$$

For $i = 1, 2, 3$

$$\begin{bmatrix} I1_A \\ I1_B \\ I1_C \end{bmatrix} = \begin{bmatrix} If_A \\ If_B \\ If_C \end{bmatrix} - \begin{bmatrix} It \\ It \\ It \end{bmatrix} = \begin{bmatrix} 81.5/\underline{-36.46} \\ 1,267.2/\underline{167.24} \\ 1.216.09/\underline{36.41} \end{bmatrix} A$$

Note from the above that each of the distribution transformer primary windings has a short-circuit current of 81.5 amps flowing. The rated currents for the three transformers are

$$Irated_i = \frac{kVA_i}{kVLN_{hi}} = \begin{bmatrix} 6.94 \\ 13.89 \\ 6.94 \end{bmatrix} A$$

Distribution Feeder Analysis

The percentage over the rated current for the transformers is

$$I_{over_i} = \frac{It}{Ihi_i} \cdot 100 = \begin{bmatrix} 1{,}173.6 \\ 586.8 \\ 1{,}173.6 \end{bmatrix} \%$$

Needless to say, the fuses on the distribution transformers are going to blow because of the back-feed current.

This method of analysis for a grounded wye – delta bank with a ground resistance can be used to simulate an ungrounded wye – delta bank by setting the grounding resistance to a very large value. For example, use a grounding resistance of 99,999.

$$Z_g = 99999$$

$$[If_{ABC}] = \begin{bmatrix} 0 \\ 1257.4/\underline{167.5} \\ 1217.0/\underline{36.1} \end{bmatrix} A$$

$$It = 0$$

$$[I_{ABC}] = \begin{bmatrix} 0 \\ 1257.4/\underline{167.5} \\ 1217.0/\underline{36.1} \end{bmatrix} A$$

Note that for this case the back-feed current from the transformer bank is zero.

being that there is a very significant back-feed current when the neutral is grounded. The back-feed current is in the range of 1,000% of the rated transformer currents, so the transformer fuses will blow for the upstream fault. It was also demonstrated that if the grounding resistance is set to a very large value, the back-feed current will be zero, thus simulating an ungrounded wye – delta transformer bank connection. The conclusion is the neutral should never be grounded either directly or through a grounding resistance.

10.3 SUMMARY

This chapter has demonstrated the application of the element models that are used in the power-flow analysis and short-circuit analysis of a distribution feeder. The modified LIT was used for the power-flow analysis. For a simple radial feeder with no laterals, the examples demonstrated that only the forward and backward sweeps were changed by adding the sweep equations for the new elements. A feeder with laterals and sublaterals will require the ladder forward and backward sweep for each lateral and sublateral. In some cases, there is a need to model a feeder with a limited number of loops. A loop flow method of modifying the ladder technique was developed, and an example was developed to demonstrate the loop flow method.

The short-circuit analysis of a feeder using the symmetrical component analysis will not work because not all possible short circuits can be modeled. Rather, a method in the phase domain for the computation of any type of short circuit was developed and demonstrated.

The back-feed, short-circuit currents due to a grounded wye – delta transformer bank were developed and demonstrated by way of an example. The final idea is to demonstrate that a grounded wye – delta transformer bank should never be used.

The examples in this chapter have been very long and should be used as a learning tool. Many of the interesting operating characteristics of a feeder can only be demonstrated through numerical examples. The examples were designed to illustrate some of these characteristics.

Armed with a computer program and using the models and techniques of this text provides the engineer with a powerful tool for solving present-day problems and long-range planning studies.

PROBLEMS

The power-flow problems in this set require the application of the modified ladder technique. Students are encouraged to write their own MATLAB scripts to solve the problems throughout this chapter. It is left to the student to determine which scripts can be modified for the problems; this will show a depth of understanding of how the programs work.

The first six problems of this set will be based upon the system in Figure 10.33.

The substation transformer is connected to an infinite bus with balanced three-phase voltages of 69 kV. The substation transformer is rated:

5000 kVA, 69 kV delta – 4.16 grounded wye, $Z = 1.5 + j8.0$ %

The phase impedance matrix for a four-wire wye line is

$$[z_{4-wire}] = \begin{bmatrix} 0.4576 + j1.0780 & 0.1560 + j0.5017 & 0.1535 + j0.3849 \\ 0.1560 + j0.5017 & 0.4666 + j1.0482 & 0.1580 + j0.4236 \\ 0.1535 + j0.3849 & 0.1580 + j0.4236 & 0.4615 + j1.0651 \end{bmatrix} \Omega/\text{mile}.$$

The secondary voltages of the infinite bus are balanced and being held at 69 kV for all power-flow problems.

The four-wire wye feeder is 0.75 miles long. An unbalanced wye-connected load is located at Node 3 and has the following values:

Phase a: 750 kVA at 0.85 lagging power factor
Phase b: 500 kVA at 0.90 lagging power factor
Phase c: 850 kVA at 0.95 lagging power factor

The load at Node 4 is zero initially.

10.1 For the system as described above and assuming that the regulators are in the neutral position,

Distribution Feeder Analysis   399

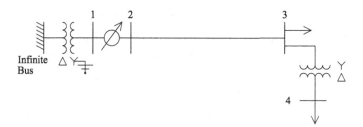

FIGURE 10.33   Wye homework system.

  (a) determine the forward and backward sweep matrices for the substation transformer and the line segment, and
  (b) use the modified ladder technique to determine the line-to-ground voltages at Node 3; use a tolerance of 0.0001 per unit, and give the voltages in actual values in volts and on a 120-volt base.
10.2 Three Type B step-voltage regulators are installed in a wye connection at the substation in order to hold the load voltages (Node 3) at a voltage level of 121 volts and a bandwidth of 2 volts.
  (a) Compute the actual equivalent line impedance between Nodes 2 and 3.
  (b) Use a potential transformer ratio of 2,400–120 volts and a current transformer ratio of 500:5 amps. Determine the R and X compensator settings calibrated in volts and ohms. The settings must be the same for all three regulators.
  (c) For the load conditions of Problem 1 and with the regulators in the neutral position, compute the voltages across the voltage relays in the compensator circuits.
  (d) Determine the appropriate tap settings for the three regulators to hold the Node 3 voltages at 121 volts in a bandwidth of 2 volts.
  (e) With the regulator's tap set, compute the actual load voltages on a 120-volt base.
10.3 A wye-connected, three-phase shunt capacitor bank of 300 kvar per phase is installed at Node 3. With the regulator compensator settings from Problem 2, determine
  (a) the new tap settings for the three regulators,
  (b) the voltages at the load on a 120-volt base, and
  (c) the voltages across the relays in the compensator circuits.
10.4 The load at Node 4 is served through an ungrounded wye – delta transformer bank. The load is connected in delta with the following values:

Phase a-b: 400 kVA at 0.9-factor power factor
Phase b-c: 150 kVA at 0.8 lagging power factor
Phase c-a: 150 kVA at 0.8 lagging power factor

The three single-phase transformers are rated as follows:

"Lighting transformer": 500 kVA, 2400 − 240 V, Z = .9 + j3.0 %
"Power transformers": 167 kVA, 2400 − 240 V, Z = 1.0 + j1.6 %

Use the original loads and the shunt capacitor bank at Node 3 and this new load at Node 4. Determine

(a) the voltages on 120-volt base at Node 3 assuming the regulators are in the neutral position,
(b) the voltages on 120-volt base at Node 4 assuming the regulators are in the neutral position,
(c) the new tap settings for the three regulators, and
(d) the Node 3 and 4 voltages on 120-volt base after the regulators have changed tap positions.

10.5 Under short-circuit conditions, the infinite bus voltage is the only voltage that is constant. The voltage regulators in the substation are in the neutral position. Determine the short-circuit currents and voltages at Nodes 1, 2, 3 for the following short circuits at Node 3:
(a) Three-phase to ground
(b) Phase b to ground
(c) Line-to-line fault on phases a-c

10.6 A line-to-line fault occurs at Node 4. Determine the currents in the fault and on the line segment between Nodes 2 and 3. Determine the voltages at Nodes 1, 2, 3 and 4.

10.7 A three-wire delta line of length 0.75 miles is serving an unbalanced delta load of:

Phase a-b: 600 kVA, 0.9 lagging power factor
Phase b-c: 800 kVA, 0.8 lagging power factor
Phase c-a: 500 kVA, 0.95 lagging power factor

The phase impedance matrix for the line is:

$$[z_{3-wire}] = \begin{bmatrix} 0.4013 + j1.4133 & 0.0953 + j0.8515 & 0.0953 + j0.7802 \\ 0.0953 + j0.8515 & 0.4013 + j1.4133 & 0.0953 + j0.7266 \\ 0.0953 + j0.7802 & 0.0953 + j0.7266 & 0.4013 + j1.4133 \end{bmatrix} \Omega/\text{mile}$$

The line is connected to a constant balanced voltage source of 4.8 kV line to line. Determine the load voltages on a 120-volt base.

10.8 Add two type B step-voltage regulators in an open delta connection using phases A-B and B-C to the system in Problem 7. The regulator should be set to hold 121 ± 1 volt. Determine the R and X settings and the final tap settings. For the open delta connection, the R and X settings will be different on the two regulators.

10.9 The three-wire line of Problem 7 is connected to a substation transformer connected delta-delta. The substation transformer is connected to a 69 kV infinite bus and is rated:

10,000 kVA, 69 kV delta – 4.8 kV delta, Z = 1.6 + j7.8 %

Determine the short-circuit currents and substation transformer secondary voltages for the following short circuits at the end of the line:
(a) Three-phase
(b) Line-to-line between phases a-b

10.10 Two three-phase systems are shown in Figure 10.34

In Figure 10.32, the solid line represents three-phase lines while the dashed lines represent phase C single-phase lines. The phase conductors are 336,400 26/7 ACSR and the neutral conductor is 1/0 ACSR. The impedances matrices are as follows:

Three-phase lines:

$$[z3] = \begin{bmatrix} 0.5396 + j1.0978 & 0.1916 + j0.5475 & 0.1942 + j0.4705 \\ 0.1916 + j0.5475 & 0.5279 + j1.1233 & 0.1884 + j0.4296 \\ 0.1942 + j0.4705 & 0.1884 + j0.4296 & 0.5330 + j1.1122 \end{bmatrix}$$

Single-phase C line:

$$[z1] = \begin{bmatrix} 0 & 0 & 0 \\ 0 & 0 & 0 \\ 0 & 0 & 0.5328 + j1.1126 \end{bmatrix} \text{Ohms/mile}$$

The two sources are 12.47 kV substations operating at rated line-to-line voltages.

The line lengths in feet are the following:

$L_{12} = 2000$, $L_{23} = 2500$, $L_{27} = 1000$, $L_{45} = 6000$, $L_{56} = 750$, $L_{58} = 5000$

The loads are as follows:

S_{3a} = 500 kW at 90 % PF S_{4a} = 500 kW at 80 % PF
S_{3b} = 600 kW at 85 % PF S_{4b} = 400 kW at 85 % PF
S_{3c} = 400 kW at 95 % PF S_{4c} = 600 kW at 90 % PF
S_{7c} = 500 kW at 80 % PF S_{4c} = 450 kW at 90 % PF

The base kVA = 5,000 and the base line-to-line kV = 12.470
The two switches are open.
(a) Determine the node voltages.

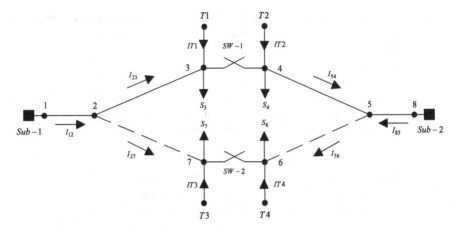

FIGURE 10.34 Looped flow system.

(b) Determine the line currents.
The two switches are closed.
(c) Determine the values of the injected currents at Nodes 3, 4, 6 and 7.
(d) With the injected currents operating, determine the node voltages and the line currents.

10.11 The system in Figure 10.35 is to be studied for steady-state and short-circuit analyses.

In Figure 10.33, the system is served from an equivalent source with balanced line-to-line voltages of 12.47 kV. The equivalent source impedance matrix is

$$[Z_{eq}] = \begin{bmatrix} 0.4311 + j4.0454 & -0.0601 - j0.1400 & -0.0600 - j0.1734 \\ -0.0601 - j0.1400 & 0.4311 + j4.0071 & -0.0600 - j0.1351 \\ -0.0600 - j0.1734 & -0.0600 - j0.1351 & 0.4309 + j4.0405 \end{bmatrix}.$$

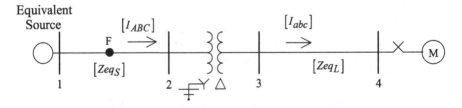

FIGURE 10.35 Small system.

Distribution Feeder Analysis

The three-phase, four-wire line impedance matrix in ohms/mile is

$$[z_{4-wire}] = \begin{bmatrix} 0.4576 + j1.0780 & 0.1560 + j0.5017 & 0.1535 + j0.3849 \\ 0.1560 + j0.5017 & 0.4666 + j1.0482 & 0.1580 + j0.4236 \\ 0.1535 + j0.3849 & 0.1580 + j0.4236 & 0.4615 + j1.0651 \end{bmatrix}.$$

The primary line is 5 miles long.
The three-phase, three-wire open wire secondary line impedance matrix in ohms/mile is

$$\frac{1}{3}[z_{sec}] = \begin{bmatrix} 1.0653 + j1.5088 & 0.0953 + j1.0468 & 0.0953 + j0.9627 \\ 0.0953 + j1.0468 & 1.0653 + j1.5088 & 0.0953 + j1.0468 \\ 0.0953 + j0.9627 & 0.0953 + j1.0468 & 1.0653 + j1.5088 \end{bmatrix}.$$

The secondary is 500 feet long.
The transformer bank is connected grounded wye – delta composed of three single-phase transformers, each rated as follows:

$$kVA = 10, \quad kVLN_{hi} = 7.2, \quad kVLL_{lo} = 0.24, \quad Zpu = 0.016 + j0.014$$

The wye-connected primary windings are connected directly to ground.
A three-phase induction motor has the following data:

150 hp 480 volt
$Z_s = 0.059 + j0.127$ pu
$Z_r = 0.046 + j0.112$ pu
$Z_m = 4.447$ pu

The motor is operating with a slip of 0.035 with balanced three-phase source voltages of 12,470 volts line to line. Determine the following:
(a) The primary and secondary line currents
(b) The line-to-line voltages at the motor
(c) The three-phase complex power at the source
    The switch to the induction motor is open when a phase line-to-ground fault occurs on phase c at the fault node (F), which is two miles from Node 1.
(d) The fault currents from the source and from the transformer bank

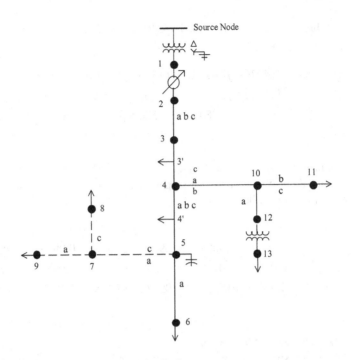

FIGURE 10.36 Unbalanced three-phase feeder.

WINDMIL ASSIGNMENT

Figure 10.36 shows the one-line diagram of an unbalanced three-phase feeder. The non-line data for the feeder is:

a. Equivalent Source
   i. Balanced 115 kV Line-to-line
   ii. $Zpos = 1.48 + j11.6$ Ohms
   iii. $Zzero = 4.73 + j21.1$ Ohms
   iv. Bus Voltage = 120
b. Substation Transformer
   i. 115 kV D – 12.47 kV grd. Y
   ii. $kVA = 10,000$
   iii. $Z = 8.026\%$, $X/R = 8$
c. Regulator
   i. CT rating = 600
   ii. % boost = 10
   iii. Step size = 0.625
   iv. Number of steps = 16
   v. Nodes: 1–2
   vi. Voltage level = ?
   vii. $R + jX = ?$
d. Single-Phase Transformer
   i. Connection: Y-D one

ii. $kVA = 100$
iii. Voltages: 7200 Y – 240 D volts
iv. $Z = 2.326\%$, $X/R = 2.1$
v. Nodes: 12–13

The line data is:
e. Three-Phase OH Lines
   i. Phase: 336,400 26/7 ACSR
   ii. Neutral: 4/0 6/1 ACSR
   iii. Phasing: a-b-c
   iv. Spacings:
      1. Position 1: $0 + j29$
      2. Position 2: $2.5 + j29$
      3. Position 3: $7 + j29$
      4. Neutral: $4 + j25$
   v. OH 1: Nodes 2–3, 2,500 feet
   vi. OH 2: Nodes 3–4, 3,000 feet
   vii. OH 3: Nodes 4–5, 2,500 feet
   viii. OH 4: Nodes 5–6, 1,000 feet
f. Two-Phase OH Line
   i. Phase: 336,400 26/7 ACSR
   ii. Neutral: 4/0 6/1 ACSR
   iii. Phasing: a-c
   iv. Spacings:
      1. Position 1: $0 + j29$
      2. Position 2: $7 + j29$
      3. Neutral: $4 + j25$
   v. OH 5: Nodes 5–7, 1,500 feet
g. Three-Phase Concentric Neutral UG
   i. CN cable: 1/0 AA, 1/3 neutral
   ii. No extra neutral
   iii. Phasing: c-b-a
   iv. Spacings:
      1. Position 1: $0 - j40$ inches
      2. Position 2: $6 - j40$ inches
      3. Position 3: $12 - j40$ inches
   v. UG 1: Nodes 4–10, 1,500 feet
h. Two-Phase Concentric Neutral UG
   i. CN cable: 1/0 AA, full neutral
   ii. No extra neutral
   iii. Phasing: b-c
   iv. Spacings:
      1. Position 1: $0 - j40$ inches
      2. Position 2: $6 - j40$ inches
   v. UG 2: Nodes 10–11, 1,000 feet
i. Single-Phase Concentric Neutral UG
   i. CN cable: 1/0 AA, full neutral

ii. No extra neutral
   iii. Phase: a
   iv. Spacings:
      1. Position 1: $0 - j40$ inches
   v. UG 3: Nodes 10–12, 500 feet
j. Single-Phase Tape Shield Cable
   i. 1/0 AA tape shield UG
   ii. Neutral: 1/0 7 strand AA
   iii. Phase c:
   iv. Spacings:
      1. Position 1: $0 - j40$ inches
      2. Neutral: $6 - j40$ inches
   v. UG 4: Nodes 7–8, 500 feet
k. Single-Phase Tape Shield Cable
   i. 1/0 AA tape shield UG
   ii. Neutral: 1/0 7 strand AA
   iii. Phase a:
   iv. Spacings:
      1. Position 1: $0 - j40$ inches
      2. Position 2: $6 - j40$ inches
      3. Neutral: $12 - j40$ inches
   v. UG 5: Nodes 7–9, 750 feet

The load data is:
Distributed Loads:

Node A	Node B	Pa	PFa %	Pb	PFb %	Pc	PFc %	Model
2	3	100	90	150	90	200	90	Y-PQ
3	4	200	90	100	90	150	90	Y-Z

Wye-Connected Spot Loads:

Node	Pa	PFa %	Pb	PFb %	Pc	PFc %	Model
3	500	85	300	95	400	90	Y-Z
4	500	80	500	80	500	90	Y-I
6	1,000	80	800	90	950	95	Y-PQ
8					200	95	Y-Z
9	350	90					
13	100	95					Y-I

Delta Connected Loads:

Node	Pab	PFab %	Pbc	PFbc %	Pca	PFca %	Model
11			350	90			D-PQ

Distribution Feeder Analysis

a. Create this system in WindMil.
b. Run voltage drop with the regulator set to "none". Do this in the Voltage Drop Analysis Manager.
c. Compute the R and X and voltage level for the voltage regulator.
   i. Hand calculation
   ii. WindMil "regulation set"
   iii. Compare settings
d. Run voltage drop with the regulators set to "step"
   i. What are the final tap positions?
   ii. Are these OK?
e. Add shunt capacitors so that the source power factor is no lower than 95% lag. Specify capacitors in multiples of 100 kvar.
f. Run voltage with the final capacitors.
   i. What is the power factor by phase at the source?
   ii. What are the final tap positions for the regulators?

REFERENCES

1. Trevino, C., Cases of Difficult Convergence in Load-Flow Problems, *IEEE Paper n.71-62-PWR*, Presented at the IEEE Summer Power Meeting, Los Angeles, 1970.
2. Kersting, W. H. and Mendive, D. L., An Application of Ladder Network Theory to the Solution of Three-Phase Radial Load-Flow Problems, *IEEE Conference Paper*, Presented at the IEEE Winter Power Meeting, New York, January 1976.
3. *Radial Test Feeders*, IEEE Distribution System Analysis Subcommittee, http://ewh.ieee.org/soc/pes/dsacom/testfeeders/index.html
4. Kersting, W.H. and Phillips, W. H., Distribution System Short-Circuit Analysis, *25th Intersociety Energy Conversion Engineering Conference*, Reno, NV, August 12–17 1990.
5. Kersting, W.H., The simulation of loop flow in radial distribution analysis programs, *IEEE Conference Paper*, Presented at the IEEE Rural Electric Power Conference, June 2014.
6. W. H. Kersting and W. Carr, "Grounded Wye-Delta Transformer Bank Backfeed Short-Circuit Currents," *IEEE Rural Electric Power Conference (REPC)*, Westminster, CO, 2016, pp. 63–68, doi: 10.1109/REPC.2016.18.

11 Center-Tapped Transformers and Secondaries

The standard method of providing three-wire service to a customer is from a center-tapped, single-phase transformer. This type of service provides the customer with two 120-V circuits and one 240-volt circuit. Two types of transformers are available for providing this service. The first is where the secondary consists of one winding that is center tapped, as shown in Figure 11.1.

For example, the secondary voltage rating of the distribution transformer in Figure 11.1 would be specified as 240/120 volts. This specifies that the full winding voltage rating is 240 volts with the center tap providing two 120-volt circuits.

A second type of transformer used to provide three-wire service is shown in Figure 11.2.

The transformer in Figure 11.2 is a three-winding transformer with the two secondary windings connected in series. The secondary on this transformer is specified as 120/240 volts. The secondary windings can be connected in series to provide the three-wire 240- and 120-volt service, or they may be connected in parallel to provide

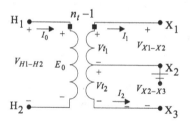

FIGURE 11.1 Center-tapped secondary winding.

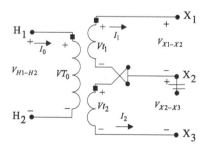

FIGURE 11.2 Three-winding transformer with secondary windings in series.

DOI: 10.1201/9781003261094-11

only 120 volts. When connected in parallel the transformer will typically be used in a three-phase connection. The secondary will be connected in wye and will provide three 120-V circuits and a three-phase, line-to-line voltage of 208 volts.

For both connections, the ideal transformer equations are as follows:

$$n_t = \frac{kVLN_{rated}}{kVLL_{rated}}$$

$$\text{example}: n_t = \frac{7200}{240} = 30 \quad (11.1)$$

$$V_{H1-H2} = 2 \cdot n_t \cdot Vt_{X1-X2}$$

$$I_0 = \frac{1}{2 \cdot n_t} \cdot (I_1 - I_2) \quad (11.2)$$

Note that in both drawings, the current I_2 is leaving the transformer secondary.

11.1 CENTER-TAPPED, SINGLE-PHASE TRANSFORMER MODEL

The model of the center-tapped transformer of Figures 11.1 and 11.2 is shown in Figure 11.3.

The model in Figure 11.3 can represent either the center-tapped secondary winding (Figure 11.1) or the two secondary windings connected in series (Figure 11.2). The impedances Z_0, Z_1, Z_2 represent the individual winding impedances.

The first step in developing the model is to determine the impedances Z_0, Z_1, Z_2. These impedances can be determined with open circuit and short circuit tests on the transformer. However, that usually is not practical. What is typically known on a transformer will be the per-unit impedance based upon the transformer rating. Unfortunately, that usually does not include the angle. When that is the case, an approximation must be made for the angle or a typical impedance value can be used. Typical values of transformer impedances per unit can be found in the text *Electric*

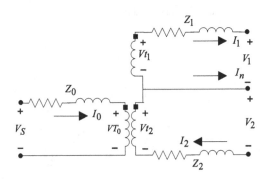

FIGURE 11.3 Center-tapped transformer model.

Center-Tapped Transformers and Secondaries

Power Distribution System Engineering, by Turan Gonen [1]. Empirical equations commonly used to convert the per-unit transformer impedance to the per-unit winding impedances of an interlaced design are given in Equation 11.3.

$$Z_0 = 0.5 \cdot R_A + j0.8 \cdot X_A$$

$$Z_1 = R_A + j0.4 \cdot X_A$$

$$Z_2 = R_A + j0.4 \cdot X_A \text{ per-unit} \tag{11.3}$$

The equations for the non-interlaced design are as follows:

$$Z_0 = 0.25 \cdot R_A - j0.6 \cdot X_A$$

$$Z_1 = 1.5 \cdot R_A + j3.3 \cdot X_A$$

$$Z_2 = 1.5 \cdot R_A + j3.1 \cdot X_A \text{ per-unit} \tag{11.4}$$

The interlaced design is the most common and should be used when in doubt. Note for this design the following:

$$Z_A = Z_0 + \frac{1}{4} \cdot (Z_1 + Z_2)$$

$$Z_A = 0.5 \cdot R_A + j0.8 \cdot X_A + \frac{1}{4} \cdot (R_A + j0.4 \cdot X_A + R_A + j0.4 \cdot X_A)$$

$$Z_A = 0.5 \cdot R_A + j0.8 \cdot X_A + \frac{1}{4} \cdot (2 \cdot R_A + j0.8 \cdot X_A)$$

$$Z_A = 0.5 \cdot R_A + j0.8 \cdot X_A + 0.5 \cdot R_A + j0.2 \cdot X_A$$

$$Z_A = R_A + jX_A \tag{11.5}$$

The per-unit impedances of the three windings must be converted to ohms based upon the transformer rating. The base impedance for (Z_0) of the primary is based upon the rated primary voltage of the transformer ($kVLN_{hi}$ or $kVLL_{hi}$). The center-tapped transformer secondary is modeled as two 120 voltage windings ($kVLN_{lo}$). The two windings in series result in the rated line-to-line voltage ($kVLL_{lo}$). The base impedances for the primary winding connected line to neutral:

$$Zbase_{hi} = \frac{kVLN_{hi}^2 \cdot 1000}{kVA_{rated}} \tag{11.6}$$

For the primary winding, they connected line to line:

$$Zbase_{hi} = \frac{kVLL_{hi}^2 \cdot 1000}{kVA_{rated}} \quad (11.7)$$

The base impedance for the two secondary windings is based upon the rated line-to-line voltage of the secondary:

$$Zbase_1 = Zbase_2 = \frac{\left(\frac{kVLL_{lo}}{2}\right)^2 \cdot 1000}{kVA_{rated}}$$

$$Zbase_1 = Zbase_2 = \frac{1}{4} \cdot Zbase_{lo}$$

$$\text{where}: Zbase_{lo} = \frac{kVLL_{lo}^2 \cdot 1000}{kVA_{rated}} \quad (11.8)$$

Example 11.1: A single-phase, center-tapped transformer is connected to the line-to-neutral system voltage.

Transformer ratings:

$$kVA = 50 \quad kVLN_{hi} = 7.2 \quad kVLL_{lo} = 0.24$$

$$Rpu_A = 0.011 \quad Xpu_A = 0.018$$

Compute primary and secondary per-unit impedances.

$$pu_0 = 0.5 \cdot Rpu_A + j0.8 \cdot Xpu_A = 0.0055 + j0.0144$$

$$Zpu_1 = Rpu_A + j0.4 \cdot Xpu_A = 0.011 + j0.0072$$

$$Zpu_2 = Zpu_1$$

$$Zpu_A = Zpu_0 + 0.25 \cdot (Zpu_1 + Zpu_2) = 0.011 + j0.018$$

Compute the transformer primary and secondary transformer impedances in ohms.

$$kVA_{base} = 50 \quad kVLN_{hi} = 7.2 \quad kVLL_{lo} = 0.24$$

$$Zbase_{hi} = \frac{kVLN_{hi}^2 \cdot 1000}{kVA_{base}} = \frac{7.2^2 \cdot 1000}{50} = 1036.8\,\Omega$$

Center-Tapped Transformers and Secondaries

$$Z_0 = Zpu_0 \cdot Zbase_{hi} = 5.7024 + j14.9299 \, \Omega$$

$$Zbase_{lo} = \frac{kVLL_{lo}^2 \cdot 1000}{kVA_{base}} = 1.152 \, \Omega$$

$$Zbase_1 = \frac{\left(\frac{kVLL_{lo}}{2}\right)^2 \cdot 1000}{kVA_{base}} = 0.288 \, \Omega$$

$$\text{Note}: Zbase_1 = \frac{Zbase_{lo}}{4} = 0.288 \, \Omega$$

$$Z_1 = Z_2 = Zpu_1 \cdot \frac{Zbase_{lo}}{4} = 0.0032 + j0.0021 \, \Omega$$

11.1.1 Matrix Equations

Referring to Figure 11.3, the ideal secondary voltages of the transformer are the following:

$$\begin{bmatrix} Vt_1 \\ Vt_2 \end{bmatrix} = \begin{bmatrix} V_1 \\ V_2 \end{bmatrix} + \begin{bmatrix} Z_1 & 0 \\ 0 & -Z_2 \end{bmatrix} \cdot \begin{bmatrix} I_1 \\ I_2 \end{bmatrix}$$

$$[Vt_{12}] = [V_{12}] + [Z_{12}] \cdot [I_{12}] \tag{11.9}$$

The ideal primary voltage as a function of the secondary ideal voltages is as follows:

$$\begin{bmatrix} E_0 \\ E_0 \end{bmatrix} = 2 \cdot n_t \cdot \begin{bmatrix} 1 & 0 \\ 0 & 1 \end{bmatrix} \cdot \begin{bmatrix} Vt_1 \\ Vt_2 \end{bmatrix}$$

$$[E_{00}] = [av] \cdot [Vt_{12}]$$

$$\text{where}: [av] = 2 \cdot n_t \cdot \begin{bmatrix} 1 & 0 \\ 0 & 1 \end{bmatrix} \tag{11.10}$$

The primary transformer current as a function of the secondary winding currents is given in Equation 11.11. The negative sign is due to the selected direction of the current I_2.

$$\begin{bmatrix} I_0 \\ I_0 \end{bmatrix} = \frac{1}{2 \cdot n_t} \cdot \begin{bmatrix} 1 & -1 \\ 1 & -1 \end{bmatrix} \cdot \begin{bmatrix} I_1 \\ I_2 \end{bmatrix}$$

$$[I_{00}] = [ai] \cdot [I_{12}]$$

$$\text{where}: [ai] = \frac{1}{2 \cdot n_t} \cdot \begin{bmatrix} 1 & -1 \\ 1 & -1 \end{bmatrix} \quad (11.11)$$

Substitute Equations 11.9 into Equations 11.10.

$$[E_{00}] = [av] \cdot ([V_{12}] + [Z_{12}] \cdot [I_{12}])$$

$$[E_{00}] = [av] \cdot [V_{12}] + [av] \cdot [Z_{12}] \cdot [I_{12}] \quad (11.12)$$

The source voltage as a function of the ideal primary voltage is as follows:

$$\begin{bmatrix} V_s \\ V_s \end{bmatrix} = \begin{bmatrix} E_0 \\ E_0 \end{bmatrix} + \begin{bmatrix} Z_0 & 0 \\ 0 & Z_0 \end{bmatrix} \cdot \begin{bmatrix} I_0 \\ I_0 \end{bmatrix}$$

$$[V_{ss}] = [E_{00}] + [Z_{00}] \cdot [I_{00}] \quad (11.13)$$

Substitute Equation 11.12 into Equation 11.13.

$$[V_{ss}] = [av] \cdot [V_{12}] + [av] \cdot [Z_{12}] \cdot [I_{12}] + [Z_{00}] \cdot [I_{00}] \quad (11.14)$$

Substitute Equation 11.11 into Equation 11.14.

$$[V_{ss}] = [av] \cdot [V_{12}] + [av] \cdot [Z_{12}] \cdot [I_{12}] + [Z_{00}] \cdot [ai] \cdot [I_{12}]$$

$$[V_{ss}] = [av] \cdot [V_{12}] + ([av] \cdot [Z_{12}] + [Z_{00}] \cdot [ai]) \cdot [I_{12}]$$

$$[V_{ss}] = [a_t] \cdot [V_{12}] + [b_t] \cdot [I_{12}]$$

$$\text{where:} [a_t] = [av] = 2 \cdot n_t \cdot \begin{bmatrix} 1 & 0 \\ 0 & 1 \end{bmatrix}$$

$$[b_t] = [av] \cdot [Z_{12}] + [Z_{00}] \cdot [ai]$$

$$[b_t] = 2 \cdot n_t \cdot \begin{bmatrix} Z_1 + \dfrac{1}{(2 \cdot n_t)^2} \cdot Z_0 & -\dfrac{1}{(2 \cdot n_t)^2} \cdot Z_0 \\ \dfrac{1}{(2 \cdot n_t)^2} \cdot Z_0 & -\left(Z_2 + \dfrac{1}{(2 \cdot n_t)^2} \cdot Z_0\right) \end{bmatrix} \quad (11.15)$$

Center-Tapped Transformers and Secondaries

Equation 11.15 is the backward sweep voltage equation for the single-phase center-tapped transformer when the secondary voltages and currents are known. The primary current as a function of the secondary voltages and currents is given by the backward sweep current equation:

$$[I_{00}] = [c_t] \cdot [V_{12}] + [d_t] \cdot [I_{12}],$$

$$\text{where } [c_t] = \begin{bmatrix} 0 & 0 \\ 0 & 0 \end{bmatrix},$$

$$[d_t] = \frac{1}{2 \cdot n_t} \cdot \begin{bmatrix} 1 & -1 \\ 1 & -1 \end{bmatrix}. \tag{11.16}$$

Equation 11.15 is used to compute the primary source voltage when the secondary terminal voltages and the secondary currents are known. It is also important to be able to compute the secondary terminal voltages when the primary source voltage and secondary currents are known (forward sweep). The forward sweep equation is derived from Equation 11.15.

$$[V_{12}] = [a_t]^{-1} \cdot ([V_{ss}] - [b_t] \cdot [I_{12}])$$

$$[V_{12}] = [a_t]^{-1} \cdot [V_{ss}] - [a_t]^{-1} \cdot [b_t] \cdot [I_{12}]$$

$$[V_{12}] = [A_t] \cdot [V_{ss}] - [B_t] \cdot [I_{12}] \tag{11.17}$$

Where

$$[A_t] = [a_t]^{-1} = \frac{1}{2 \cdot n_t} \cdot \begin{bmatrix} 1 & 0 \\ 0 & 1 \end{bmatrix}$$

$$[B_t] = [a_t]^{-1} \cdot [b_t] = \frac{1}{2 \cdot n_t} \cdot b_t = \begin{bmatrix} \left(Z_1 + \frac{1}{(2 \cdot n_t)^2} \cdot Z_0\right) & -\frac{1}{(2 \cdot n_t)^2} \cdot Z_0 \\ \frac{1}{(2 \cdot n_t)^2} \cdot Z_0 & -\left(Z_2 + \frac{1}{(2 \cdot n_t)^2} \cdot Z_0\right) \end{bmatrix}$$

$$\tag{11.18}$$

Example 11.2: The 50 kVA center-tapped transformer of Example 11.1 serves constant impedance loads, as shown in Figure 11.4.

Transformer Rating: 50 kVA, 7200 – 240/120 V, R_A = 0.011 pu, X_A = 0.018 pu.

Loads:

S_1 = 10 kVA at 95 % lagging power factor
S_2 = 15 kVA at 90% lagging power factor
S_3 = 25 kVA at 85% lagging power factor
Source Voltage: 7200/0 V
Determine: $[A_t]$, $[B_t]$, and $[d_t]$ matrices

(1) Load voltages, secondary currents, and load currents
(2) Primary current

The winding impedances from Example 11.1 are as follows:

$$Z_0 = 5.7024 + j14.9299$$

$$Z_1 = Z_2 = 0.0032 + j0.0021 \text{ ohms}$$

Compute the turns ratio: $n_t = \dfrac{7200}{240} = 30.$
Compute the matrices:

$$[A_t] = \dfrac{1}{2 \cdot n_t} \begin{bmatrix} 1 & 0 \\ 0 & 1 \end{bmatrix} = \begin{bmatrix} 0.0167 & 0 \\ 0 & 0.0167 \end{bmatrix},$$

$$[B_t] = \begin{bmatrix} Z_1 + \dfrac{Z_0}{(2 \cdot n_t)^2} & -\dfrac{Z_0}{(2 \cdot n_t)^2} \\ \dfrac{Z_0}{(2 \cdot n_t)^2} & -\left(Z_2 + \dfrac{Z_0}{(2 \cdot n_t)^2}\right) \end{bmatrix} = \begin{bmatrix} 0.0048 + j0.0062 & -0.0016 - j0.0041 \\ 0.0016 + j0.0041 & -0.0048 - j0.0062 \end{bmatrix},$$

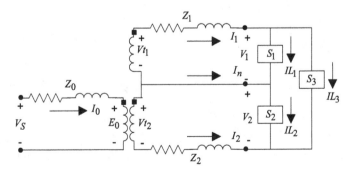

FIGURE 11.4 Center-tapped transformer serving constant impedance loads.

Center-Tapped Transformers and Secondaries

$$[d_t] = \frac{1}{2 \cdot n_t} \cdot \begin{bmatrix} 1 & -1 \\ 1 & -1 \end{bmatrix} = \begin{bmatrix} 0.0167 & -0.0167 \\ 0.0167 & -0.0167 \end{bmatrix}$$

The first forward sweep is computed by setting the secondary line currents to zero.

$$[V_{12}] = [A_t] \cdot [V_{SS}] - [B_t] \cdot [I_{12}] = \begin{bmatrix} 120\underline{/0} \\ 120\underline{/0} \end{bmatrix} V$$

where : $[V_{SS}] = \begin{bmatrix} 7200\underline{/0} \\ 7200\underline{/0} \end{bmatrix} V$

$$[I_{12}] = \begin{bmatrix} 0 \\ 0 \end{bmatrix}$$

The three load voltages are

$$[V_{ld}] = \begin{bmatrix} V_1 \\ V_2 \\ V_1 + V_2 \end{bmatrix} = \begin{bmatrix} 120\underline{/0} \\ 120\underline{/0} \\ 240\underline{/0} \end{bmatrix} V$$

The load currents are

$i = 1$ to 3,

$$Id_i = \left(\frac{SL_i \cdot 1000}{Vld_i} \right)^* = \begin{bmatrix} 83.3\underline{/-18.2} \\ 125.0\underline{/-25.8} \\ 104.2\underline{/-31.8} \end{bmatrix} A$$

The secondary line currents are given by

$$[I_{12}] = \begin{bmatrix} 1 & 0 & 1 \\ 0 & -1 & -1 \end{bmatrix} \cdot \begin{bmatrix} IL_1 \\ IL_2 \\ IL_3 \end{bmatrix} = \begin{bmatrix} 186.2\underline{/-25.8} \\ 228.9\underline{/151.5} \end{bmatrix} A$$

Define $[DI] = \begin{bmatrix} 1 & 0 & 1 \\ 0 & -1 & 1 \end{bmatrix}$.

The current in the neutral is

$$I_n = (Id_2 - Id_1) = 43.8\underline{/-40.5} A$$

The backward sweep computes the primary current:

$$\begin{bmatrix} I_0 \\ I_0 \end{bmatrix} = [d_t] \cdot [I_{12}] = \begin{bmatrix} 6.9156\underline{/-27.3} \\ 6.9156\underline{/-27.3} \end{bmatrix} A$$

Using the computed secondary line currents, the second forward sweep is

$$[V_{12}] = [A_t] \cdot [V_{SS}] - [B_t] \cdot [I_{12}] = \begin{bmatrix} 117.9\underline{/-0.64} \\ 117.8\underline{/-0.63} \end{bmatrix} V$$

The calculations above demonstrate the first and second forward sweeps and the first backward sweep. This process can continue, but it is much easier to write a Mathcad program to compute the final load voltages. The program is shown in Figure 11.5. The initial values are

$$[start] = \begin{bmatrix} 0 \\ 0 \end{bmatrix} \quad Tol = 0.00001$$

The Mathcad program follows the same general steps that all programs will follow.

1. Initialize
2. Set loop

$$X := \begin{vmatrix} I_{12} \leftarrow start \\ V_{old} \leftarrow start \\ \text{for } n \in 1..20 \\ \quad \begin{vmatrix} V_{12} \leftarrow A_t \cdot V_{SS} - B_t \cdot I_{12} \\ V_{ld} \leftarrow \begin{pmatrix} V_{12_1} \\ V_{12_2} \\ V_{12_1} + V_{12_2} \end{pmatrix} \\ \text{for } i \in 1..2 \\ \quad Error_i \leftarrow \dfrac{||V_{12_i}| - |V_{old_i}||}{120} \\ \text{break if } \max(Error) < Tol \\ \text{for } i \in 1..3 \\ \quad IL_i \leftarrow \dfrac{SL_i \cdot 1000}{V_{ld_i}} \\ I_{12} \leftarrow DI \cdot IL \\ I_{00} \leftarrow d_t \cdot I_{12} \\ V_{old} \leftarrow V_{12} \end{vmatrix} \\ Out_1 \leftarrow V_{ld} \\ Out \end{vmatrix}$$

FIGURE 11.5 Mathcad Program for Example 11.2.

Center-Tapped Transformers and Secondaries

3. Forward sweep
4. Check for convergence
   a. If converged output results
   b. If not converged, continue
5. Compute new load and line currents
6. Backward sweep
7. End of loop

After four iterations, the final load voltages are

$$[V_{ld}] = \begin{bmatrix} 117.88/-0.64 \\ 117.71/-0.63 \\ 235.60/-0.64 \end{bmatrix} V$$

11.1.2 Center-Tapped Transformer Serving Loads through a Triplex Secondary

Shown in Figure 11.6 is a center-tapped transformer serving a load through a triplex secondary.

Before the system of Figure 11.6 can be modeled, the impedance matrix for the triplex secondary must be determined. The impedances of the triplex are computed using Carson's equations and the Kron reduction, as described in Chapter 4. Applying Carson's equations will result in a 3 × 3 matrix. The Kron reduction method is used to "fold" the impedance of the neutral conductor into that of the two-phase conductors. A triplex secondary consisting of two insulated conductors and one un-insulated neutral conductor is shown in Figure 11.7.

The spacings between conductors that are applied in Carson's equations are given by the following:

$$D_{12} = \frac{dia + 2 \cdot T}{12}$$

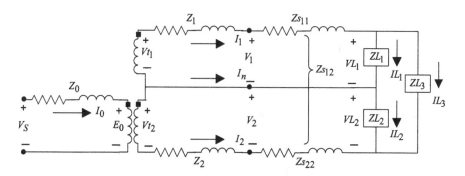

FIGURE 11.6 Center-tapped transformer with secondary.

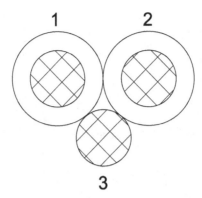

FIGURE 11.7 Triplex secondary.

$$D_{13} = \frac{dia + T}{12}$$

$$D_{23} = \frac{dia + T}{12} \text{ ft} \quad (11.19)$$

Where

dia = diameter of conductor in inches,
T = thickness of insulation in inches.

Applying Carson's equations:

$$zp_{ii} = r_i + 0.09530 + j0.12134 \cdot \left(\ln \frac{1}{GMR_i} + 7.93402 \right)$$

$$zp_{ij} = 0.09530 + j0.12134 \cdot \left(\ln \frac{1}{D_{ij}} + 7.93402 \right) \quad (11.20)$$

Where
r_i = conductor resistance in ohms/mile,
GMR_i = conductor geometric mean radius in feet,
D_{ij} = distance in ft between conductors i and j.

The secondary voltage equation in matrix form is

$$\begin{bmatrix} v_1 \\ v_2 \\ v_n \end{bmatrix} = \begin{bmatrix} V_{1g} \\ V_{2g} \\ V_{ng} \end{bmatrix} - \begin{bmatrix} VL_{1g} \\ VL_{2g} \\ VL_{ng} \end{bmatrix} = \begin{bmatrix} zp_{11} & zp_{12} & zp_{13} \\ zp_{21} & zp_{22} & zp_{23} \\ zp_{31} & zp_{32} & zp_{33} \end{bmatrix} \cdot \begin{bmatrix} I_1 \\ I_2 \\ I_n \end{bmatrix} \quad (11.21)$$

Center-Tapped Transformers and Secondaries

When the neutral is grounded at the transformer and the load, then

$$v_n = V_{ng} - VL_{ng} = 0. \tag{11.22}$$

This leads to the Kron reduction equation in partitioned form:

$$\begin{bmatrix} [v_{12}] \\ [0] \end{bmatrix} = \begin{bmatrix} [zp_{ii}] & [zp_{in}] \\ [zp_{nj}] & [zp_{nn}] \end{bmatrix} \cdot \begin{bmatrix} [I_{12}] \\ [I_n] \end{bmatrix} \tag{11.23}$$

Solving Equation 11.22 for the neutral current:

$$[I_n] = -[zp_{nn}]^{-1} \cdot [zp_{ni}] \cdot [I_{12}]$$

$$[I_n] = [t_n] \cdot [I_{12}]$$

$$\text{where}: [t_n] = -[zp_{nn}]^{-1} \cdot [zp_{ni}] \tag{11.24}$$

The Kron reduction gives the 2 × 2 phase impedance matrix:

$$[zs] = [zp_{ij}] - [zp_{in}] \cdot [zp_{nn}]^{-1} \cdot [zp_{nj}] \tag{11.25}$$

For a secondary length L,

$$[Zs] = [zs] \cdot L = \begin{bmatrix} Zs_{11} & Zs_{12} \\ Zs_{21} & Zs_{22} \end{bmatrix} \tag{11.26}$$

Referring to Figure 11.6, the voltage backward sweep for the secondary is given by the following:

$$\begin{bmatrix} V_1 \\ V_2 \end{bmatrix} = \begin{bmatrix} VL_1 \\ VL_2 \end{bmatrix} + \begin{bmatrix} Zs_{11} & Zs_{12} \\ Zs_{21} & Zs_{22} \end{bmatrix} \cdot \begin{bmatrix} I_1 \\ I_2 \end{bmatrix}$$

$$[V_{12}] = [a_{sec}] \cdot [VL_{12}] + [b_{sec}] \cdot [I_{12}] \tag{11.27}$$

Because of the short length of the secondary, the line currents leaving the transformer are equal to the line currents at the load, so no current backward sweep is needed for the secondary. In order to remain consistent for the general analysis of a feeder, the matrix $[d_{sec}]$ is defined as follows:

$$[I_{12}] = [d_{sec}] \cdot [I_{12}]$$

where: $[d_{sec}] = [U] = \begin{bmatrix} 1 & 0 \\ 0 & 1 \end{bmatrix}$ (11.28)

The voltage forward sweep equation for the secondary is determined by solving for the load voltages in Equation 11.27.

$$[VLVL_{12}] = [A_{sec}] \cdot [VL_{12}] - [B_{sec}] \cdot [I_{12}]$$

where: $[A_{sec}] = [a_{sec}]^{-1}$

$$[B_{sec}] = [A_{sec}] \cdot [b_{sec}]$$ (11.29)

Example 11.3: The secondary in Figure 11.7 is 100 feet of 1/0 AA triplex. Determine the phase impedance matrix for the triplex secondary.

From the table for 1/0 AA: $GMR = 0.111$ ft., $Diameter = 0.368"$, $r = 0.973$ Ω/mile
The insulation thickness of the phase conductors is 80 mil = 0.08 in.
The distance matrix with the diagonal terms equal to the GMR is computed to be

$$[D] = \begin{bmatrix} 0.0111 & 0.0440 & 0.0373 \\ 0.0440 & 0.0111 & 0.0373 \\ 0.0373 & 0.0373 & 0.0111 \end{bmatrix} \text{ft}$$

Applying Carson's equations, the primitive impedance matrix is

$$[zp] = \begin{bmatrix} 1.0683 + j1.5088 & 0.0953 + j1.3417 & 0.0953 + j1.3617 \\ 0.0953 + j1.3417 & 1.0683 + j1.5088 & 0.0953 + j1.3617 \\ 0.0953 + j1.3617 & 0.0953 + j1.3617 & 1.0683 + j1.5088 \end{bmatrix} \text{Ω/mile}$$

Define:

$$[zp_{ii}] = \begin{bmatrix} 1.0683 + j1.5088 & 0.0953 + j1.3417 \\ 0.0953 + j1.3417 & 1.0683 + j1.5088 \end{bmatrix} \text{Ω}$$

$$[zp_{in}] = \begin{bmatrix} 0.0953 + j1.3617 \\ 0.0953 + j1.3617 \end{bmatrix} \text{Ω}$$

$$[zp_{ni}] = [0.0953 + j1.3617 \quad 0.0953 + j1.3617] \text{Ω}$$

$$[zp_{nn}] = [1.0683 + j1.5088]$$

Center-Tapped Transformers and Secondaries

The Kron reduction:

$$[zs] = [zp_{ii}] - [zp_{in}] \cdot [zp_{nn}]^{-1} \cdot [zp_{ni}] \ \Omega$$

$$[zs] = \begin{bmatrix} 1.5304 + j0.6132 & 0.5574 + j0.4461 \\ 0.5574 + j0.4461 & 1.5304 + j0.6132 \end{bmatrix} \Omega$$

The secondary impedance matrix for a length of 100 feet is

$$[Zs] = \begin{bmatrix} 0.0290 + j0.0116 & 0.0106 + j0.0084 \\ 0.0106 + j0.0084 & 0.0290 + j0.0116 \end{bmatrix} \Omega.$$

The forward and backward sweep equations for the secondary are as follows:

$$[U] = \begin{bmatrix} 1 & 0 \\ 0 & 1 \end{bmatrix}$$

$$[A_{sec}] = [U]$$

$$[Zs] = \begin{bmatrix} 0.0290 + j0.0116 & 0.0106 + j0.0084 & 0.0290 + j0.0116 & 0.0106 + j0.0084 \\ 0.0106 + j0.0084 & 0.0290 + j0.0116 & 0.0106 + j0.0084 & 0.0290 + j0.0116 \end{bmatrix}$$

$$[d_{sec}] = [U]$$

The Mathcad program of Example 11.2 is modified so that
Forward Sweep:

$$[V_{12}] = [A_t] \cdot [V_{ss}] - [B_t] \cdot [I_{12}]$$

$$[VL_{12}] = [A_{sec}[\ ][V_{12}][B_{sec}[\ ][I_{12}]]]$$

In the program, the load voltages are

$$[V_{ld}] = \begin{bmatrix} VL_1 \\ VL_2 \\ VL_1 + VL_2 \end{bmatrix}$$

The remainder of the program stays the same. After five iterations, the final voltages are

$$\begin{bmatrix} V_1 \\ V_2 \end{bmatrix} = \begin{bmatrix} 117.89/\underline{-0.64} \\ 117.75/\underline{-0.62} \end{bmatrix} V,$$

$$\begin{bmatrix} VL_1 \\ VL_2 \\ VL_3 \end{bmatrix} = \begin{bmatrix} 114.98/\underline{-0.19} \\ 122.30/\underline{-1.31} \\ 237.27/\underline{-0.77} \end{bmatrix} V$$

Note that the voltage VL_2 is great than V_2, indicating a voltage rise on that phase. This is not uncommon when the line currents are very unbalanced.

The secondary line currents are

$$\begin{bmatrix} I_1 \\ I_2 \end{bmatrix} = \begin{bmatrix} 190.9/-26.2 \\ 227.8/150.4 \end{bmatrix} A$$

The primary line current is

$$I_0 = 6.97/-28.1 \, A$$

Using the neutral current transform matrix of Equation 11.23, the current flowing in the neutral conductor is

$$I_n = [t_n] \cdot [I_{12}] = 28.8/-15.8 \, A$$

The current flowing in ground is

$$I_g = -(I_n + I_1 + I_2) = 20.8/-93.1 \, A$$

It is always good to check the validity of the results. This is particularly true because there should be some questions about the voltage rise on phase 2. The check can be done by using basic circuit and transformer theory to compute the source voltage using the load voltages and line currents output from the program.

$$[V_{12}] = [VL_{12}] + [Z_s] \cdot [I_{12}] = \begin{bmatrix} 117.89/-0.64 \\ 117.75/-0.62 \end{bmatrix} V$$

$$[Vt_{12}] = [V_{12}] + [Z_{12}] \cdot [I_{12}] = \begin{bmatrix} 118.61/-0.59 \\ 118.61/-0.59 \end{bmatrix} V$$

This is the first indication that the solution is correct since the two ideal voltages on the secondary are equal. That is a must. Knowing the ideal voltages and the secondary line currents, the primary voltage and line current can be computed.

$$E_0 = 2 \cdot n_t \cdot Vt_1 = 7116.3/-0.59 \, A$$

$$I_0 = \frac{1}{2 \cdot n_t} \cdot (I_1 - I_2) = 6.97/-28.1 \, A$$

$$V_s = E_0 + Z_0 \cdot I_0 = 7200/\underline{0} \, V$$

Since the original source voltage has been computed, the results of the program have been shown to be correct. Whenever there is a question about the validity of a program solution, it is good to use basic circuit and transformer theory to prove the results are correct. Don't ever assume the results are correct because they came from a computer program.

Center-Tapped Transformers and Secondaries

11.2 UNGROUNDED WYE – DELTA TRANSFORMER BANK WITH CENTER-TAPPED TRANSFORMER

The most common transformer connection for providing service to a combination of three-phase and single-phase loads is the ungrounded wye-delta. In order to provide the usual three-wire service for the single-phase loads, one of the three transformers, the "lighting" transformer will have a center tap. The other two transformers are referred to as the "power" transformers.

The connection diagram for the standard 30° ungrounded wye – delta center tap transformer on phase A connection is shown in Figure 11.8. The derivations will be in terms of primary phases A-B-C and secondary phases a-b-c-n.

11.2.1 Basic Transformer Equations

The turns ratios for all transformers are given by

$$n_t = \frac{kVLN_{Rated\ Primary}}{kVLL_{Rated\ Secondary}} \quad (11.30)$$

The basic transformer equations for the center-tapped transformer are the following:

$$Vt_{an} = Vt_{nb} = \frac{1}{2 \cdot n_t} \cdot VT_{AN}$$

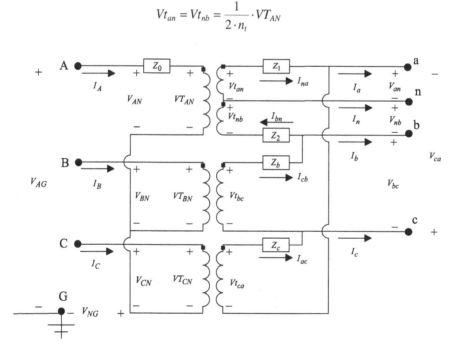

FIGURE 11.8 Ungrounded wye – delta transformer center-tapped connection.

$$VT_{AN} = 2 \cdot n_t \cdot Vt_{an}$$

$$I_A = \frac{1}{2 \cdot n_t} \cdot (I_{na} + I_{bn}) \tag{11.31}$$

For the transformer bank, the basic "ideal" transformer voltage equations as a function of the turns ratio are as follows:

$$\begin{bmatrix} VT_{AN} \\ VT_{BN} \\ VT_{CN} \end{bmatrix} = n_t \cdot \begin{bmatrix} 2 & 0 & 0 & 0 \\ 0 & 0 & 1 & 0 \\ 0 & 0 & 0 & 1 \end{bmatrix} \cdot \begin{bmatrix} Vt_{an} \\ Vt_{nb} \\ Vt_{bc} \\ Vt_{ca} \end{bmatrix}$$

$$[VTLN_{ABC}] = [AV] \cdot [Vt_{anbc}]$$

where: $[AV] = n_t \cdot \begin{bmatrix} 2 & 0 & 0 & 0 \\ 0 & 0 & 1 & 0 \\ 0 & 0 & 0 & 1 \end{bmatrix}$ \tag{11.32}

$$\begin{bmatrix} Vt_{an} \\ Vt_{nb} \\ Vt_{bc} \\ Vt_{ca} \end{bmatrix} = \frac{1}{n_t} \cdot \begin{bmatrix} 0.5 & 0 & 0 \\ 0.5 & 0 & 0 \\ 0 & 1 & 0 \\ 0 & 0 & 1 \end{bmatrix} \cdot \begin{bmatrix} VT_{AN} \\ VT_{BN} \\ VT_{CN} \end{bmatrix}$$

$$[Vt_{anbc}] = [BV] \cdot [VTLN_{ABC}]$$

where: $[BV] = \dfrac{1}{n_t} \cdot \begin{bmatrix} 0.5 & 0 & 0 \\ 0.5 & 0 & 0 \\ 0 & 1 & 0 \\ 0 & 0 & 1 \end{bmatrix}$ \tag{11.33}

The basic "ideal" transformer current equations as a function of the turn's ratio are the following:

$$\begin{bmatrix} I_A \\ I_B \\ I_C \end{bmatrix} = \frac{1}{n_t} \cdot \begin{bmatrix} 0.5 & 0.5 & 0 & 0 \\ 0 & 0 & 1 & 0 \\ 0 & 0 & 0 & 1 \end{bmatrix} \cdot \begin{bmatrix} I_{na} \\ I_{bn} \\ I_{cb} \\ I_{ac} \end{bmatrix}$$

$$[I_{ABC}] = [AI] \cdot [ID_{anbc}]$$

Center-Tapped Transformers and Secondaries

$$\text{where}: [AI] = \frac{1}{n_t} \cdot \begin{bmatrix} 0.5 & 0.5 & 0 & 0 \\ 0 & 0 & 1 & 0 \\ 0 & 0 & 0 & 1 \end{bmatrix}$$

(11.34)

11.2.2 Forward Sweep

Refer to Figure 11.8. In the forward sweep, the line-to-ground voltages at the terminals of the transformer bank will be known.

$$[VLG_{ABC}] = \begin{bmatrix} V_{AG} \\ V_{BG} \\ V_{CG} \end{bmatrix}$$

(11.35)

In order to determine the voltages across the transformer, it is necessary to first determine the "ideal" primary voltages, defined as:

$$[VTLN_{ABC}] = \begin{bmatrix} VT_{AN} \\ VT_{BN} \\ VT_{CN} \end{bmatrix}$$

(11.36)

The first step is to determine the voltages of the "idea" transformer to ground.

$$[VTLG_{ABC}] = [VLG_{ABC}] - [ZT_0] \cdot [I_{ABC}]$$

$$\text{where}: [VTLG_{ABC}] = \begin{bmatrix} VT_{AG} \\ VT_{BG} \\ VT_{CG} \end{bmatrix}$$

$$[ZT_0] = \begin{bmatrix} Z_0 & 0 & 0 \\ 0 & 0 & 0 \\ 0 & 0 & 0 \end{bmatrix}$$

$$[I_{ABC}] = \begin{bmatrix} I_A \\ I_B \\ I_C \end{bmatrix}$$

(11.37)

The line-to-line "ideal" voltages are as follows:

$$[VTLL_{ABC}] = [Dv] \cdot [VTLG_{ABC}]$$

where: $[VTLL_{ABC}] = \begin{bmatrix} VT_{AB} \\ VT_{BC} \\ VT_{CA} \end{bmatrix}$

$$[Dv] = \begin{bmatrix} 1 & -1 & 0 \\ 0 & 1 & -1 \\ -1 & 0 & 1 \end{bmatrix}$$ (11.38)

With reference to Figure 11.8, the primary "ideal" voltages to ground as a function of the primary line-to-ground voltages are the following:

$$VT_{AN} = VT_{AG} - V_{NG}$$

$$VT_{BN} = VT_{BG} - V_{NG}$$

$$VT_{CN} = VT_{CG} - V_{NG}$$

$$[VTLN_{ABC}] = [VTLG_{ABC}] - [VNG]$$

where: $[VNG] = \begin{bmatrix} V_{NG} \\ V_{NG} \\ V_{NG} \end{bmatrix}$ (11.39)

The primary line-to-line voltages across the "ideal" transformer windings are

$$[VTLL_{ABC}] = [Dv] \cdot [VTLN_{ABC}]$$ (11.40)

Substitute Equation 11.39 into Equation 11.39.

$$[VTLL_{ABC}] = [Dv] \cdot [VTLG_{ABC}] - [Dv] \cdot [VNG]$$

however: $[Dv] \cdot [VNG] = [0]$

therefore: $[VTLL_{ABC}] = [Dv] \cdot [VTLG_{ABC}]$ (11.41)

In Equation 11.41, the "ideal" line-to-line voltages are known. The "ideal" line-to-neutral voltages are needed to continue the forward sweep. In Equation 11.40, it appears that the line-to-neutral voltages can be computed by using the inverse of the matrix [Dv]. Unfortunately, that matrix is singular. Two of the equations in Equation

Center-Tapped Transformers and Secondaries 429

11.40 can be used, but a third independent equation is needed. The two equations from 11.40 that will be used are

$$VT_{BC} = VT_{BN} - VT_{CN}$$

$$VT_{CA} = VT_{CN} - VT_{AN} \quad (11.42)$$

The third independent equation comes from writing KVL around the delta secondary. The sum of the secondary voltages around the delta must equal zero. With a reference to Figure 11.8,

$$V_{an} + V_{nb} + V_{bc} + V_{ca} = 0,$$

$$Vt_{an} - Z_1 \cdot I_{na} + Vt_{nb} - Z_2 \cdot I_{bn} + Vt_{bc} - Z_b \cdot I_{cb} + Vt_{bc} - Z_c \cdot I_{ac} = 0,$$

$$Vt_{an} + Vt_{nb} + Vt_{bc} + Vt_{ca} = Z_1 \cdot I_{na} + Z_2 \cdot I_{bn} + Z_b \cdot I_{cb} + Z_c \cdot I_{ac},$$

$$Vt_{an} + Vt_{nb} + Vt_{bc} + Vt_{ca} = [ZD_{anbc}] \cdot [ID_{anbc}],$$

$$\text{where} [ZD_{anbc}] = [Z_1, Z_2, Z_b, Z_c],$$

$$[ID_{anbc}] = \begin{bmatrix} I_{na} \\ I_{bn} \\ I_{cb} \\ I_{ac} \end{bmatrix},$$

$$\text{but}: Vt_{an} + Vt_{nb} + Vt_{bc} + Vt_{ca} = \frac{1}{n_t} \cdot \left(\frac{VT_{AN}}{2} + \frac{VT_{AN}}{2} + VT_{BN} + VT_{CN} \right) =$$

$$\frac{1}{n_t} \cdot (.5 \cdot VT_{AN} + .5 \cdot VT_{AN} + VT_{BN} + VT_{CN}) = [ZD_{anbc}] \cdot [ID_{anbc}]$$

$$VT_{AN} + VT_{BN} + VT_{CN} = n_t \cdot [ZD_{anbc}] \cdot [ID_{anbc}] = X$$

$$\text{where}: \ X = n_t \cdot [ZD_{anbc}] \cdot [ID_{anbc}] \quad (11.43)$$

It is important to know that in Equation 11.43, the secondary transformer currents will be set to zero during the first forward sweep. After that, the most recent secondary currents from the backward sweep will be used.

Equations 11.42 and 11.43 are combined in matrix form:

$$\begin{bmatrix} X \\ VT_{BC} \\ VT_{CA} \end{bmatrix} = \begin{bmatrix} 1 & 1 & 1 \\ 0 & 1 & -1 \\ -1 & 0 & 1 \end{bmatrix} \cdot \begin{bmatrix} VT_{AN} \\ VT_{BN} \\ VT_{CN} \end{bmatrix}$$

$$[VXLL_{ABC}] = [DX] \cdot [VTLN_{ABC}] \tag{11.44}$$

The ideal primary voltages are computed by taking the inverse of $[DX]$:

$$[VTLN_{ABC}] = [Dx] \cdot [VXLL_{ABC}]$$

$$\text{where}: [Dx] = [DX]^{-1} = \frac{1}{3} \cdot \begin{bmatrix} 1 & -1 & -2 \\ 1 & 2 & 1 \\ 1 & -1 & 1 \end{bmatrix} \tag{11.45}$$

With the "ideal" line-to-neutral voltages known, the forward sweep continues with the computation of the secondary "ideal" voltages:

$$\begin{bmatrix} Vt_{an} \\ Vt_{nb} \\ Vt_{bc} \\ Vt_{ca} \end{bmatrix} = \frac{1}{n_t} \cdot \begin{bmatrix} 0.5 & 0 & 0 \\ 0.5 & 0 & 0 \\ 0 & 1 & 0 \\ 0 & 0 & 1 \end{bmatrix} \cdot \begin{bmatrix} VT_{AN} \\ VT_{BN} \\ VT_{CN} \end{bmatrix}$$

$$[Vt_{anbc}] = [BV] \cdot [VTLN_{ABC}] \tag{11.46}$$

The secondary transformer terminal voltages are given by

$$\begin{bmatrix} V_{an} \\ V_{nb} \\ V_{bc} \\ V_{ca} \end{bmatrix} = \begin{bmatrix} Vt_{an} \\ Vt_{nb} \\ Vt_{bc} \\ Vt_{ca} \end{bmatrix} - \begin{bmatrix} Z_1 & 0 & 0 & 0 \\ 0 & Z_2 & 0 & 0 \\ 0 & 0 & Z_y & 0 \\ 0 & 0 & 0 & Z_z \end{bmatrix} \begin{bmatrix} I_{na} \\ I_{bn} \\ I_{cb} \\ I_{ac} \end{bmatrix},$$

$$[V_{anbc}] = [Vt_{anbc}] - [Zt_{anbc}] \cdot [ID_{anbc}] \tag{11.47}$$

In the first forward sweep, the secondary delta currents are assumed to be zero.

On the first backward sweep, the secondary line currents will be known. To determine the currents in the delta as a function of the line currents, only three KCL

Center-Tapped Transformers and Secondaries

equations can be used. The fourth independent equation comes from recognizing that the sum of the primary line currents must equal zero. The three KCL equations to use are

$$I_a = I_{na} - I_{ac},$$

$$I_b = I_{cb} - I_{bn},$$

$$I_c = I_{ac} - I_{cb} \tag{11.48}$$

Since the sum of the line currents must equal zero the fourth equation is given by the following:

$$I_A + I_B + I_C = 0 = \frac{1}{2 \cdot n_t} \cdot (I_{na} + I_{bn}) + \frac{1}{n_t} \cdot (I_{cb} + I_{ac})$$

$$I_A + I_B + I_C = 0 = \frac{1}{2 \cdot n_t} \cdot (I_{na} + I_{bn} + 2 \cdot I_{cb} + 2 \cdot I_{ac})$$

$$2 \cdot n_t \cdot (I_A + I_B + I_C) = 0 = I_{na} + I_{bn} + 2 \cdot I_{cb} + 2 \cdot I_{ac}$$

$$0 = I_{na} + I_{bn} + 2 \cdot I_{cb} + 2 \cdot I_{ac} \tag{11.49}$$

Combine Equations 11.48 and 11.49 into matrix form:

$$\begin{bmatrix} I_a \\ I_b \\ I_c \\ 0 \end{bmatrix} = \begin{bmatrix} 1 & 0 & 0 & -1 \\ 0 & -1 & 1 & 0 \\ 0 & 0 & -1 & 1 \\ 1 & 1 & 2 & 2 \end{bmatrix} \cdot \begin{bmatrix} I_{na} \\ I_{bn} \\ I_{ca} \\ I_{ac} \end{bmatrix}$$

$$[I_{abc0}] = [X1] \cdot [ID_{anbc}] \tag{11.50}$$

The delta currents can now be computed by taking the inverse of $[X1]$:

$$[ID_{anbc}] = [X1]^{-1} \cdot [I_{abc0}],$$

$$[ID_{anbc}] = [x1] \cdot [I_{abc0}],$$

$$\begin{bmatrix} I_{na} \\ I_{bn} \\ I_{ca} \\ I_{ac} \end{bmatrix} = \frac{1}{6} \cdot \begin{bmatrix} 5 & 1 & 3 & 1 \\ -1 & -5 & -3 & 1 \\ -1 & 1 & -3 & 1 \\ -1 & 1 & 3 & 1 \end{bmatrix} \cdot \begin{bmatrix} I_a \\ I_b \\ I_c \\ 0 \end{bmatrix},$$

$$[ID_{anbc}] = [x1] \cdot [I_{abc0}],$$

$$\text{where } [x1] = [X1]^{-1} = \frac{1}{6} \cdot \begin{bmatrix} 5 & 1 & 3 & 1 \\ -1 & -5 & -3 & 1 \\ -1 & 1 & -3 & 1 \\ -1 & 1 & 3 & 1 \end{bmatrix} \qquad (11.51)$$

Note in Equation 11.51 that the fourth column of the inverse isn't needed since the fourth term in the current vector is zero. Therefore, Equation 11.51 is modified to

$$\begin{bmatrix} I_{na} \\ I_{bn} \\ I_{cb} \\ I_{ac} \end{bmatrix} = \frac{1}{6} \cdot \begin{bmatrix} 5 & 1 & 3 \\ -1 & -5 & -3 \\ -1 & 1 & -3 \\ -1 & 1 & 3 \end{bmatrix} \cdot \begin{bmatrix} I_a \\ I_b \\ I_c \end{bmatrix},$$

$$[ID_{anbc}] = [Dd] \cdot [I_{abc}],$$

$$\text{where } [Dd] = \frac{1}{6} \cdot \begin{bmatrix} 5 & 1 & 3 \\ -1 & -5 & -3 \\ -1 & 1 & -3 \\ -1 & 1 & 3 \end{bmatrix} \qquad (11.52)$$

Substitute Equation 11.52 into Equation 11.34.

$$[I_{ABC}] = [AI] \cdot [ID_{anbc}]$$

$$[I_{ABC}] = [AI] \cdot [Dd] \cdot [I_{abc}]$$

$$\text{Define}: [d_t] = [AI] \cdot [Dd]$$

$$[I_{ABC}] = [d_t] \cdot [I_{abc}] \qquad (11.53)$$

Equation 11.53 is the necessary equation used in the backward sweep to compute the primary line currents.

With the primary line currents known, the primary line-to-neutral voltages are computed as

$$[VTLN_{ABC}] = [VLN_{ABC}] - [ZT_0] \cdot [I_{ABC}],$$

$$\text{but } [I_{ABC}] = [d_t] \cdot [I_{abc}],$$

Center-Tapped Transformers and Secondaries

$$[VTLN_{ABC}] = [VLN_{ABC}] - [ZT_0] \cdot [d_t] \cdot [I_{abc}]. \quad (11.54)$$

The secondary transformer voltages are computed by

$$[V_{anbc}] = [Vt_{anbc}] - [Zt_{anbc}] \cdot [ID_{anbc}],$$

$$[ID_{anbc}] = [Dd] \cdot [I_{abc}],$$

$$[V_{anbc}] = [Vt_{anbc}] - [Zt_{anbc}] \cdot [Dd] \cdot [I_{abc}]. \quad (11.55)$$

Substitute Equation 11.46 into Equation 11.55.

$$[V_{anbc}] = [Vt_{anbc}] - [Zt_{anbc}] \cdot [Dd] \cdot [I_{abc}]$$

$$[Vt_{anbc}] = [BV] \cdot [VTLN_{ABC}]$$

$$[V_{anbc}] = [BV] \cdot [VTLN_{ABC}] - [Zt_{anbc}] \cdot [Dd] \cdot [I_{abc}] \quad (11.56)$$

Substitute Equation 11.54 into Equation 11.56.

$$[V_{anbc}] = [BV] \cdot [VTLN_{ABC}] - [Zt_{anbc}] \cdot [Dd] \cdot [I_{abc}]$$

$$[VTLN_{ABC}] = [VLN_{ABC}] - [ZT_0] \cdot [d_t] \cdot [I_{abc}]$$

$$[V_{anbc}] = [BV] \cdot ([VLN_{ABC}] - [ZT_0] \cdot [d_t] \cdot [I_{abc}]) - [Zt_{anbc}] \cdot [Dd] \cdot [I_{abc}]$$

$$[V_{anbc}] = [BV] \cdot [VLN_{ABC}] - ([BV] \cdot [ZT_0] \cdot [d_t] + [Zt_{anbc}] \cdot [Dd]) \cdot [I_{abc}]$$

Define : $[A_t] = [BV]$

$$[B_t] = [BV] \cdot [ZT_0] \cdot [d_t] + [Zt_{anbc}] \cdot [Dd]$$

$$[V_{anbc}] = [A_t] \cdot [VLN_{ABC}] - [B_t] \cdot [I_{abc}] \quad (11.57)$$

11.2.3 Backward Sweep

The terminal line-to-neutral voltages are

$$[VLN_{ABC}] = [VTLN_{ABC}] + [ZT_0] \cdot [I_{ABC}],$$

but $[VTLN_{ABC}] = [AV] \cdot [Vt_{anbc}],$

and $[I_{ABC}] = [d_t] \cdot [I_{abc}]$;

therefore, $[VLN_{ABC}] = [AV] \cdot [Vt_{anbc}] + [ZT_0] \cdot [d_t] \cdot [I_{abc}]$ (11.58)

The "ideal" secondary voltages as a function of the secondary terminal voltages are

$$[Vt_{anbc}] = [V_{anbc}] + [Zt_{anbc}] \cdot [ID_{abc}],$$

$$[Vt_{anbc}] = [V_{anbc}] + [Zt_{anbc}] \cdot [Dd] \cdot [I_{abc}]$$ (11.59)

Substitute Equation 11.59 into Equation 11.58.

$$[VLN_{ABC}] = [AV] \cdot ([V_{anbc}] + [Zt_{anbc}] \cdot [Dd] \cdot [I_{abc}]) + [d_t] \cdot [I_{abc}]$$

$$[VLN_{ABC}] = [AV] \cdot [V_{anbc}] + ([AV] \cdot [Zt_{anbc}] \cdot [Dd] + [ZT_0] \cdot [d_t]) \cdot [I_{abc}]$$

Therefore, $[VLN_{ABC}] = [a_t] \cdot [V_{anbc}] + [b_t] \cdot [I_{abc}]$,

where $[a_t] = [AV]$,

$$[b_t] = [AV] \cdot [Zt_{anbc}] \cdot [Dd] + [ZT_0] \cdot [d_t]$$ (11.60)

It is important to know that normally on the backward sweep, the node voltages are not computed using Equation 11.60. Only the currents are calculated back to the source. However, as a check to confirm the results of the power-flow program, Equation 11.60, using the computed secondary voltages and currents, is used to confirm that the source voltages are the same as were used in the forward sweep.

11.2.4 Summary

It is important to note that the turns ratio is given by

$$n_t = \frac{kVLN_{hi}}{kVLL_{lo}}.$$ (11.61)

In the derivation of the forward and backward matrices, it was found that all of the matrices can be defined by the combination of matrices based upon basic circuit theory. The definitions are as follows:

$$[a_t] = [AV]$$

Center-Tapped Transformers and Secondaries

Example 11.4 Figure 11.9 shows an ungrounded wye – delta transformer bank servicing 120/240-volt, single-phase loads and a three-phase induction motor.

The single-phase loads are rated:

$SL_1 = 3\,kVA, 120\,volts, 0.95\,lagging\,power\,factor$
$SL_2 = 5\,kVA, 120\,volts, 0.90\,lagging\,power\,factor$
$SL_3 = 8\,kVA, 240\,volts, 0.85\,lagging\,power\,factor$

The load vector is

for $i = 1$ to 3,

$$SL_i = kVA_i \angle acos(PF_i) = \begin{bmatrix} 2.85 + j0.9367 \\ 4.5 + j2.1794 \\ 6.8 + j4.2143 \end{bmatrix} kVA$$

The three-phase induction motor data are as follows:
25 hp 240 volts, impedances:

$Zs = 0.0774 + j0.1843\,\Omega$

$Zr = 0.0908 + j0.1843\,\Omega$

$Zm = 0 + j4.8385\,\Omega$

The motor is operating at a slip of 0.035.
The transformer data are as follows:
Lighting Transformer: 25 kVA, 7,200–240/120 volts, $ZL_{pu} = 0.012 + j0.017$
Power Transformers: 10 kVA, 7,200–240 volts, $ZP_{pu} = 0.016 + j0.014$
Source Voltages: Balanced line-to-neutral 7,200 volts

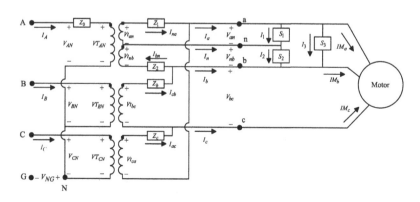

FIGURE 11.9 Ungrounded wye – delta bank serving combination loads.

Determine:
a. Transformer impedances in ohms for the model in Figure 11.9
b. Transformer forward and backward sweep matrices
c. Motor phase admittance matrix
d. Operating currents
  i. Single-phase loads
  ii. Motor
e. Operating voltages
i. Single-phase loads
ii. Motor

Compute the winding per-unit impedances for the lighting transformer.

$$Zpu_0 = 0.5 \cdot \text{Re}(Z_L) + j0.8 \cdot \text{Im}(Z_L) = 0.006 + j0.0136$$

$$Zpu_1 = Zpu_2 = \text{Re}(Z_L) + j0.4 \cdot \text{Im}(Z_L) = 0.012 + j0.0068$$

Convert the lighting transformer impedances to ohms.

$$Zbase_{hi} = \frac{kVLN_{hi}^2 \cdot 1000}{kVA} = \frac{7.2^2 \cdot 1000}{25} = 2073.6\,\Omega$$

$$Z_0 = 2073.6 \cdot (0.006 + j0.0136) = 12.446 + j28.201\,\Omega$$

$$Zbase_{lo} = \frac{kVLL_{lo}^2 \cdot 1000}{kVA_P} = \frac{0.24^2 \cdot 1000}{25} = 2.304\,\Omega$$

$$Z_1 = Z_2 = \frac{2.304}{4} \cdot (0.012 + j0.0068) = 0.0069 + j0.0039\,\Omega$$

Convert the per-unit impedances of the power transformers to ohms.

$$Zbase_{lo} = \frac{0.24^2 \cdot 1000}{10} = 5.76\,\Omega$$

$$Z_b = Z_c = 5.76 \cdot (0.016 + j0.014) = 0.0922 + j0.0806\,\Omega$$

Compute the turns ratio.

$$n_t = \frac{7200}{240} = 30$$

Compute the forward sweep matrices.

$$[A_t] = [BV] = \begin{bmatrix} 0.0167 & 0 & 0 \\ 0.0167 & 0 & 0 \\ 0 & 0.0333 & 0 \\ 0 & 0 & 0.0333 \end{bmatrix}$$

$$[B_t] = [BV] \cdot [ZT_0] \cdot [d_t] + [Zt_{anbc}] \cdot [Dd]$$

Center-Tapped Transformers and Secondaries

$$[B_t] = \begin{bmatrix} 0.0081+j0.0085 & -0.0012-j0.0046 & 0.0035+j0.002 \\ 0.0012+j0.0046 & -0.0081-j0.0085 & -0.0035-j0.002 \\ -0.0154-j0.0134 & 0.0154+j0.0134 & -0.0461-j0.0403 \\ -0.0154-j0.0134 & 0.0154+j0.0134 & 0.0461+j0.0403 \end{bmatrix}$$

Compute the backward sweep matrices.

$$\text{Where } [a_t] = [AV] = n_t \cdot \begin{bmatrix} 2 & 0 & 0 & 0 \\ 0 & 0 & 1 & 0 \\ 0 & 0 & 0 & 1 \end{bmatrix} = \begin{bmatrix} 60 & 0 & 0 & 0 \\ 0 & 0 & 30 & 0 \\ 0 & 0 & 0 & 30 \end{bmatrix}$$

$$[b_t] \cdot \frac{n_t}{6} \cdot \begin{bmatrix} 5 \cdot Z_1 + 4 \cdot \frac{Z_0}{n_t^2} & Z_1 - 4 \cdot \frac{Z_0}{n_t^2} & 3 \cdot Z_1 \\ -0.5 \cdot Z_b & 0.5 \cdot Z_b & -1.5 \cdot Z_b \\ -0.5 \cdot Z_c & 0.5 \cdot Z_c & 1.5 \cdot Z_c \end{bmatrix}$$

$$[b_t] = [AV] \cdot [Zt_{anbc}] \cdot [Dd] + [ZT_0] \cdot [d_t]$$

$$[b_t] = \begin{bmatrix} 0.4838+j0.5092 & -0.0691-j0.2742 & 0.2074+j0.1175 \\ -0.4608-j0.4032 & 0.4608+j0.4032 & -1.3824-j1.2096 \\ -0.4608-j0.4032 & 0.4608+j0.4032 & 1.3824+j1.2096 \end{bmatrix}$$

$$[d_t] = [AI] \cdot [Dd] = \begin{bmatrix} 0.0111 & -0.0111 & 0 \\ -0.0056 & 0.0056 & -0.0167 \\ -0.0056 & 0.0056 & 0.0167 \end{bmatrix}$$

Motor admittance matrix:
Define the positive and negative sequence slips.

$$s_1 = 0.035$$

$$s_2 = 2 - s_1 = 1.965$$

Compute the sequence load resistances and input sequence impedances.

for $k = 1$ and 2

$$RL_k = \frac{1-s_k}{s_k} \cdot R_r = \begin{bmatrix} 2.5035 \\ -0.0446 \end{bmatrix}$$

$$ZM_k = Zs + \frac{Zm \cdot (Zr + RL_k)}{Zm + Zr + RL_k} = \begin{bmatrix} 1.9778+j1.3434 \\ 0.1203+j0.3622 \end{bmatrix}$$

Compute the input sequence admittances.

$$YM_k = \frac{1}{ZM_k} = \begin{bmatrix} 0.3460 - j0.2350 \\ 0.8256 - j2.4865 \end{bmatrix}$$

The sequence admittance matrix is

$$[YM_{012}] = \begin{bmatrix} 1 & 0 & 0 \\ 0 & t^* \cdot YM_1 & 0 \\ 0 & 0 & t \cdot YM_2 \end{bmatrix} = \begin{bmatrix} 1 & 0 & 0 \\ 0 & 0.1052 - j0.2174 & 0 \\ 0 & 0 & 1.1306 - j1.0049 \end{bmatrix},$$

$$\text{where } t = \frac{1}{\sqrt{3}}/\underline{30}.$$

The phase admittance matrix is

$$[YM_{abc}] = [A] \cdot [YM_{012}] \cdot [A]^{-1}$$
$$= \begin{bmatrix} 0.7543 - j0.4074 & -0.1000 - j0.0923 & 0.3347 + j0.4997 \\ 0.3547 + j0.4997 & 0.7453 - j0.4074 & -0.1000 - j0.0923 \\ -0.1000 - j0.0923 & 0.3547 + 0.4997 & 0.7453 - j0.4074 \end{bmatrix},$$

where $a = 1/\underline{120}$,

$$[A] = \begin{bmatrix} 1 & 1 & 1 \\ 1 & a^2 & a \\ 1 & a & a^2 \end{bmatrix}$$

Set the source voltage vector.

$$[VLG_{ABC}] = \begin{bmatrix} 7200/\underline{0} \\ 7200/\underline{-120} \\ 7200/\underline{120} \end{bmatrix} V$$

The Mathcad program to compute the voltages and currents is shown in Figure 11.10. The starting matrices and the KCL current matrix and tolerance are defined as

$$[istart] = \begin{bmatrix} 0 \\ 0 \\ 0 \end{bmatrix} [vstart] = \begin{bmatrix} 0 \\ 0 \\ 0 \\ 0 \end{bmatrix} Tol = 0.00001$$

Note in this program that at the start of each iteration, the transformer bank line-to-neutral voltages must first be computed. This is necessary since the primary

Center-Tapped Transformers and Secondaries

$$X := \begin{vmatrix} I_{abc} \leftarrow istart \\ ID_{anbc} \leftarrow Dd \cdot I_{abc} \\ I_{ABC} \leftarrow istart \\ V_{old} \leftarrow vstart \\ \text{for } n \in 1..10 \\ \quad \begin{vmatrix} VTLG_{ABC} \leftarrow VLG_{ABC} - ZT_{,0} \cdot I_{ABC} \\ VTLL_{ABC} \leftarrow Dv \cdot VTLG_{ABC} \\ XX \leftarrow n_{,t} \cdot ZD_{anbc} \cdot ID_{anbc} \\ VXLL_{ABC} \leftarrow \begin{pmatrix} XX \\ VTLL_{ABC_2} \\ VTLL_{ABC_3} \end{pmatrix} \\ VTLN_{ABC} \leftarrow Dx\,VXLL_{ABC} \\ VLN_{ABC} \leftarrow VTLN_{ABC} + ZT_{,0} \cdot I_{ABC} \\ V_{anbc} \leftarrow A_{,t} \cdot VLN_{ABC} - B_{,t} \cdot I_{abc} \\ \text{for } j \in 1..4 \\ \quad Error_j \leftarrow \left|\left|V_{anbc_j}\right| - \left|V_{old_j}\right|\right| \\ \text{break if max(Error)} < Tol \\ V_{ld} \leftarrow \begin{pmatrix} V_{anbc_1} \\ V_{anbc_2} \\ V_{anbc_1} + V_{anbc_2} \end{pmatrix} \\ VM \leftarrow \begin{pmatrix} V_{ld_3} \\ V_{anbc_3} \\ V_{anbc_4} \end{pmatrix} \\ \text{for } i \in 1..3 \\ \quad IL_i \leftarrow \dfrac{\overline{SL_i \cdot 1000}}{V_{ld_i}} \\ IM \leftarrow YM_{abc} \cdot VM \\ I_{abc} \leftarrow DI \cdot IL + IM \\ ID_{anbc} \leftarrow Dd \cdot I_{abc} \\ I_{ABC} \leftarrow d_{,t} \cdot I_{abc} \\ V_{old} \leftarrow V_{anbc} \end{vmatrix} \\ Out_1 \leftarrow V_{anbc} \\ Out \end{vmatrix}$$

FIGURE 11.10 Program for Example 11.4.

of the wye connection is open and not grounded. The program in Figure 11.10 only shows the secondary voltages being output. That is done here to conserve space. The output can be increased by adding the other voltages and currents of interest to the list at the end of the program.

After four iterations, the results are as follows:

Currents:

$$[IL] = \begin{bmatrix} 25.6/\underline{-18.6} \\ 42.6/\underline{-26.2} \\ 34.1/\underline{-32.2} \end{bmatrix} A$$

$$[IM] = \begin{bmatrix} 56.3/\underline{-65.6} \\ 56.1/\underline{176.6} \\ 58.1/\underline{54.6} \end{bmatrix} A$$

$$[I_{abc}] = \begin{bmatrix} 108.8/\underline{-45.5} \\ 129.6/\underline{161.9} \\ 58.1/\underline{54.6} \end{bmatrix} A$$

$$[I_{ABC}] = \begin{bmatrix} 2.57/\underline{-30.6} \\ 1.66/\underline{-175.0} \\ 1.56/\underline{111.1} \end{bmatrix} A$$

Voltages:

$$[VLN_{ABC}] = \begin{bmatrix} 7140.8/\underline{-0.02} \\ 7231.7/\underline{-120.4} \\ 7227.8/\underline{120.4} \end{bmatrix} V$$

$$[V_{anbc}] = \begin{bmatrix} 117.4/\underline{-0.39} \\ 117.3/\underline{-0.38} \\ 235.1/\underline{-120.1} \\ 236.1/\underline{119.7} \end{bmatrix} V$$

$$[V_{ld}] = \begin{bmatrix} 117.4/\underline{-0.39} \\ 117.3/\underline{-0.38} \\ 234.7/\underline{-0.39} \end{bmatrix} V$$

$$[VM] = \begin{bmatrix} 234.7/\underline{-0.39} \\ 235.1/\underline{-120.1} \\ 236.1/\underline{119.7} \end{bmatrix} V$$

For a check of the accuracy of the results, the backward sweep, using the computed secondary voltages and currents, is used to compute the primary terminal line-to-neutral voltages.

$$[VLN_{ABC}] = [a_t] \cdot [V_{anbc}] + [b_t] \cdot [I_{abc}]$$

Center-Tapped Transformers and Secondaries

$$[VLN_{ABC}] = \begin{bmatrix} 7140.8/\underline{-0.02} \\ 7231.7/\underline{-120.4} \\ 7227.8/\underline{120.4} \end{bmatrix} V$$

It is noted that these voltages exactly match the initially computed terminal line-to-neutral voltages.

The motor voltage and current unbalances are computed to be:

$$V_{unbalance} = 0.3382\%$$

$$I_{unbalance} = 2.2205\%$$

$$[b_t] = [AV] \cdot [Zt_{anbc}] \cdot [Dd] + [ZT_0] \cdot [d_t]$$

$$[d_t] = [AI] \cdot [Dd]$$

$$[A_t] = [BV]$$

$$[B_t] = [BV] \cdot [ZT_0] \cdot [d_t] + [Zt_{anbc}] \cdot [Dd] \qquad (11.62)$$

The individual matrices in Equation 11.62 define the relationship between parameters by the following:

$$[ID_{anbc}] = [Dd] \cdot [I_{abc}]$$

$$[VTLN_{ABC}] = [AV] \cdot [Vt_{anbc}]$$

$$[I_{ABC}] = [AI] \cdot [ID_{anbc}]$$

$$[Vt_{anbc}] = [BV] \cdot [VT_{ABC}] \qquad (11.63)$$

The definitions of Equations 11.62 and 11.63 will be used to develop the models for the open wye – open delta connections.

11.3 OPEN WYE – OPEN DELTA TRANSFORMER CONNECTIONS

Quite often, an open wye – open delta transformer consisting of one lighting transformer and one power transformer will be used to serve combination single-phase and three-phase loads. For this connection, the neutral of the primary wye-connected windings must be grounded.

11.3.1 THE LEADING OPEN WYE – OPEN DELTA CONNECTION

In the "leading" connection, the voltage applied to the lighting transformer will lead the voltage applied to the power transformer by 120°. The leading open wye – open delta connection is shown in Figure 11.11.

The voltage phasors at no-load for the leading connection in Figure 11.11 are shown in Figure 11.12:

Notice in Figure 11.12 that there are three line-to-line voltages. Two of those voltages come directly from the primary voltages applied to the lighting and power transformers. The third voltage is a result of the three line-to-line voltages that must equal zero.

11.3.2 THE LAGGING OPEN WYE – OPEN DELTA CONNECTION

In the "lagging" connection, the voltage applied to the lighting transformer will lag the voltage applied to the power transformer by 120°. The lagging open wye – open delta connection is shown in Figure 11.13.

The voltage phasors at no load for the lagging connection in Figure 11.13 are shown in Figure 11.14:

It is very important to note that for both connections, the phase sequence on the secondary is a-b-c. That will always be the assumption but great care must be taken

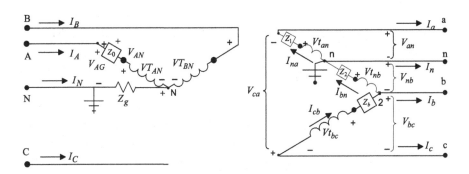

FIGURE 11.11 Leading open wye – open delta connection.

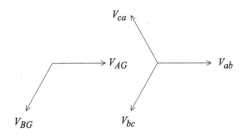

FIGURE 11.12 Leading open wye – open delta voltage phasors.

Center-Tapped Transformers and Secondaries

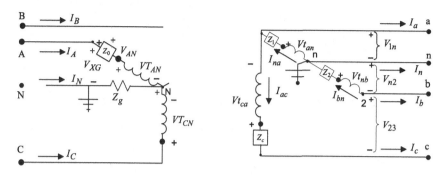

FIGURE 11.13 Lagging open wye – open delta connection.

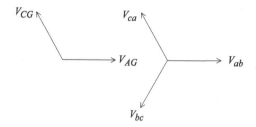

FIGURE 11.14 Lagging open wye – open delta voltage phasors.

to assure that the labeling of the phases will result in the correct a-b-c sequence. Note for both connections, the primary neutral is grounded through an impedance Z_g.

11.3.3 Forward Sweep

There will be a slight difference in the matrices for the leading and lagging connections. In order to define the matrices, the subscript L will be used on various matrices.

$$L = 1 \text{ leading connection}$$
$$L = 2 \text{ lagging connection} \tag{11.64}$$

The "ideal" primary transformer voltages for both connections are

$$\begin{bmatrix} VT_{AN} \\ VT_{BN} \\ VT_{CN} \end{bmatrix} = \begin{bmatrix} V_{AG} \\ V_{BG} \\ V_{CG} \end{bmatrix} - \begin{bmatrix} Z_0 + Z_g & Z_g & Z_g \\ Z_g & Z_g & Z_g \\ Z_g & Z_g & Z_g \end{bmatrix} \cdot \begin{bmatrix} I_A \\ I_B \\ I_C \end{bmatrix},$$

$$[VTLN_{ABC}] = [VTLG_{ABC}] - [ZOG] \cdot [I_{ABC}],$$

$$\text{where } [ZOG] = \begin{bmatrix} Z_0 + Z_g & Z_g & Z_g \\ Z_g & Z_g & Z_g \\ Z_g & Z_g & Z_g \end{bmatrix} \quad (11.65)$$

The "ideal" secondary transformer voltages are

$$[Vt_{anbc}]_L = [BV]_L \cdot [VTLN_{ABC}]$$

Leading Connection:

$$[Vt_{anbc}]_1 = \begin{bmatrix} Vt_{an} \\ Vt_{nb} \\ Vt_{ab} \end{bmatrix} \quad [BV]_1 = \frac{1}{n_t} \cdot \begin{bmatrix} 0.5 & 0 & 0 \\ 0.5 & 0 & 0 \\ 0 & 1 & 0 \end{bmatrix}$$

Lagging Connection

$$[Vt_{anbc}]_2 = \begin{bmatrix} Vt_{an} \\ Vt_{nb} \\ Vt_{ca} \end{bmatrix} \quad [BV]_2 = \frac{1}{n_t} \cdot \begin{bmatrix} 0.5 & 0 & 0 \\ 0.5 & 0 & 0 \\ 0 & 0 & 1 \end{bmatrix} \quad (11.66)$$

Substitute Equation 11.65 into Equation 11.66.

$$[Vt_{anbc}]_L = [BV]_L \cdot ([VLG_{ABC}] - [ZOG] \cdot [I_{ABC}])$$

$$[Vt_{anbc}]_L = [BV]_L \cdot [VLG_{ABC}] - [BV]_L \cdot [ZOG] \cdot [I_{ABC}] \quad (11.67)$$

The transformer secondary currents are defined as follows:

$$\text{Leading connection}: [ID_{anbc}]_1 = \begin{bmatrix} I_{na} \\ I_{bn} \\ I_{cb} \end{bmatrix}$$

$$\text{Lagging connection}: [ID_{anbc}]_2 = \begin{bmatrix} I_{na} \\ I_{bn} \\ I_{ac} \end{bmatrix} \quad (11.68)$$

Center-Tapped Transformers and Secondaries

The primary line currents as a function of the secondary open delta currents are

$$[I_{ABC}] = [AI]_L \cdot [ID_{anbc}]_L,$$

$$\text{where } [AI]_1 = \frac{1}{n_t} \cdot \begin{bmatrix} 0.5 & 0.5 & 0 \\ 0 & 0 & 1 \\ 0 & 0 & 0 \end{bmatrix},$$

$$[AI]_2 = \frac{1}{n_t} \cdot \begin{bmatrix} 0.5 & 0.5 & 0 \\ 0 & 0 & 0 \\ 0 & 0 & 1 \end{bmatrix} \quad (11.69)$$

The secondary open delta currents as a function of the secondary line currents are

$$[ID_{anbc}]_L = [Dd]_L \cdot [I_{abcn}],$$

$$\text{where } [I_{abcn}] = \begin{bmatrix} I_a \\ I_b \\ I_c \\ I_n \end{bmatrix},$$

$$[Dd]_1 = \begin{bmatrix} 1 & 0 & 0 & 0 \\ 0 & -1 & -1 & 0 \\ 0 & 0 & -1 & 0 \end{bmatrix},$$

$$[Dd]_2 = \begin{bmatrix} 1 & 0 & 1 & 0 \\ 0 & -1 & 0 & 0 \\ 0 & 0 & 1 & 0 \end{bmatrix} \quad (11.70)$$

Substitute Equations 11.69 and 11.70 into Equation 11.67.

$$[ID_{anbc}]_L = [Dd]_L \cdot [I_{abcn}]$$

$$[I_{ABC}] = [AI]_L \cdot [ID_{anbc}]_L$$

$$[I_{ABC}] = [AI]_L \cdot [Dd]_L \cdot [I_{abcn}]$$

$$[Vt_{anbc}]_L = [BV]_L \cdot [VLG_{ABC}] - [BV]_L \cdot [Z0G] \cdot [I_{ABC}]$$

$$[Vt_{anbc}]_L = [BV]_L \cdot [VLG_{ABC}] - [BV]_L \cdot [Z0G] \cdot [AI]_L \cdot [Dd]_L \cdot [I_{abcn}] \quad (11.71)$$

The transformer bank secondary voltages are

$$[V_{abc}]_L = [Vt_{anbc}]_L - [Zt_{\sec}\ ][ID_{anbc}]_L,$$

$$\text{where } [V_{abc}]_1 = \begin{bmatrix} V_{an} \\ V_{nb} \\ V_{bc} \end{bmatrix} \quad [V_{abc}]_2 = \begin{bmatrix} V_{an} \\ V_{nb} \\ V_{ca} \end{bmatrix},$$

$$\text{but } [ID_{anbc}]_L = [Dd]_L \cdot [I_{abcn}],$$

$$[V_{abc}]_L = [Vt_{anbc}]_L - [Zt_{\sec}\ ][Dd]_L [I_{abcn}]]. \quad (11.72)$$

Substitute Equation 11.71 into Equation 11.72.

$$[Vt_{anbc}]_L = [BV]_L \cdot [VLG_{ABC}] - [BV]_L \cdot [Z0G] \cdot [AI]_L \cdot [Dd]_L \cdot [I_{abcn}]$$

$$[V_{abc}]_L = [Vt_{anbc}]_L - [Zt_{anbc}] \cdot [Dd]_L \cdot [I_{abcn}]$$

$$[V_{abc}]_L = \big([BV]_L \cdot [VLG_{ABC}] - [BV]_L \cdot [Z0G] \cdot [AI]_L \cdot [Dd]_L \cdot [I_{abcn}]\big)$$

$$- [Zt_{anbc}] \cdot [Dd]_L \cdot [I_{abcn}]$$

$$[V_{abc}]_L = [BV]_L \cdot [VLG_{ABC}] - \big([BV]_L \cdot [Z0G] \cdot [AI]_L + [Zt_{anbc}]\big) \cdot [Dd_L] \cdot [I_{abcn}]$$
(11.73)

The secondary line voltages are

$$\begin{bmatrix} V_{an} \\ V_{nb} \\ V_{bc} \\ V_{ca} \end{bmatrix} = [V_{anbc}] = [CV]_L \cdot [V_{abc}]_L,$$

$$\text{where } [CV]_1 = \begin{bmatrix} 1 & 0 & 0 \\ 0 & 1 & 0 \\ 0 & 0 & 1 \\ -1 & -1 & -1 \end{bmatrix},$$

Center-Tapped Transformers and Secondaries

$$[CV]_2 = \begin{bmatrix} 1 & 0 & 0 \\ 0 & 1 & 0 \\ -1 & -1 & -1 \\ 0 & 0 & 1 \end{bmatrix} s \qquad (11.74)$$

Substitute Equation 11.73 into Equation 11.74.

$$[V_{abc}]_L = [BV]_L \cdot [VLG_{ABC}] - ([BV]_L \cdot [Z0G] \cdot [AI]_L + [Zt_{anbc}]) \cdot [Dd_L] \cdot [I_{abcn}]$$

$$[V_{anbc}] = [CV]_L \cdot [V_{abc}]_L$$

$$[V_{anbc}] = [CV]_L \cdot [BV]_L \cdot [VLG_{ABC}]$$

$$- [CV]_L ([BV]_L \cdot [Z0G] \cdot [AI]_L + [Zt_{anbc}]) \cdot [Dd_L] \cdot [I_{abcn}]$$

Define: $[A_t]_L = [CV]_L \cdot [BV]_L$

$$[B_t]_L = [CV]_L ([BV]_L \cdot [Z0G] \cdot [AI]_L + [Zt_{anbc}]) \cdot [Dd_L]$$

therefore: $[V_{anbc}] = [A_t]_L \cdot [VLG_{ABC}] - [B_t]_L \cdot [I_{abcn}]$ \qquad (11.75)

11.3.4 Backward Sweep

Substitute Equation 11.

$$[ID_{anbc}]_L = [Dd]_L \cdot [I_{abcn}]$$

$$[I_{ABC}] = [AI]_L \cdot [ID_{anbc}]_L$$

$$[I_{ABC}] = [AI]_L \cdot [Dd]_L \cdot [I_{abcn}]$$

$$[I_{ABC}] = [d_t]_L \cdot [I_{abcn}]$$

where: $[d_t]_L = [AI]_L \cdot [Dd]_L$ \qquad (11.76)

Example 11.5: For the system of Example 11.4, the transformer bank is changed to open wye – open delta. Analyze the system using the leading and lagging connections. The Mathcad program for the analyses of the leading and lagging connections is shown in Figure 11.15

Leading Connection with $L = 1$:
  Appling Equations 11.75 and 11.76, the forward and backward sweep matrices are

$$X := \begin{vmatrix} I_{.abcn} \leftarrow vstart \\ ID_{.anbc} \leftarrow Dd_L \cdot I_{.abcn} \\ I_{.ABC} \leftarrow istart \\ V_{.old} \leftarrow vstart \\ \text{for } n \in 1..20 \\ \quad \begin{vmatrix} V_{.anbc} \leftarrow A_{.t} \cdot VLG_{ABC} - B_{.t} \cdot I_{.abcn} \\ \text{for } j \in 1..4 \\ \quad Error_j \leftarrow ||V_{.anbc_j}| - |V_{.old_j}|| \\ \text{break if } \max(Error) < Tol \\ V_{.ld} \leftarrow \begin{pmatrix} V_{.anbc_1} \\ V_{.anbc_2} \\ V_{.anbc_1} + V_{.anbc_2} \end{pmatrix} \\ VM \leftarrow \begin{pmatrix} V_{.ld_3} \\ V_{.anbc_3} \\ V_{.anbc_4} \end{pmatrix} \\ \text{for } i \in 1..3 \\ \quad IL_i \leftarrow \dfrac{\overline{SL_i \cdot 1000}}{V_{.ld_i}} \\ IM \leftarrow YM_{abc} \cdot VM \\ I_{.abcn} \leftarrow AIM \cdot IM + AIL_L \cdot IL \\ ID_{.anbc} \leftarrow Dd_L \cdot I_{.abcn} \\ I_{.ABC} \leftarrow d_{.t} \cdot I_{.abcn} \\ V_{.old} \leftarrow V_{.anbc} \\ Out_1 \leftarrow V_{.anbc} \\ Out \end{vmatrix} \end{vmatrix}$$

FIGURE 11.15 Program for the leading and lagging connections.

Center-Tapped Transformers and Secondaries

$$[A_t] = \begin{bmatrix} 0.0167 & 0 & 0 \\ 0.0167 & 0 & 0 \\ 0 & 0.0333 & 0 \\ -0.0333 & -0.0333 & 0 \end{bmatrix},$$

$$[B_t] = \begin{bmatrix} 0.0118+j0.0118 & -0.0048-j0.0078 & -0.0076-j0.0078 & 0 \\ -0.0048-j0.0078 & -0.0118-j0.0118 & -0.145-j0.0118 & 0 \\ 0.0028 & -0.0028 & -0.1005-j0.0806 & 0 \\ -0.0194-0.0196 & 0.0194+j0.0196 & 0.1227+j0.1002 & 0 \end{bmatrix},$$

$$[d_t] = \begin{bmatrix} 0.0167 & -0.0167 & -0.0167 & 0 \\ 0 & 0 & -0.0333 & 0 \\ 0 & 0 & 0 & 0 \end{bmatrix}$$

The secondary load voltages are

$$[V_{anbc}] = \begin{bmatrix} 117.1/-0.13 \\ 117.0/-0.12 \\ 234.2/-121.1 \\ 230.8/119.4 \end{bmatrix} V$$

The secondary line currents are

$$[I_{abcn}] = \begin{bmatrix} 110.6/-43.3 \\ 133.4/160.9 \\ 53.2/53.7 \\ 17.7/-37.1 \end{bmatrix} A$$

The primary line currents are

$$[I_{ABC}] = \begin{bmatrix} 3.98/-42.9 \\ 1.77/-126.3 \\ 0 \end{bmatrix} A$$

The motor voltage unbalance is computed to be 0.96%.
The motor current unbalance is computed to be 5.47%.
The operating kVA of the lighting and power transformers are
for $i = 1$ to 3

$$kVA_i = \frac{VLG_{ABC_i} \cdot I_{ABC_i}^*}{1000} = \begin{bmatrix} 28.7 \\ 12.8 \\ 0 \end{bmatrix}$$

Lagging Connections $L = 2$:
Applying Equations 11.75 and 11.76, the forward and backward sweep matrices are

$$[A_t] = \begin{bmatrix} 0.0167 & 0 & 0 \\ 0.0167 & 0 & 0 \\ -0.0333 & 0 & -0.0333 \\ 0 & 0 & 0.0333 \end{bmatrix},$$

$$[B_t] = \begin{bmatrix} 0.0118+j0.0118 & -0.0048-j0.0078 & -0.0145-j0.0118 & 0 \\ 0.0048+j0.0078 & -0.0118-j0.0118 & 0.0076+j0.0078 & 0 \\ -0.0194-0.0196 & 0.0194+0.0196 & -0.1227-j0.1002 & 0 \\ 0.0028 & -0.0028 & 0.1005+j0.0806 & 0 \end{bmatrix},$$

$$[d_t] = \begin{bmatrix} 0.0167 & -0.0167 & 0.0167 & 0 \\ 0 & 0 & 0 & 0 \\ 0 & 0 & 0.0333 & 0 \end{bmatrix}.$$

The secondary load voltages are

$$[V_{anbc}] = \begin{bmatrix} 117.4/\underline{-0.82} \\ 117.3/\underline{-0.81} \\ 228.9/\underline{-120.3} \\ 233.8/\underline{120.7} \end{bmatrix} V$$

The secondary line currents are

$$[I_{abcn}] = \begin{bmatrix} 112.9/\underline{-47.1} \\ 126.0/\underline{159.5} \\ 56.3/\underline{61.2} \\ 17.6/\underline{-37.8} \end{bmatrix} A$$

The primary line currents are

$$[I_{ABC}] = \begin{bmatrix} 3.92/\underline{-19.2} \\ 0 \\ 1.88/\underline{61.2} \end{bmatrix} A$$

The motor voltage unbalance is computed to be 1.52%.
The motor current unbalance is computed to be 8.32%.

Center-Tapped Transformers and Secondaries 451

The operating kVA of the lighting and power transformers are

for $i = 1$ to 3,

$$kVA_i = \frac{VLG_{ABC_i} \cdot I_{ABC_i}{}^*}{1000} = \begin{bmatrix} 28.2 \\ 0 \\ 13.5 \end{bmatrix}$$

Note that the voltage and current unbalances for the lagging connection are greater than the voltage and current unbalances for the leading connection.

11.4 FOUR-WIRE SECONDARY

Typically, the combination single-phase and three-phase loads will not be directly connected to the transformer but rather will be connected through a length of open four-wire secondary or a quadruplex cable secondary. This is depicted in Figure 11.16.

The first step in modeling the open four-wire or quadruplex cable secondary is to compute the self and mutual impedances. As always, Carson's equations are used to compute the 4 × 4 primitive impedance matrix. Since the secondary neutral is grounded at both ends, the Kron reduction method is used to eliminate the fourth row and column, which results in the 3 × 3 phase impedance matrix. Chapter 4 gives the details on the application of Carson's equations and the Kron reduction.

The 3 × 3 phase impedance matrix gives the self-impedance of the three line conductors and the mutual impedance between those conductors. The voltage drops on the three line conductors are as follows:

$$v_a = Zs_{aa} \cdot I_a + Zs_{ab} \cdot I_b + Zs_{ac} \cdot I_c$$

$$v_b = Zs_{ba} \cdot I_a + Zs_{bb} \cdot I_b + Zs_{bc} \cdot I_c$$

$$v_c = Zs_{ca} \cdot I_a + Zs_{cb} \cdot I_b + Zs_{cc} \cdot I_c \qquad (11.77)$$

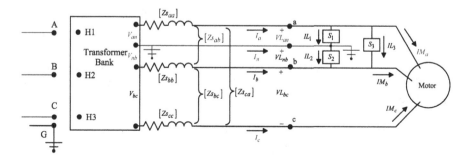

FIGURE 11.16 Four-wire secondary serving combination loads.

The model of the four-wire secondary will again be in terms of the abcd and AB generalized matrices. The first step in developing the model is to write KVL around the three "window" loops and the outside loop in Figure 11.16.

$$V_{an} = VL_{an} + v_a$$

$$V_{nb} = VL_{nb} - v_b$$

$$V_{bc} = VL_{bc} + v_b - v_c$$

$$V_{ca} = VL_{ca} + v_c - v_a \qquad (11.78)$$

Substitute Equations 11.77 into Equations 11.78.

$$\begin{bmatrix} V_{an} \\ V_{nb} \\ V_{bc} \\ V_{ca} \end{bmatrix} = \begin{bmatrix} VL_{an} \\ VL_{nb} \\ VL_{bc} \\ VL_{ca} \end{bmatrix} + \begin{bmatrix} Zs_{aa} & Zs_{ab} & Zs_{ac} \\ -Zs_{ba} & -Zs_{bb} & -Zs_{bc} \\ Zs_{ba} - Zs_{ca} & Zs_{bb} - Zs_{cb} & Zs_{bc} - Zs_{cc} \\ Zs_{ca} - Zs_{aa} & Zs_{cb} - Zs_{ab} & Zs_{cc} - Zs_{ac} \end{bmatrix} \cdot \begin{bmatrix} I_a \\ I_b \\ I_c \end{bmatrix}$$

$$[V_{anbc}] = [VL_{anbc}] + [Zs_{abc}] \cdot [I_{abc}] \qquad (11.79)$$

Equation 11.79 is in the form of

$$[V_{anbc}] = [a_s] \cdot [VL_{anbc}] + [b_s] \cdot [I_{abc}],$$

$$\text{where } [a_s] = \begin{bmatrix} 1 & 0 & 0 & 0 \\ 0 & 1 & 0 & 0 \\ 0 & 0 & 1 & 0 \\ 0 & 0 & 0 & 1 \end{bmatrix},$$

$$[b_s] = \begin{bmatrix} Zs_{aa} & Zs_{ab} & Zs_{ac} \\ -Zs_{ba} & -Zs_{bb} & -Zs_{bc} \\ Zs_{ba} - Zs_{ca} & Zs_{bb} - Zs_{cb} & Zs_{bc} - Zs_{cc} \\ Zs_{ca} - Zs_{aa} & Zs_{cb} - Zs_{ab} & Zs_{cc} - Zs_{ac} \end{bmatrix} \qquad (11.80)$$

Since there are no shunt devices between the transformer and the loads, the currents leaving the transformers are equal to the line currents serving the loads. Therefore,

$$[d_s] = \begin{bmatrix} 1 & 0 & 0 \\ 0 & 1 & 0 \\ 0 & 0 & 1 \end{bmatrix} \qquad (11.81)$$

Center-Tapped Transformers and Secondaries

The matrices for the forward sweep are

$$[A_s] = [a_s],$$

$$[B_s] = [b_s]. \quad (11.82)$$

Example 11.6: The configuration of a quadruplex secondary cable is shown in Figure 11.17.

The phase conductors for the quadruplex cable are 1/0 AA, and the grounded neutral conductor is 1/0 ACSR. The insulation thickness of the conductors is 80 mil.

Determine the phase impedance matrix and the a, b, d, and A and B matrices for the quadruplex cable where the length L is 100 feet.

From Appendix A:

1/0 AA: GMR = 0.0111 ft., Diameter = 0.368 inches, Resistance = 0.97 Ω/mile

1/0 ACSR: = 0.00446 ft., Diameter = 0.398 inches, Resistance = 1.12 Ω/mile

The spacing matrix for this configuration with the GMRs on the diagonal is

$$[D] = \begin{bmatrix} 0.0111 & 0.0440 & 0.0440 & 0.0386 \\ 0.0440 & 0.0111 & 0.0440 & 0.0698 \\ 0.0440 & 0.0440 & 0.0111 & 0.0386 \\ 0.0386 & 0.0698 & 0.0386 & 0.0045 \end{bmatrix}$$

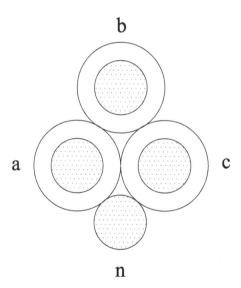

FIGURE 11.17 1/0 quadruplex.

Plugging these spacings into Carson's equations yields the primitive impedance matrix:

$$[zp] = \begin{bmatrix} 1.0653 + j1.5088 & 0.0953 + j1.3417 & 0.0953 + j1.3417 & 0.0953 + j1.3577 \\ 0.0953 + j1.3417 & 1.0653 + j1.5088 & 0.0953 + j1.3417 & 0.0953 + j1.2857 \\ 0.0953 + j1.3417 & 0.0953 + j1.3417 & 1.0653 + j1.5088 & 0.0953 + j1.3577 \\ 0.0953 + j1.3577 & 0.0953 + j1.2857 & 0.0953 + j1.3577 & 1.2153 + j1.6195 \end{bmatrix}$$

The Kron reduction yields the phase impedance matrix:

$$[z_{abc}] = \begin{bmatrix} 1.5068 + j0.7076 & 0.5106 + j0.5811 & 0.5368 + j0.5405 \\ 0.5106 + j0.5811 & 1.4558 + j0.7868 & 0.5106 + j0.5811 \\ 0.5368 + j0.5405 & 0.5106 + j0.5811 & 1.5068 + j0.7076 \end{bmatrix}$$

The matrices for 100 feet of this quadruplex are as follows:

$$Zs = z_{abc} \cdot \frac{L}{5280} = \begin{bmatrix} 0.0258 + j0.0134 & 0.0097 + j0.0110 & 0.0102 + j0.0102 \\ 0.0097 + j0.0110 & 0.0276 + j0.0149 & 0.0097 + j0.0110 \\ 0.0102 + j0.0102 & 0.0097 + j0.0110 & 0.0285 + j0.0134 \end{bmatrix}$$

$$[a_q] = [A_q] = \begin{bmatrix} 1 & 0 & 0 & 0 \\ 0 & 1 & 0 & 0 \\ 0 & 0 & 1 & 0 \\ 0 & 0 & 0 & 1 \end{bmatrix}$$

$$[b_q] = [B_q] = \begin{bmatrix} 0.0285 + j0.01314 & 0.0097 + j0.0110 & 0.0102 + j0.0102 \\ -0.0097 - j0.0110 & -0.0276 - j0.0149 & -0.0097 - j0.0110 \\ -0.0005 + j0.0008 & 0.0179 + j0.0039 & -0.0189 - j0.0024 \\ -0.0184 - j0.0032 & 0 & 0.0184 + j0.0032 \end{bmatrix}$$

$$[d_q] = \begin{bmatrix} 1 & 0 & 0 \\ 0 & 1 & 0 \\ 0 & 0 & 1 \end{bmatrix}$$

11.5 PUTTING IT ALL TOGETHER

Shown in Figure 11.18 is the IEEE 4 node test feeder [2], which will be used to study each of the three-phase wye – delta (closed and open) transformer connections developed in this chapter.

11.5.1 UNGROUNDED WYE – DELTA CONNECTION

The IEEE 4 node text feeder consists of an infinite 12.47 source connected to a 5-mile-long primary overhead line serving a three-phase transformer bank. The

Center-Tapped Transformers and Secondaries

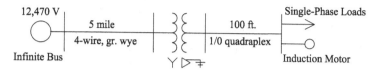

FIGURE 11.18 IEEE 4 node test feeder.

secondary is 100 feet long and is a four-wire quadruplex cable serving single-phase 120- and 240-volt loads and a three-phase induction motor. With the known source voltage, a complete analysis of the feeder is desired. This will include the voltages at all nodes and the currents flowing on the primary and secondary lines.

Based upon the techniques presented in this text, the steps in the analysis are as follows:

1. Determine the forward and backward sweep matrices [A], [B], and [d] for the primary and secondary lines and the transformer bank.
2. Mold the induction motor using the equivalent motor admittance matrix. The matrix [YM_{abc}] should be computed based upon the slip.

The matrices for the overhead line are developed in Chapter 4. The matrices for the transformer bank and quadruplex cable have been developed in this chapter.

Example 11.7: The system of Figure 11.18 is to be analyzed with the following data.

A 5-mile-long overhead three-phase line is between Nodes 1 and 2. This overhead line is the same as the line described in Problem 11.1. The generalized matrices for the primary line are computed to be

$$[a_p] = [d_p] = [A_p] = \begin{bmatrix} 1 & 0 & 0 \\ 0 & 1 & 0 \\ 0 & 0 & 1 \end{bmatrix} [c_p] = \begin{bmatrix} 0 & 0 & 0 \\ 0 & 0 & 0 \\ 0 & 0 & 0 \end{bmatrix},$$

$$[b_p] = [B_p] = \begin{bmatrix} 1.6873 + j5.2391 & 0.7674 + j1.9247 & 0.7797 + j2.5084 \\ 0.7674 + j1.9247 & 1.7069 + j5.1742 & 0.7900 + j2.1182 \\ 0.7797 + j2.5084 & 0.7900 + j2.1182 & 1.7326 + j5.0897 \end{bmatrix}$$

The transformer bank between Nodes 2 and 3 is an ungrounded wye – delta and is the same as in Example 11.4 where the parameter matrices are computed as follows:

$$[A_t] = \begin{bmatrix} 0.0167 & 0 & 0 \\ 0.0167 & 0 & 0 \\ 0 & 0.0333 & 0 \\ 0 & 0 & 0.0333 \end{bmatrix}$$

$$[B_t] = \begin{bmatrix} 0.0081+j0.0085 & -0.0012-j0.0046 & 0.0035+j0.002 \\ 0.0012+j0.0046 & -0.0081-j0.0085 & -0.0035-j0.002 \\ -0.0154-j0.0134 & 0.0154+j0.0134 & -0.0461-j0.0403 \\ -0.0154-j0.0134 & 0.0154+j0.0134 & 0.0461+j0.0403 \end{bmatrix}$$

$$[d_t] = \begin{bmatrix} 0.0111 & -0.0111 & 0 \\ -0.0056 & 0.0056 & -0.0167 \\ -0.0056 & 0.0056 & 0.0167 \end{bmatrix}$$

The 100-foot quadruplex secondary is the same as in Example 11.6 where the parameter matrices are computed as follows:

$$[A_q] = \begin{bmatrix} 1 & 0 & 0 & 0 \\ 0 & 1 & 0 & 0 \\ 0 & 0 & 1 & 0 \\ 0 & 0 & 0 & 1 \end{bmatrix}$$

$$[B_q] = \begin{bmatrix} 0.0285+j0.01314 & 0.0097+j0.0110 & 0.0102+j0.0102 \\ -0.0097-j0.0110 & -0.0276-j0.0149 & -0.0097-j0.0110 \\ -0.0005+j0.0008 & 0.0179+j0.0039 & -0.0189-j0.0024 \\ -0.0184-j0.0032 & 0 & 0.01840+j0.0032 \end{bmatrix}$$

$$[d_q] = \begin{bmatrix} 1 & 0 & 0 \\ 0 & 1 & 0 \\ 0 & 0 & 1 \end{bmatrix}$$

The single-phase loads are at Node 4:

$$S_1 = 3.0\,\text{kVA},\ 0.95\,\text{lag},\ 120\,\text{volts}$$

$$S_2 = 5.0\,\text{kVA},\ 0.90\,\text{lag},\ 120\,\text{volts}$$

$$S_3 = 8.0\,\text{kVA},\ 0.85\,\text{lag},\ 240\,\text{volts}$$

The three-phase induction motor is the same as in Example 11.4. With a slip of 0.035, the shunt admittance matrix was computed to be

$$[YM_{abc}] = \begin{bmatrix} 0.7543-j0.4074 & -0.1000-j0.0923 & 0.3347+j0.4997 \\ 0.3547+j0.4997 & 0.7453-j0.4074 & -0.1000-j0.0923 \\ -0.1000-j0.0923 & 0.3547+0.4997 & 0.7453-j0.4074 \end{bmatrix} S$$

The Mathcad program to perform the analysis is shown in Figure 11.19.

Center-Tapped Transformers and Secondaries

$$X := \begin{array}{|l} I_{abc} \leftarrow \text{istart} \\ ID_{anbc} \leftarrow Dd \cdot I_{abc} \\ I_{ABC} \leftarrow \text{istart} \\ V_{old} \leftarrow \text{vstart} \\ \text{for } n \in 1..10 \\ \quad \begin{array}{|l} VLG_{ABC} \leftarrow A_{.p} \cdot ELG_{ABC} - B_{.p} \cdot I_{ABC} \\ VTLG_{ABC} \leftarrow VLG_{ABC} - ZT_{.0} \cdot I_{ABC} \\ VTLL_{ABC} \leftarrow Dv \cdot VTLG_{ABC} \\ XX \leftarrow n_{.t} \cdot ZD_{.anbc} \cdot ID_{.anbc} \\ VXLL_{ABC} \leftarrow \begin{pmatrix} XX \\ VTLL_{ABC_2} \\ VTLL_{ABC_3} \end{pmatrix} \\ VTLN_{ABC} \leftarrow Dx \, VXLL_{ABC} \\ VLN_{ABC} \leftarrow VTLN_{ABC} + ZT_{.0} \cdot I_{ABC} \\ V_{.anbc} \leftarrow A_{.t} \cdot VLN_{ABC} - B_{.t} \cdot I_{abc} \\ VL_{.anbc} \leftarrow A_{.q} \cdot V_{.anbc} - B_{.q} \cdot I_{abc} \\ \text{for } j \in 1..4 \\ \quad Error_j \leftarrow \left| \left| V_{.anbc_j} \right| - \left| V_{.old_j} \right| \right| \\ \text{break if max(Error)} < \text{Tol} \\ VL \leftarrow \begin{pmatrix} VL_{.anbc_1} \\ VL_{.anbc_2} \\ VL_{.anbc_1} + VL_{.anbc_2} \end{pmatrix} \\ VM \leftarrow \begin{pmatrix} VL_3 \\ VL_{.anbc_3} \\ VL_{.anbc_4} \end{pmatrix} \\ \text{for } i \in 1..3 \\ \quad IL_i \leftarrow \dfrac{SL_i \cdot 1000}{VL_i} \\ IM \leftarrow YM_{abc} \cdot VM \\ I_{.abc} \leftarrow DI \cdot IL + IM \\ ID_{.anbc} \leftarrow Dd \cdot I_{abc} \\ I_{.ABC} \leftarrow d_{.t} \cdot I_{abc} \\ V_{.old} \leftarrow V_{.anbc} \\ Out_i \leftarrow V_{.anbc} \\ Out \end{array} \end{array}$$

FIGURE 11.19 Program for ungrounded wye-delta connection.

The source at Node 1 is an ideal source of 12.47 kV line to line. The specified source line-to-ground voltages are

$$[ELG_{ABC}] = \begin{bmatrix} 7200\underline{/0} \\ 7200\underline{/-120} \\ 7200\underline{/120} \end{bmatrix} V$$

After the forward and backward sweep matrices are computed for each of the components, a Mathcad program is used to analyze the system. The Mathcad program of Example 11.4 is modified to include the primary line and the secondary quadruplex secondary voltage drops. After eight iterations, the voltage errors are less than the desired tolerance of 0.00001. The final motor and load voltages are

$$[VM_{abc}] = \begin{bmatrix} 230.5/-0.1 \\ 232.8/-119.5 \\ 233.6/119.7 \end{bmatrix} V,$$

$$[VL] = \begin{bmatrix} 115.9/0.2 \\ 114.5/-0.4 \\ 230.5/-0.1 \end{bmatrix} V$$

The final motor and load currents are

$$[IM_{abc}] = \begin{bmatrix} 53.8/-66.0 \\ 55.8/178.8 \\ 58.7/54.7 \end{bmatrix} A,$$

$$[IL_{abc}] = \begin{bmatrix} 25.9/-18.0 \\ 43.7/-26.2 \\ 34.7/-31.9 \end{bmatrix} A$$

The final transformer terminal line-to-neutral voltages and currents are

$$[VLN_{ABC}] = \begin{bmatrix} 7135.6/-0.1 \\ 7226.6/-120.4 \\ 7224.5/120.4 \end{bmatrix} V,$$

$$[I_{ABC}] = \begin{bmatrix} 2.5643/-29.7 \\ 1.6869/-174.4 \\ 1.5355/110.9 \end{bmatrix} A$$

The percent motor voltage unbalance is as follows:

$$V_{avg} = \sum_{k=1}^{3} |VM_k| = 232.3$$

$$dV = \||VM_i| - V_{avg}\| = \begin{bmatrix} 1.8345 \\ 0.5258 \\ 1.3087 \end{bmatrix}$$

Center-Tapped Transformers and Secondaries

$$V_{unbalance} = \frac{\max(dV)}{V_{avg}} \cdot 100 = \frac{1.8345}{232.3} \cdot 100 = 0.7897\%$$

The percent motor current unbalance is as follows:

$$I_{avg} = \sum_{k=1}^{3} |VM_k| = 56.1305$$

$$dI = \left\| |IM_i| - I_{avg} \right\| = \begin{bmatrix} 2.3035 \\ 0.2867 \\ 2.5902 \end{bmatrix}$$

$$I_{unbalance} = \frac{\max(dI)}{I_{avg}} \cdot 100 = \frac{2.5902}{56.1305} \cdot 100 = 4.6146\%$$

The operating kVA of each of the transformers is as follows:

for : $i = 1, 2, 3$

$$kVA = \frac{VLN_{ABC_i} \cdot (I_{ABC_i})^*}{1000} = \begin{bmatrix} 18.3\underline{/29.6} \\ 12.2\underline{/54.0} \\ 11.1\underline{/9.5} \end{bmatrix} kVA$$

If the input or output power of the motor had been specified instead of the slip, after each convergence of the modified ladder method, a new value of slip would have to be computed for the motor. This is a double iterative process that works. The first step would be to use the initial motor voltages after the first forward sweep to compute the necessary slip. Once the slip has been determined, the backward sweep begins. The forward/backward sweeps would continue until convergence. The converged motor voltages would be used to compute the new required slip. Again the forward/backward sweeps are used. This process continues until both the specified motor power and the specified source voltages are matched.

Example 11.7 is intended to demonstrate how the ladder forward/backward sweep iterative method will work. The example used an ungrounded wye – delta transformer bank.

11.5.2 Open Wye – Delta Connections

The same routine used in Example 11.7 can be used for the leading and lagging open wye – open delta connections by using the *A*, *B*, and *d* matrices for each connection. Also, the terminal line-to-neutral voltages are computed directly from the transformer bank terminal line-to-ground voltages. That eliminates the method of computing the line-to-neutral voltages in Example 11.7.

The modified Mathcad program is shown in Figure 11.20.

Note in Figure 11.20 that $L = 1$ for the leading connection and $L = 2$ for the lagging connection.

$$X := \begin{vmatrix} I_{abcn} \leftarrow vstart \\ ID_{anbc} \leftarrow Dd_L \cdot I_{abcn} \\ I_{ABC} \leftarrow istart \\ V_{old} \leftarrow vstart \\ \text{for } n \in 1..20 \\ \quad \begin{vmatrix} VLG_{ABC} \leftarrow A_p \cdot ELG_{ABC} - B_p \cdot I_{ABC} \\ VLN_{ABC} \leftarrow VLG_{ABC} - Z0G \cdot I_{ABC} \\ V_{anbc} \leftarrow A_t \cdot VLG_{ABC} - B_t \cdot I_{abcn} \\ VL_{anbc} \leftarrow A_q \cdot V_{anbc} - B_q \cdot I_{abcn} \\ \text{for } j \in 1..4 \\ \quad Error_j \leftarrow \left| |V_{anbc_j}| - |V_{old_j}| \right| \\ \text{break if } \max(Error) < Tol \\ VL \leftarrow \begin{pmatrix} VL_{anbc_1} \\ VL_{anbc_2} \\ VL_{anbc_1} + VL_{anbc_2} \end{pmatrix} \\ VM \leftarrow \begin{pmatrix} VL_3 \\ V_{anbc_3} \\ V_{anbc_4} \end{pmatrix} \\ \text{for } i \in 1..3 \\ \quad IL_i \leftarrow \dfrac{SL_i \cdot 1000}{VL_i} \\ IM \leftarrow YM_{abc} \cdot VM \\ I_{abcn} \leftarrow AIM \cdot IM + AIL_L \cdot IL \\ ID_{anbc} \leftarrow Dd_L \cdot I_{abcn} \\ I_{ABC} \leftarrow d_t \cdot I_{abcn} \\ V_{old} \leftarrow V_{anbc} \end{vmatrix} \\ Out_1 \leftarrow V_{anbc} \\ Out_2 \leftarrow VL_{anbc} \\ Out_3 \leftarrow IL \\ Out_4 \leftarrow IM \\ Out_5 \leftarrow I_{abcn} \\ Out_6 \leftarrow VLN_{ABC} \\ Out_7 \leftarrow VLG_{ABC} \\ Out_8 \leftarrow I_{ABC} \\ Out_9 \leftarrow n \\ Out_{10} \leftarrow VM \\ Out \end{vmatrix}$$

FIGURE 11.20 Program for open wye – open delta connections.

The only matrix changes from the closed wye – delta connection are the matrices associated with the transformer connection.

11.5.3 Comparisons of Voltage and Current Unbalances

It is interesting to compare the induction motor voltage and current unbalances for the closed and open wye – delta connections where for all connections the lighting transformer is rated 25 kVA and the power transformer is rated 10 kVA.

Center-Tapped Transformers and Secondaries 461

Example 11.8: Compute the node voltages and line currents for the leading and lagging open wye – delta connections. The grounding impedance is 5 ohms.

Leading Connection $L = 1$:

$$[A_t] = \begin{bmatrix} 0.0167 & 0 & 0 \\ 0.0167 & 0 & 0 \\ 0 & 0.0333 & 0 \\ -0.0333 & -0.0333 & 0 \end{bmatrix}$$

$$[B_t] = \begin{bmatrix} 0.0118+j0.0118 & -0.0048-j0.0078 & -0.0076-j0.0078 & 0 \\ 0.0048+j0.0078 & -0.0118-j0.0118 & -0.0145-j0.0118 & 0 \\ 0.0028 & -0.0028 & -0.1005 & 0 \\ -0.0194-0.0196 & 0.0194+j0.0196 & 0.1227+j0.1002 & 0 \end{bmatrix}$$

$$[d_t] = \begin{bmatrix} 0.0167 & -0.0167 & -0.0167 & 0 \\ 0 & 0 & -0.0333 & 0 \\ 0 & 0 & 0 & 0 \end{bmatrix}$$

After six iterations, the voltage errors are less than the desired tolerance of 0.00001. The final motor and load voltages are

$$[VM_{abc}] = \begin{bmatrix} 229.3/\underline{0.1} \\ 234.2/\underline{-121.2} \\ 230.5/\underline{119.2} \end{bmatrix} V,$$

$$[VL] = \begin{bmatrix} 115.4/\underline{0.4} \\ 113.8/\underline{-0.1} \\ 229.3/\underline{0.1} \end{bmatrix} V$$

The final motor and load currents are

$$[IM_{abc}] = \begin{bmatrix} 52.4/\underline{-62.4} \\ 61.4/\underline{175.3} \\ 53.9/\underline{52.7} \end{bmatrix} A,$$

$$[IL_{abc}] = \begin{bmatrix} 26.0/\underline{-17.8} \\ 43.9/\underline{-26.0} \\ 34.9/\underline{-31.7} \end{bmatrix} A$$

The final transformer terminal line-to-neutral voltages and currents are

$$[VLN_{ABC}] = \begin{bmatrix} 7059.2/\underline{-0.3} \\ 7189.7/\underline{-120.3} \\ 7226.1/\underline{120.1} \end{bmatrix} V,$$

$$[I_{ABC}] = \begin{bmatrix} 3.9488/-41.9 \\ 1.7965/-127.33 \\ 0 \end{bmatrix} V$$

The percent motor voltage unbalance is as follows:

$$V_{avg} = \sum_{k=1}^{3} |VM_k| = 231.3 \, V$$

$$dV = \left\| |VM_i| - V_{avg} \right\| = \begin{bmatrix} 2.0688 \\ 2.9048 \\ 0.8371 \end{bmatrix} \Omega$$

$$V_{unbalance} = \frac{\max(dV)}{V_{avg}} \cdot 100 = \frac{2.9048}{231.3} \cdot 100 = 1.2557\%$$

The percent motor current unbalance is as follows:

$$I_{avg} = \frac{1}{3} \cdot \sum_{k=1}^{3} |IM_k| = 55.9 \, A$$

$$dI_i = \left\| |IM_i| - I_{avb} \right\| = \begin{bmatrix} 3.4737 \\ 5.4682 \\ 1.9945 \end{bmatrix} A$$

$$I_{unbal} = \frac{\max(dI)}{I_{avg}} \cdot 100 = 9.7840\%$$

The operating kVA of each of the transformers is as follows:

for : $i = 1, 2, 3$

$$kVA = \frac{VLN_{ABC_i} \cdot (I_{ABC_i})^*}{1000} = \begin{bmatrix} 27.9/41.6 \\ 12.9/73.0 \\ 0 \end{bmatrix} kVA$$

Lagging Connection $L = 2$:

$$[A_t] = \begin{bmatrix} 0.0167 & 0 & 0 \\ 0.0167 & 0 & 0 \\ -0.0333 & 0 & -0.0333 \\ 0 & 0 & 0.0333 \end{bmatrix}$$

Center-Tapped Transformers and Secondaries

$$[B_t] = \begin{bmatrix} 0.0118 + j0.0118 & -0.0048 - j0.0078 & 0.0145 + j0.0188 & 0 \\ 0.0048 + j0.0078 & -0.0118 - j0.0118 & 0.0076 + j0.0078 & 0 \\ -0.0194 - 0.0196 & 0.0194 + j0.0196 & -0.1227 - j0.1002 & 0 \\ 0.0028 & -0.0028 & 0.1005 + j0.0806 & 0 \end{bmatrix}$$

$$[d_t] = \begin{bmatrix} 0.0167 & -0.0167 & -0.0167 & 0 \\ 0 & 0 & 0 & 0 \\ 0 & 0 & 0.0333 & 0 \end{bmatrix}$$

After six iterations, the voltage errors are less than the desired tolerance of 0.00001. The final motor and load voltages are

$$[VM_{abc}] = \begin{bmatrix} 230.4 / -0.6 \\ 227.7 / -120.4 \\ 233.3 / 120.7 \end{bmatrix} V,$$

$$[VL] = \begin{bmatrix} 115.9 / -0.3 \\ 114.5 / -0.9 \\ 230.4 / -0.6 \end{bmatrix} V$$

The final motor and load currents are

$$[IM_{abc}] = \begin{bmatrix} 58.0 / -68.2 \\ 52.8 / 173.7 \\ 56.6 / 61.6 \end{bmatrix} V,$$

$$[IL_{abc}] = \begin{bmatrix} 25.9 / -18,5 \\ 43.7 / -26.7 \\ 34.7 / -32.4 \end{bmatrix} V$$

The final transformer terminal line-to-neutral voltages and currents are

$$[VLN_{ABC}] = \begin{bmatrix} 7086.9 / -0.9 \\ 7222.7 / -120.1 \\ 7195.7 / 120.2 \end{bmatrix} V,$$

$$[I_{ABC}] = \begin{bmatrix} 3.9237 / -18.6 \\ 0 \\ 1.8851 / 61.6 \end{bmatrix} A$$

The percent motor unbalance is as follows:

$$V_{avg} = \sum_{k=1}^{3} |VM_k| = 230.4 \, V$$

$$dV = \left| |VM_i| - V_{avg} \right| = \begin{bmatrix} 0.0550 \\ 2.7646 \\ 2.8196 \end{bmatrix} V$$

$$V_{unbalance} = \frac{\max(dV)}{V_{avg}} \cdot 100 = \frac{2.8196}{230.4} \cdot 100 = 1.2236\%$$

The percent motor unbalance is as follows:

$$I_{avg} = \frac{1}{3} \cdot \sum_{k=1}^{3} |IM_k| = 55.8 \, A$$

$$dI_i = \left| |IM_i| - I_{avb} \right| = \begin{bmatrix} 2.2390 \\ 2.9952 \\ 0.7562 \end{bmatrix} A$$

$$I_{unbal} = \frac{\max(dI)}{I_{avg}} \cdot 100 = 5.3682\%$$

The operating kVA of each of the transformers is as follows:

for : $i = 1, 2, 3$

$$kVA = \frac{VLN_{ABC_i} \cdot (I_{ABC_i})^*}{1000} = \begin{bmatrix} 27.8/\underline{17.7} \\ 0 \\ 13.5645/\underline{58.6} \end{bmatrix} kVA$$

Table 11.1 demonstrates why the lagging connection should be selected if an open connection is going to be used. It has also been shown that the 25 kVA lighting transformer and the 10 kVA power transformer in the open connection lead to overloading for both connections. If the open connection is changed to the lighting transformer being 37.5 kVA and the power transformer 15 kVA, the overloading is avoided but the voltage and current unbalances increase. The comparison is shown in Table 11.2.

11.6 SUMMARY

This chapter has developed the models for the single-phase center-tapped transformer and for the three-phase banks using the center-tapped transformer. Examples have demonstrated how the models can be analyzed. The most important feature is demonstrated in Example 11.7 where not only is the transformer bank modeled but also the primary and secondary lines, along with the admittance matrix model of the induction motor.

The primary purpose of this chapter is to bring the total concept of distribution analysis to the forefront. Every element of a distribution feeder can be modeled using

Center-Tapped Transformers and Secondaries

TABLE 11.1
Voltage and Current Motor Unbalances

Connection	Voltage Unbalance %	Current Unbalance %
Closed Wye – Delta	0.3382	2.2205
Leading Open Wye – Delta	1.2557	9.7840
Lagging Open Wye – Delta	1.2236	5.3684

TABLE 11.2
Voltage and Current Motor Unbalances

Connection	Voltage Unbalance %	Current Unbalance %
Closed Wye – Delta	0.3382	2.2205
Leading Open Wye – Delta	1.3132	9.4592
Lagging Open Wye – Delta	1.2559	3.5248

the generalized matrices. When all of the matrices are known, the modified ladder forward/backward sweep iterative routine is used to compute all node voltages and line currents in the system. As demonstrated in the examples, a Mathcad program can be developed to do the analyses of a simple radial feeder with no laterals. For complex systems, a commercial program, such as WindMil, should be used.

PROBLEMS

11.1 A 25 kVA, centered-tapped, single-phase transformer is rated:

25 kVA, 7200-$240/120$, $R_A = 0.012$ pu, $X_A = 0.017$ pu.

The transformer serves the following constant PQ loads:

5 kVA, 0.95 PF lag at nominal 120 volts
8 kVA, 0.90 PF lag at nominal 120 volts
10 kVA, 0.85 PF lag at nominal 120 volts

Determine the following when the primary transformer voltage is 6,900 volts:
(a) The forward and backward matrices $[A_t]$, $[B_t]$ and $[d_t]$
(b) Load voltages, secondary transformer currents, and load currents
(c) Primary current

11.2 The transformer of Problem 11.1 is connected to the same loads through 200 feet of three-wire, open-wire secondary. The conductors are 1/0 AA, and the spacings between conductors are

$D_{12} = 6$ inches, $D_{23} = 6$ inches, $D_{13} = 13$ inches.

(a) Determine the secondary impedances and matrices.
(b) The primary source voltage is 7,350 volts, determine the load voltages.
(c) Determine the primary, secondary, and load currents.

11.3 Combination single-phase loads and a three-phase induction motor are served from an ungrounded wye – delta transformer bank, as shown in Figure 11.9. The single-phase loads are to be modeled as constant impedance:

$S_1 = 15$ kVA, 0.95 lag, $S_2 = 10$ kVA, 0.90 lag, $S_3 = 25$ kVA, 0.85 lag.

The three-phase induction motor has the following parameters:

25 kVA, 240 volts
$R_s = 0.035$ pu, $R_r = 0.0375$ pu, $X_s = X_r = 0.10$ pu, $X_m = 3.0$ pu
$Slip = 0.035$

The transformers are rated:

Lighting transformer: 50 kVA, 7,200 – 240/120 volts, $Z = 0.011 + j0.018$ pu
Power transformers: 25 kVA, 7,200 – 240 volts, $Z = 0.012 + j0.017$ pu

The loads are served through 100 feet of quadruplex consisting of three 2/0 AA insulated conductors and one 2/0 ACSR conductor. The insulation thickness is 80 mil.

The transformer bank is connected to a balanced 12.47 kV (line-to-line) source.

Determine the following:
(a) The forward and backward sweep matrices for the transformer connection
(b) The forward and backward sweep matrices for the quadruplex
(c) The constant impedance values of the three-phase loads
(d) The single-phase load voltages
(e) The line-to-line motor voltages
(f) The primary and secondary line currents

11.4 Repeat Problem 11.3 if the loads are being served from a "leading" open wye – open delta transformer bank. The transformers are rated:

Lighting transformer: 75 kVA, 7200 – 240.120 volt, $Z = 0.010 + j0.021$ pu
Power transformer: 37.5 kVA, 7200 – 240 volt, $Z = 0.013 + j0.019$ pu

11.5 Repeat Problem 11.3, only rather than the slip being specified, the input real power to the motor is to be 20 kW. This will require a double iterative process.
(a) Determine the slip.
(b) Determine the same voltages and currents as in Problem 11.4.
(c) Determine the input kVA and power factor of the motor.

Center-Tapped Transformers and Secondaries 467

WINDMIL ASSIGNMENT

Use the system that was developed in Chapter 10.

1. Add the single-phase, center-tapped transformer from Example 11.1 to Node 8. The transformer serves single-phase loads through 100 feet of triplex as defined in Example 11.2. The loads are as follows:
   a. $S1$: 10 kW at 95% power factor lagging
   b. $S2$: 15 kW at 90% power factor lagging
   c. $S12$: 25 kW at 85% power factor lagging
2. Add the transformer, secondary, single-phase, and motor loads from Example 5 to Node 9.
   a. Specify a slip of 3.5% for the motor.
3. Add an open wye – open delta transformer bank to Node 11. The transformers are as follows:
   a. Lighting: 50 kVA, 7,200 – 120/240 center tap, $Z = 2.11$, $X/R = 1.6364$
   b. Power: 25 kVA, 7,200, 240, $Z = 2.08$, $X/R = 1.4167$
   c. The lighting transformer is connected to phase b
   d. The power transformer is connected to phase c
   e. The loads are served by 150 feet of 1/0 quadruplex, as defined in part 1 above
   f. The motor is the same as part 2 above. The motor is to operate at 20 kW
4. At Node 10, add a three-phase, delta – delta transformer
   a. $kVA = 500$
   b. Voltage = 12.47 kV line to line, –0.480 kV line to line
   c. $Z = 1.28$ %, $X/R = 1.818$
   d. Connect a "swing" generator to the transformer
      i. Supply 350 kW
      ii. Hold voltage at 1.02 per unit
5. Make whatever changes are necessary to satisfy all of the following conditions:
   a. Phase power factor at the source to be not less than 95% lagging
   b. The load voltages must not be less than:
      i. Node: 120 volts
      ii. Transformer secondary terminal: 114 volts
   c. The voltage unbalance at either of the motors to not exceed 3 %
   d. Regulator must not be at tap 16 on any phase

REFERENCES

1. Gonen, T, *Electric Power Distribution System Engineering*, CRC Press, Boca Raton, FL, 2007.
2. *Radial Test Feeders*, IEEE Distribution System Analysis Subcommittee, https://site.ieee.org/pes-testfeeders/resources/

Appendix A

Conductor Data

Size	Stranding	Material	DIAM	GMR	RES	Capacity
			Inches	Feet	Ω/Mile	Amps
1		ACSR	0.355	0.00418	1.38	200
1	7 STRD	Copper	0.328	0.00992	0.765	270
1	CLASS A	AA	0.328	0.00991	1.224	177
2	6/1	ACSR	0.316	0.00418	1.69	180
2	7 STRD	Copper	0.292	0.00883	0.964	230
2	7/1	ACSR	0.325	0.00504	1.65	180
2	AWG SLD	Copper	0.258	0.00836	0.945	220
2	CLASS A	AA	0.292	0.00883	1.541	156
3	6/1	ACSR	0.281	0.0043	2.07	160
3	AWG SLD	Copper	0.229	0.00745	1.192	190
4	6/1	ACSR	0.25	0.00437	2.57	140
4	7/1	ACSR	0.257	0.00452	2.55	140
4	AWG SLD	Copper	0.204	0.00663	1.503	170
4	CLASS A	AA	0.232	0.007	2.453	90
5	6/1	ACSR	0.223	0.00416	3.18	120
5	AWG SLD	Copper	0.1819	0.0059	1.895	140
6	6/1	ACSR	0.198	0.00394	3.98	100
6	AWG SLD	Copper	0.162	0.00526	2.39	120
6	CLASS A	AA	0.184	0.00555	3.903	65
7	AWG SLD	Copper	0.1443	0.00468	3.01	110
8	AWG SLD	Copper	0.1285	0.00416	3.8	90
9	AWG SLD	Copper	0.1144	0.00371	4.6758	80
10	AWG SLD	Copper	0.1019	0.00330	5.9026	75
12	AWG SLD	Copper	0.0808	0.00262	9.3747	40
14	AWG SLD	Copper	0.0641	0.00208	14.8722	20
16	AWG SLD	Copper	0.0508	0.00164	23.7262	10
18	AWG SLD	Copper	0.0403	0.00130	37.6726	5
19	AWG SLD	Copper	0.0359	0.00116	47.5103	4
20	AWG SLD	Copper	0.032	0.00103	59.684	3
22	AWG SLD	Copper	0.0253	0.00082	95.4835	2
24	AWG SLD	Copper	0.0201	0.00065	151.616	1
1/0		ACSR	0.398	0.00446	1.12	230
1/0	7 STRD	Copper	0.368	0.01113	0.607	310
1/0	CLASS A	AA	0.368	0.0111	0.97	202
2/0		ACSR	0.447	0.0051	0.895	270
2/0	7 STRD	Copper	0.414	0.01252	0.481	360
2/0	CLASS A	AA	0.414	0.0125	0.769	230
3/0	12 STRD	Copper	0.492	0.01559	0.382	420
3/0	6/1	ACSR	0.502	0.006	0.723	300
3/0	7 STRD	Copper	0.464	0.01404	0.382	420
3/0	CLASS A	AA	0.464	0.014	0.611	263
3/8	INCH STE	Steel	0.375	0.00001	4.3	150
4/0	12 STRD	Copper	0.552	0.0175	0.303	490
4/0	19 STRD	Copper	0.528	0.01668	0.303	480
4/0	6/1	ACSR	0.563	0.00814	0.592	340

Appendix A

Size	Stranding	Material	DIAM Inches	GMR Feet	RES Ω/Mile	Capacity Amps
4/0	7 STRD	Copper	0.522	0.01579	0.303	480
4/0	CLASS A	AA	0.522	0.0158	0.484	299
250,000	12 STRD	Copper	0.6	0.01902	0.257	540
250,000	19 STRD	Copper	0.574	0.01813	0.257	540
250,000	CON LAY	AA	0.567	0.0171	0.41	329
266,800	26/7	ACSR	0.642	0.0217	0.385	460
266,800	CLASS A	AA	0.586	0.0177	0.384	320
300,000	12 STRD	Copper	0.657	0.0208	0.215	610
300,000	19 STRD	Copper	0.629	0.01987	0.215	610
300,000	26/7	ACSR	0.68	0.023	0.342	490
300,000	30/7	ACSR	0.7	0.0241	0.342	500
300,000	CON LAY	AA	0.629	0.0198	0.342	350
336,400	26/7	ACSR	0.721	0.0244	0.306	530
336,400	30/7	ACSR	0.741	0.0255	0.306	530
336,400	CLASS A	AA	0.666	0.021	0.305	410
350,000	12 STRD	Copper	0.71	0.0225	0.1845	670
350,000	19 STRD	Copper	0.679	0.0214	0.1845	670
350,000	CON LAY	AA	0.679	0.0214	0.294	399
397,500	26/7	ACSR	0.783	0.0265	0.259	590
397,500	30/7	ACSR	0.806	0.0278	0.259	600
397,500	CLASS A	AA	0.724	0.0228	0.258	440
400,000	19 STRD	Copper	0.726	0.0229	0.1619	730
450,000	19 STRD	Copper	0.77	0.0243	0.1443	780
450,000	CON LAG	AA	0.77	0.0243	0.229	450
477,000	26/7	ACSR	0.858	0.029	0.216	670
477,000	30/7	ACSR	0.883	0.0304	0.216	670
477,000	CLASS A	AA	0.795	0.0254	0.216	510
500,000	19 STRD	Copper	0.811	0.0256	0.1303	840
500,000	37 STRD	Copper	0.814	0.026	0.1303	840
500,000	CON LAY	AA	0.813	0.026	0.206	483
556,500	26/7	ACSR	0.927	0.0313	0.1859	730
556,500	30/7	ACSR	0.953	0.0328	0.1859	730
556,500	CLASS A	AA	0.858	0.0275	0.186	560
600,000	37 STRD	Copper	0.891	0.0285	0.1095	940
600,000	CON LAY	AA	0.891	0.0285	0.172	520
605,000	26/7	ACSR	0.966	0.0327	0.172	760
605,000	54/7	ACSR	0.953	0.0321	0.1775	750
636,000	27/7	ACSR	0.99	0.0335	0.1618	780
636,000	30/19	ACSR	1.019	0.0351	0.1618	780
636,000	54/7	ACSR	0.977	0.0329	0.1688	770
636,000	CLASS A	AA	0.918	0.0294	0.163	620
666,600	54/7	ACSR	1	0.0337	0.1601	800
700,000	37 STRD	Copper	0.963	0.0308	0.0947	1040
700,000	CON LAY	AA	0.963	0.0308	0.148	580
715,500	26/7	ACSR	1.051	0.0355	0.1442	840
715,500	30/19	ACSR	1.081	0.0372	0.1442	840
715,500	54/7	ACSR	1.036	0.0349	0.1482	830
715,500	CLASS A	AA	0.974	0.0312	0.145	680
750,000	37 STRD	AA	0.997	0.0319	0.0888	1090
750,000	CON LAY	AA	0.997	0.0319	0.139	602
795,000	26/7	ACSR	1.108	0.0375	0.1288	900

Appendix A

Size	Stranding	Material	DIAM Inches	GMR Feet	RES Ω/Mile	Capacity Amps
795,000	30/19	ACSR	1.14	0.0393	0.1288	910
795,000	54/7	ACSR	1.093	0.0368	0.1378	900
795,000	CLASS A	AA	1.026	0.0328	0.131	720

Appendix B

Concentric Neutral 15 kV Cable

Conductor Size AWG or kcmil	Diameter over Insulation Inches	Diameter over Screen Inches	Outside Diameter Inches	Copper Neutral No. x AWG	Ampacity UG Duct Amps
FULL NEUTRAL					
2(7×)	0.78	0.85	0.98	10 × 14	120
1(19×)	0.81	0.89	1.02	13 × 14	135
1/0(19×)	0.85	0.93	1.06	16 × 14	155
2/0(19×)	0.90	0.97	1.13	13 × 12	175
3/0(19×)	0.95	1.02	1.18	16 × 12	200
4/0(19×)	1.01	1.08	1.28	13 × 10	230
250(37×)	1.06	1.16	1.37	16 × 10	255
350(37×)	1.17	1.27	1.47	20 × 10	300
1/3 NEUTRAL					
2(7×)	0.78	0.85	0.98	6 × 14	135
1(19×)	0.81	0.89	1.02	6 × 14	155
1/0(19×)	0.85	0.93	1.06	6 × 14	175
2/0(19×)	0.90	0.97	1.10	7 × 14	200
3/0(19×)	0.95	1.02	1.15	9 × 14	230
4/0(19×)	1.01	1.08	1.21	11 × 14	240
250(37×)	1.06	1.16	1.29	13 × 14	260
350(37×)	1.17	1.27	1.39	18 × 14	320
500(37×)	1.29	1.39	1.56	16 × 12	385
750(61×)	1.49	1.59	1.79	15 × 10	470
1,000(61×)	1.64	1.77	1.98	20 × 10	550

Tape Shielded 15 kV Cable
Tape Thickness = 5 mils

Conductor Size AWG or kcmil	Diameter over Insulation Inches	Diameter over Screen Inches	Jacket Thickness mils	Outside Diameter Inches	Ampacity in UG Duct Amps
1/0	0.82	0.88	80	1.06	165
2/0	0.87	0.93	80	1.10	190
3/0	0.91	0.97	80	1.16	215
4/0	0.96	1.02	80	1.21	245
250	1.01	1.08	80	1.27	270
350	1.11	1.18	80	1.37	330
500	1.22	1.30	80	1.49	400
750	1.40	1.48	110	1.73	490
1,000	1.56	1.66	110	1.91	565

Index

A

Allocation factor (AF), 23
Approximate line segment model, 120–126
Average demand, 11

B

Backfeed ground fault currents, 384–397
  one downstream transformer bank, 385–387
  three-phase circuit analysis, 387–392

C

Capacitor
  delta-connected capacitor bank, 296–297
  wye-connected capacitor bank, 296
Carson's equations
  concentric neutral cable, 74–80
  overhead lines, 64–71
  parallel distribution lines, 71–73
  triplex secondary, 419–422
Center-tapped transformers
  four-wire secondary 451–454
  lagging open wye–open delta connection, 442
  leading open wye–open delta transformer connection, 442
  open wye–delta connections, 441
  ungrounded wye–delta connection, 454
Closed delta–connected regulators, 194–197
Concentric neutral cable
  series impedance of, 74–80
  shunt admittance of, 102–106
Concentric neutral 15 kV cable, 473
Conductors, 469–471

D

Demand, 11
Demand factor, 15
Distribution feeder analysis
  power-flow analysis, 335–374
  short-circuit studies, 375–397
Distribution feeder map, 6
Distribution substations, 2–4, 23
Distribution system
  feeder electrical characteristics, 6
  power system components, 1
  radial feeders, 4–5
Distribution transformer loading, 12
  demand factor, 16
  diversified demand, 13
  diversity factor, 15
  load diversity, 17
  load duration curve, 14
  maximum diversified demand, 14
  maximum noncoincident demand, 14
  utilization factor, 17

E

Electrically parallel lines, 140
Electric vehicle
  level 1 charger, 335
  level 2 charger, 29, 335
  level 3 charger, 335
  load model, 335–340
  24–hour demand curve, 29
Equivalent T circuit, three-phase induction machine, 325
Exact line segment model, 111–120

F

Feeder
  demand curve, 17
  load allocation, 18
  voltage drop calculations, 19–21

G

General voltage drop equation, 38–42

H

High-side and low-side switching, 2

I

Induction generator, 325
Induction motor, 306–325

L

Ladder iterative technique, 42–52, 128–131
Line drop compensator, 175–182
Load
  delta-connected, 293–295
  single phase, 292
  two-phase, 295
  wye-connected loads, 287–293
Load diversity, 17
Load factor, 12

475

M

Maximum demand, 11
Maximum diversified demand, 14
Maximum noncoincident demand, 14
Maximum system voltage, 151
Metering, distribution substation, 3
Modified line model, 120–126

N

Neutral and ground currents, computation of, 122–126
Nominal system voltage, 151
Nominal utilization voltage, 151

O

Overhead lines
   series impedance of, 55–73
   shunt admittance of, 94–102

P

Parallel lines, general matrices for
   electrically parallel lines, 140–145
   physically parallel lines, 135–140
Parallel overhead distribution lines, 71–73
Parallel underground distribution lines, 83–87
Phase impedance matrix, 64–71
Power system components, 1
Primitive impedance matrix, for overhead lines, 63
Protection, distribution substation, 3

R

Radial feeders, 4–6

S

Service voltage, 154–155
Shunt capacitors, *see* capacitor
Smart grid, 3, 8, 173, 178
Standard voltage ratings, 151–153
Step-voltage regulators
   closed delta–connected regulators, three phase, 194–197
   open delta–connected regulators, three phase, 197–210
   single-phase step-voltage regulators, 171–175
   Type A step-voltage regulator, 171–172
   Type B step-voltage regulator, 172–175
   wye-connected regulators, three phase, 182–194
System voltage, 151

T

Tape-shielded cables
   series impedance of, 80–83
   shunt admittance of, 106–107
Three-phase induction machine, 297–325
Transformer models
   delta–delta connection, 264–274
   delta–grounded wye step-down connection, 219–234
   delta–grounded wye step-up connection, 234–236
   generalized matrices, 218–219
   grounded wye–delta step-down connection, 247–255
   grounded wye–grounded wye connection, 261–264
   open delta–open delta, 274–278
   two-winding transformer theory, 153–158
   ungrounded wye–delta step-down connection, 236–246
   ungrounded wye–delta step-up connection, 246–247
Transformer load management program, 22
Transposed three-phase lines, 37, 40, 52, 56
Two-winding autotransformer, 158–169
Two-winding transformer theory, 153–158

U

Underground cables, *see* concentric neutral cable, tape-shielded cables
Uniformly distributed loads, 346–349
Utilization factor, 17
Utilization voltage, 151–152

V

Voltage drop, 24, 38–42, 94
Voltage regulator, *see* step-voltage regulator
Voltage unbalance, 312–314

Printed in the United States
by Baker & Taylor Publisher Services